Handbook of Sensors

Volume I

Handbook of Sensors
Volume I

Edited by **Marvin Heather**

New Jersey

Published by Clanrye International,
55 Van Reypen Street,
Jersey City, NJ 07306, USA
www.clanryeinternational.com

Handbook of Sensors: Volume I
Edited by Marvin Heather

International Standard Book Number: 978-1-63240-290-5 (Hardback)

Printed in the United States of America.

Contents

Preface

A sensor can be defined as a device that can detect changes in physical stimuli and turn them into signals which can be measured and recorded. Sensors are thus based on the simple foundational principle that the physical properties of sensors must be able to change by external stimuli or environmental changes causing that characteristic to either produce electric signals or modulate external electric signals. Since there are many different types of stimuli, they can be measured by using quite different and distinct physical phenomena, and therefore subsequently, different sensors as well. There is a huge variety of sensors in the market and advancements are being made in their quality on a regular basis. Thus selecting a sensor might depend on various factors like cost, environmental conditions, availability and power consumption. Sensors are devices that are used every day in our daily lives in practical applications such as in tactile sensors in doors or lamps which dim or brighten based on certain stimuli like touch and sound. Other functions include cars, medicine, manufacturing, aerospace and even robotics. The demand for new and specialized sensors is increasing every day and thus the market for research in this field is wide open for innovations.

This book is an attempt to compile, collate and understand the research being done in the field of sensors and the data found in the advancements in this field. I am grateful to those who put in their hard work and efforts into this field as well as those who supported us in this endeavor. I would like to thank my publisher for giving me this incredible opportunity of working with international experts. Lastly, I would like to thank my family for their endless support.

Editor

1

Robust Modeling of Low-Cost MEMS Sensor Errors in Mobile Devices Using Fast Orthogonal Search

M. Tamazin,[1] A. Noureldin,[1,2] and M. J. Korenberg[1]

[1] *Electrical and Computer Engineering Department, Queen's University, Kingston, ON, Canada K7L 3N6*
[2] *Electrical and Computer Engineering Department, Royal Military College of Canada, Kingston, ON, Canada K7K 7B4*

Correspondence should be addressed to M. Tamazin; tamazin@gmail.com

Academic Editor: Xiaoji Niu

Accessibility to inertial navigation systems (INS) has been severely limited by cost in the past. The introduction of low-cost microelectromechanical system-based INS to be integrated with GPS in order to provide a reliable positioning solution has provided more wide spread use in mobile devices. The random errors of the MEMS inertial sensors may deteriorate the overall system accuracy in mobile devices. These errors are modeled stochastically and are included in the error model of the estimated techniques used such as Kalman filter or Particle filter. First-order Gauss-Markov model is usually used to describe the stochastic nature of these errors. However, if the autocorrelation sequences of these random components are examined, it can be determined that first-order Gauss-Markov model is not adequate to describe such stochastic behavior. A robust modeling technique based on fast orthogonal search is introduced to remove MEMS-based inertial sensor errors inside mobile devices that are used for several location-based services. The proposed method is applied to MEMS-based gyroscopes and accelerometers. Results show that the proposed method models low-cost MEMS sensors errors with no need for denoising techniques and using smaller model order and less computation, outperforming traditional methods by two orders of magnitude.

1. Introduction

Presently, GPS-enabled mobile devices offer various positioning capabilities to pedestrians, drivers, and cyclists. GPS provides absolute positioning information, but when signal reception is attenuated and becomes unreliable due to multipath, interference, and signal blockage, augmentation of GPS with inertial navigation systems (INS) or the like is needed. INS is inherently immune to the signal jamming, spoofing, and blockage vulnerabilities of GPS, but the accuracy of INS is significantly affected by the error characteristics of the inertial sensors it employs [1].

GPS/INS integrated navigation systems are extensively used [2], for example, in mobile devices that require low-cost microelectromechanical System (MEMS) inertial sensors (gyroscopes and accelerometers) due to their low cost, low power consumption, small size, and portability. The inadequate long-term performance of most commercially available MEMS-based INS limits their usefulness in providing reliable navigation solutions. MEMSs are challenging in any consumer navigation system because of their large errors, extreme stochastic variance, and quickly changing error characteristics.

According to [3], the inertial sensor errors of a low-cost INS consist of two parts: a deterministic part and a random part. The deterministic part includes biases and scale factors, which are determined by calibration and then removed from the raw measurements. The random part is correlated over time and is basically due to the variations in the INS sensor bias terms. These errors are mathematically integrated during the INS mechanization process, which results in increasingly inaccurate position and attitude over time. Therefore, these errors must be modeled.

The fusion of INS and GPS data is a highly synergistic coupling as INS can provide reliable short-term positioning information during GPS outages, while GPS can correct for longer-term INS errors [1]. INS and GPS integration (i.e., data fusion) is typically achieved through an optimal estimation technique, such as the Kalman filter or Particle filter [4].

Despite having an INS/GPS integration algorithm to correct for INS errors, it is still advantageous to have an accurate INS solution before the data fusion process. This requires preprocessing (i.e., prefiltering or denoising) each of the inertial sensor (gyroscope and accelerometer) signals before they are used to compute position, velocity, and attitude. This paper offers a robust method based on fast orthogonal search (FOS) to model the stochastic errors of low-cost MEMS sensors for smart mobile phones.

Orthogonal search [5] is a technique developed for identifying difference equation and functional expansion models by orthogonalizing over the actual data record. It mainly utilizes Gram-Schmidt orthogonalization to create a series of orthogonal functions from a given set of arbitrary functions. This enables signal representation by a functional expansion of arbitrary functions and therefore provides a wider selection of candidate functions that can be used to represent the signal. FOS is a variant of orthogonal search [6] where one major difference is that FOS achieves orthogonal identification without creating orthogonal functions at any stage of the process. As a result FOS is many times faster and less memory storage intensive than the earlier technique, while equally as accurate and robust [5–7].

Many techniques have been used previously to denoise and stochastically model the inertial sensor errors [3, 8, 9]. For example, several levels of wavelet decomposition have been used to denoise the raw INS data and eliminate high-frequency disturbances [3, 8, 9]. Modeling inertial sensor errors using autoregressive (AR) models was performed in [3], where the Yule-Walker, the covariance, and Burg AR methods were used. The AR model parameters were estimated after reducing the INS sensor measurements noise using wavelet denoising techniques.

FOS has been applied before in several applications [5–7, 10–12]. In [13], FOS was used to augment a Kalman filter (KF) to enhance the accuracy of a low-cost 2D MEMS-based navigation system by modeling only the azimuth error. FOS is used in this paper to model the raw MEMS gyroscope and accelerometer measurement errors in the time domain. In this paper, the performance of FOS is compared to linear modeling techniques such as Yule-Walker, the covariance and Burg AR methods in terms of mean-square errors (MSEs) and computational time.

2. Problem Statement

It is generally accepted that the long-term errors are modeled as correlated random noise. Correlated random noise is typically characterized by an exponentially decaying autocorrelation function with a finite correlation time. When the autocorrelation function of some of the noise sequences of MEMS measurements is studied, it has been shown that a first-order Gauss-Markov (GM) process may not be adequate in all cases to model such noise behavior. The shape of the autocorrelation sequence is often different from that of a first-order GM process, which is represented by a decaying exponential as shown in Figure 1. The GM process is characterized by an autocorrelation function of the form $R_{xx}(\tau) = \sigma^2 e^{-\beta|\tau|}$, where σ^2 is variance of the process and the correlation time

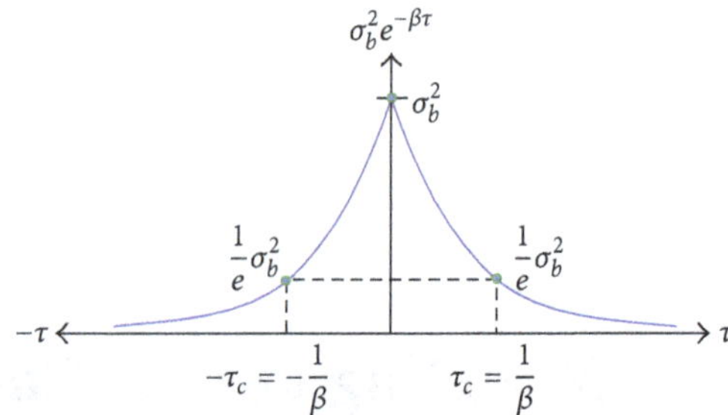

FIGURE 1: The autocorrelation sequence of a first-order Gauss-Markov (GM) process.

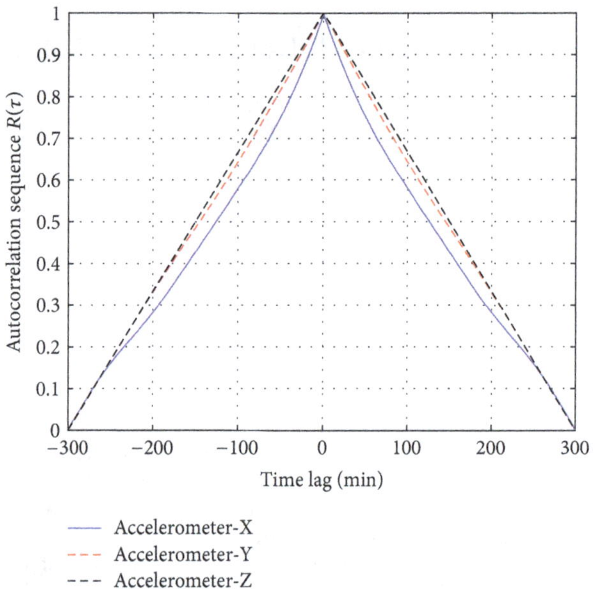

FIGURE 2: The computed autocorrelation sequence for an MEMS accelerometer data.

($1/e$ point) is given by $1/\beta$. The autocorrelation function approaches zero as $\tau \rightarrow \infty$, as depicted in Figure 1, indicating that the process gradually becomes less and less correlated as the time separation between samples increases [1].

Most of the computed autocorrelation sequences follow higher-order GM processes. An example of such computed autocorrelation sequences for one hour of static data of an MEMS accelerometer is shown in Figure 2. It clearly represents a higher-order GM process. These higher-order GM processes can be modeled using an autoregressive (AR) process of an appropriate order. In [3] it has been decided to model the randomness of the inertial sensor measurements using an AR process of order higher than one. With the present computational efficiency of microprocessor systems, efficient modeling of MEMS residual biases can be realized, and, thus, accurate prediction and estimation of such errors can be provided.

3. Modeling Methods of AR Processes

The autoregressive moving average (ARMA) modeling is based on the mathematical modeling of a time series of

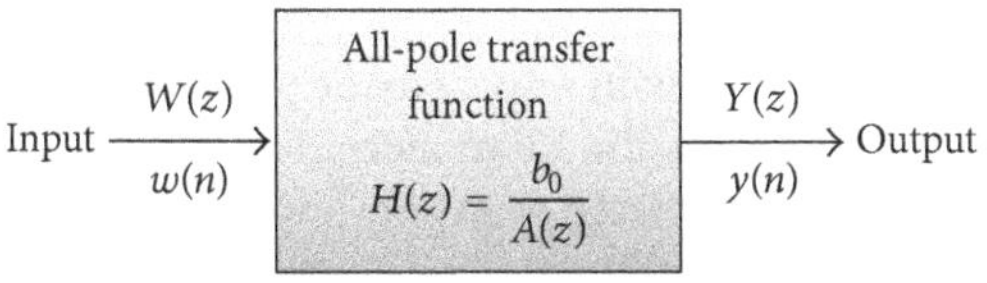

Input $\xrightarrow[w(n)]{W(z)}$ All-pole transfer function $H(z) = \frac{b_0}{A(z)}$ $\xrightarrow[y(n)]{Y(z)}$ Output

FIGURE 3: The input-output relationship of an autoregressive (AR) process.

measurements assuming that each value of such series is dependent on (a) a weighted sum of the "previous" values of the same series (AR part) and (b) a weighted sum of the "present and previous" values of a different time series (MA part) [14]. The ARMA process can be described using a pole-zero (ARMA) transfer function system $H(z)$ as follows:

$$H(z) = \frac{Y(z)}{W(z)} = \frac{B(z)}{A(z)} = \frac{\sum_{k=0}^{q} b_k z^{-k}}{1 + \sum_{k=1}^{p} a_k z^{-k}}, \tag{1}$$

where $W(z)$ is the z-transform of the input $w(n)$, $Y(z)$ is the z-transform of the output $y(n)$, p is the order of the AR process, q is the order of the MA process, and $a_1, a_2, \ldots, a_p$ and $b_1, b_2, \ldots, b_q$ are the AR and MA process parameters (weights), respectively. The AR process is a special case of an ARMA process, where q in (1) will be zero and thus $H(z)$ will be only an all-pole transfer function of the form

$$H(z) = \frac{Y(z)}{W(z)} = \frac{B(z)}{A(z)} = \frac{b_0}{1 + \sum_{k=1}^{p} a_k z^{-k}}. \tag{2}$$

Therefore, the name "Autoregressive" comes from the fact that each signal sample is regressed on (or predicted from) the previous values of itself [3]. In the time domain, the previous AR transfer function relationship can be obtained after applying the inverse z-transform for (2).

The resultant equation is written as [14]

$$\begin{aligned} y(n) &= -\sum_{k=1}^{p} a_k y(n-k) + b_0 w(n) \\ y(n) &= -a_1 y(n-1) - a_2 y(n-2) \\ &\quad - \cdots - a_p y(n-p) + b_0 w(n). \end{aligned} \tag{3}$$

The previous input-output relationship in both frequency and time domains is shown in Figure 3.

The problem in this case is to determine the values of the AR model parameters (predictor coefficients) a_k that optimally represent the random part of the inertial sensor biases. This is performed by minimizing the error $e(n)$ between the original signal $y(n)$ represented by the "AR process" of (3) and the estimated signal $\hat{y}(n)$, which is estimated by an "AR model" of the form [8]

$$\hat{y}(n) = -\sum_{k=1}^{p} a_k y(n-k). \tag{4}$$

The cost function for this minimization problem is the energy E of $e(n)$, which is given as

$$\begin{aligned} E &= \sum_{n=1}^{N} e^2(n) = \sum_{n=1}^{N} [y(n) - \hat{y}(n)]^2 \\ &= \sum_{n=1}^{N} \left[-\sum_{k=1}^{p} a_k y(n-k) + b_0 w(n) + \sum_{k=1}^{p} a_k y(n-k) \right]^2 \\ &= \sum_{n=1}^{N} b_0^2 w^2(n) = \min, \end{aligned} \tag{5}$$

where N is the total number of data samples. In this case, $w(n)$ is a sequence of stationary uncorrelated sequences (white noise) with zero mean and unity variance.

Therefore, the resultant energy from (5) $[\sum_{n=1}^{N} b_0^2 w^2(n)]$ will be b_0^2. Therefore, b_0^2 represents the estimated variance σ_w^2 of the white noise input to the AR model or, more generally, the prediction mean-square error σ_e^2. This is due to the fact that the AR model order p is completely negligible with respect to the MEMS data sample size N.

Several methods have been reported to estimate the a_k parameter values by fitting an AR model to the input data. Three AR methods are considered in this paper, namely, the Yule-Walker method, the covariance method, and Burg's method. In principle, all of these estimation techniques should lead to approximately the same parameter values if fairly large data samples are used [3].

3.1. The Yule-Walker Method. The Yule-Walker method, which is also known as the autocorrelation method, determines first the autocorrelation sequence $R(\tau)$ of the input signal (inertial sensor residual bias in our case). Then, the AR model parameters are optimally computed by solving a set of linear normal equations. These normal equations are obtained using the formula [15]

$$\frac{\partial E}{\partial a_k} = 0, \tag{6}$$

which leads to the following set of normal equations:

$$Ra = -r \longleftrightarrow a = -R^{-1} r, \tag{7}$$

where

$$a = \begin{bmatrix} a_1 \\ a_2 \\ \vdots \\ a_p \end{bmatrix} \tag{8a}$$

$$r = \begin{bmatrix} R(1) \\ R(2) \\ \vdots \\ R(p) \end{bmatrix}, \tag{8b}$$

$$R = \begin{bmatrix} R(0) & R(1) & \cdots & R(p-1) \\ R(1) & R(0) & \cdots & R(p-2) \\ \vdots & \vdots & \cdots & \vdots \\ R(p-1) & R(p-2) & \cdots & R(0) \end{bmatrix}. \tag{8c}$$

If the mean-square error σ_e^2 is also required, it can be determined by

$$\begin{bmatrix} R(0) & R(1) & \dots & R(p-1) \\ R(1) & R(0) & \cdots & R(p-2) \\ \vdots & \vdots & \cdots & \vdots \\ R(p-1) & R(p-2) & \dots & R(0) \end{bmatrix} \begin{bmatrix} 1 \\ a_1 \\ \vdots \\ a_p \end{bmatrix} = \begin{bmatrix} \sigma_e^2 \\ 0 \\ \vdots \\ 0 \end{bmatrix}. \quad (9a)$$

Equations (7) and (9a) are known as the Yule-Walker equations [7–9, 13]. Instead of solving (9a) directly (i.e., by first computing R^{-1}), it can efficiently be solved using the Levinson-Durbin (LD) algorithm which proceeds recursively to compute $a_1, a_2, \dots, a_p$, and σ_e^2. The LD algorithm is an iterative technique that computes the next prediction coefficient (AR parameter) from the previous one. This LD recursive procedure can be summarized in the following [9]:

$$E_0 = R(0) \quad (9b)$$

$$\gamma_k = -\frac{R(k) + \sum_{i=1}^{k-1} a_{i,k-1} R(k-i)}{E_{k-1}}, \quad 1 \le k \le p \quad (9c)$$

$$a_{k,k} = \gamma_k \quad (9d)$$

$$a_{i,k} = a_{i,k-1} + \gamma_k a_{k-i,k-1}, \quad 1 \le i \le k-1 \quad (9e)$$

$$E_k = \left(1 - \gamma_k^2\right) E_{k-1}. \quad (9f)$$

Equations (9b)–(9f) are solved recursively for $k = 1, 2, \dots, p$ and the final solution for the AR parameters is provided by

$$a_i = a_{i,p}, \quad 1 \le i \le p. \quad (9g)$$

Therefore, the values of the AR prediction coefficients in the Yule-Walker method are provided directly based on minimizing the forward prediction error $e_f(n)$ in the least-squares sense. The intermediate quantities γ_k represented by (9c) are known as the reflection coefficients. In (9f), both energies E_k and E_{k-1} are positive, and, thus, the magnitude of γ_k should be less than one to guarantee the stability of the all-pole filter.

However, the Yule-Walker method performs adequately only for long data records [15]. The inadequate performance in case of short data records is usually due to the data windowing applied by the Yule-Walker algorithm. Moreover, the Yule-Walker method may introduce a large bias in the AR-estimated coefficients since it does not guarantee a stable solution of the model [16].

3.2. The Covariance Method. The covariance method is similar to the Yule-Walker method in that it minimizes the forward prediction error in the least-squares sense, but it does not consider any windowing of the data. Instead, the windowing is performed with respect to the prediction error to be minimized. Therefore, the AR model obtained by this method is typically more accurate than the one obtained from the Yule-Walker method [17].

Furthermore, it uses the covariance $C(\tau_i, \tau_j)$ instead of $R(\tau)$. In this case, the Toeplitz structure of the normal equations used in the autocorrelation method is lost, and hence the LD algorithm cannot be used for the computations. To achieve an efficient C^{-1} in this case, Cholesky factorization is usually utilized [15].

The method provides more accurate estimates than the Yule-Walker method especially for short data records. However, the covariance method may lead to unstable AR models since the LD algorithm is not used for solving the covariance normal equations [18].

3.3. Burg's Method. Burg's method was introduced in 1967 to overcome most of the drawbacks of the other AR modeling techniques by providing both stable models and high resolution (i.e., more accurate estimates) for short data records [19]. Burg's method tries to make the maximum use of the data by defining both a forward and a backward prediction error terms, $e_f(n)$ and $e_b(n)$. The energy to be minimized in this case (E_{Burg}) is the sum of both the forward and backward prediction error energies; that is,

$$E_{\text{Burg}} = \sum_{n=1}^{N} \left[e_f^2(n) + e_b^2(n)\right] = \min, \quad (10)$$

where e_f and e_b are defined as

$$e_f = y(n) + a_1 y(n-1) + a_2 y(n-2) + \cdots + a_p y(n-p) \quad (11a)$$

$$e_b = y(n-p) + a_1 y(n-p+1) + a_2 y(n-p+2) + \cdots + a_p y(n). \quad (11b)$$

The forward and backward prediction error criteria are the same, and, hence, they have the same optimal solution for the model coefficients [20]. Considering the energies in (9f) to be E_{Burg}, the forward and backward prediction errors can, therefore, be expressed recursively as

$$e_{f^k}(n) = e_{f^{k-1}}(n) + \gamma_k e_{b^{k-1}}(n-1) \quad (12a)$$

$$e_{b^k}(n) = e_{b^{k-1}}(n-1) + \gamma_k e_{f^{k-1}}(n). \quad (12b)$$

These recursion formulas form the basis of what is called Lattice (or Ladder) realization of a prediction error filtering (see Figure 4).

As has been shown for the Yule-Walker method, the accuracy of the estimated parameters $a_1, a_2, \dots, a_p$, and σ_e^2 depends mainly on accurate estimates of the autocorrelation sequence $R(\tau)$. However, this can be rarely achieved due to the prewindowing of data [17] or the existence of large

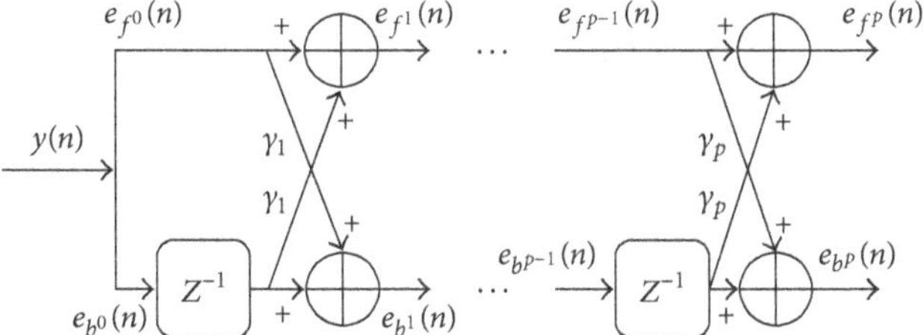

FIGURE 4: The forward-backward predication error lattice filter general structure.

measurement noise [21]. To avoid the difficulties of the computation of the autocorrelation sequences, Burg in his method estimates first the reflection coefficients γ_k using another formula instead of (9c). This formula is derived by substituting (12a) and (12b) into (13) and setting the derivative of E_{Burg} with respect to γ_k (instead of a_k in the Yule-Walker and covariance methods) to zero.

This leads to the form

$$\gamma_k = \frac{-2\sum_{n=1}^{N}\left[e_{f^{k-1}}(n)\, e_{b^{k-1}}(n-1)\right]}{\sum_{n=1}^{N} e_{f^{k-1}}^2(n) + \sum_{n=1}^{N} e_{b^{k-1}}^2(n-1)} \tag{13}$$

which shows clearly that the magnitude of γ_k is forced (guaranteed) to be less than one, and thus the obtained model is guaranteed to be stable. Equations (12a), (12b), and (13) form the recursive structure of Burg's lattice filter, which is shown in Figure 4 with the initial conditions of $e_f^0(n) = e_b^0(n) = y(n)$. Finally, the prediction coefficients a_k are obtained by constraining them to satisfy (9e) in the LD algorithm.

Therefore, the utilization of (9e) and (13) together will always ensure the stability of Burg's method solution [22]. Moreover, the utilization of both forward and backward prediction errors minimization usually yields better estimation results than using the forward prediction approach used in the previous two methods. Finally, it has been reported by [23] that Burg's method generally provides better residual estimates than the Yule-Walker method [19].

4. Fast Orthogonal Search (FOS) Method

FOS [5–7, 10–12] is a general purpose modeling technique, which can be applied to spectral estimation and time-frequency analysis. The algorithm uses an arbitrary set of nonorthogonal candidate functions $p_m(n)$ and finds a functional expansion of an input $y(n)$ in order to minimize the mean square error (MSE) between the input and the functional expansion.

The functional expansion of the input $y(n)$ in terms of the arbitrary candidate functions $p_m(n)$ is given by

$$y(n) = \sum_{m=0}^{M} a_m P_m(n) + e(n), \tag{14}$$

where a_m is the set of weights of the functional expansion, $P_0(n) = 1$, and the $P_m(n)$ are the model terms selected from the set of candidate functions, and $e(n)$ is the modeling error.

These model terms can involve the system input x and output y and cross-products and powers thereof:

$$P_m(n) = y(n-l_1)\cdots y(n-l_i)\, x(n-k_1)\cdots x(n-k_j),$$
$$m \ge 1, \quad i \ge 0, \quad j \ge 0, \quad \forall i: 1 \le l_i \le L, \quad \forall j: 0 \le k_j \le K. \tag{15}$$

By choosing non-orthogonal candidate functions, there is no unique solution for (14). However, FOS may model the input with fewer model terms than an orthogonal functional expansion [11]. For the FFT to model a frequency that does not have an integral number of periods in the record length, energy is spread into all the other frequencies, which is a phenomenon known as spectral leakage [24]. By using candidate functions that are non-orthogonal, FOS may be able to model this frequency between two FFT bins with a single term resulting in many fewer weighting terms in the model [5, 25].

FOS begins by creating a functional expansion using orthogonal basis functions such that

$$y(n) = \sum_{m=0}^{M} g_m w_m(n) + e(n), \tag{16}$$

where $w_m(n)$ is a set of orthogonal functions derived from the candidate functions $p_m(n)$, g_m is the weight, and $e(n)$ is an error term. The orthogonal functions $w_m(n)$ are derived from the candidate functions $p_m(n)$ using the Gram-Schmidt (GS) orthogonalization algorithm. The orthogonal functions $w_m(n)$ are implicitly defined by the Gram-Schmidt coefficients α_{mr} and do not need to be computed point-by-point.

The Gram-Schmidt coefficients α_{mr} and the orthogonal weights g_m can be found recursively using the equations [11]

$$w_0 = p_0(n) \tag{17}$$

$$D(m,0) = \overline{p_m(n) p_0(n)} \tag{18}$$

$$D(m,r) = \overline{p_m(n) p_r(n)} - \sum_{i=0}^{r-1} \alpha_{ri} D(m,i) \tag{19}$$

$$\alpha_{mr} = \frac{\overline{p_m(n) w_r(n)}}{\overline{w_r^2(n)}} = \frac{D(m,r)}{D(r,r)} \tag{20}$$

$$C(0) = \overline{y(n) p_0(n)} \tag{21}$$

$$C(m) = \overline{y(n)\, p_m(n)} - \sum_{r=0}^{m-1} \alpha_{mr} C(r) \tag{22}$$

$$g_m = \frac{C(m)}{D(m,m)}. \tag{23}$$

In its last stage, FOS calculates the weights of the original functional expansion a_m (6), from the weights of the orthogonal series expansion g_m and Gram-Schmidt coefficients α_{mr}. The value of a_m can be found recursively using

$$a_m = \sum_{i=m}^{M} g_i v_i, \tag{24}$$

where $v_m = 1$ and

$$v_i = -\sum_{r=m}^{i-1} \alpha_{ir} v_r, \qquad i = m+1,\ m+2, \ldots, M. \tag{25}$$

FOS requires the calculation of the correlation between the candidate functions and the calculation of the correlation between the input and the candidate functions. The correlation between the input and the candidate function $\overline{y(n)p_m(n)}$ is typically calculated point-by-point once at the start of the algorithm and then stored for later quick retrieval.

The MSE of the orthogonal function expansion has been shown to be [5, 6, 11]

$$\overline{\varepsilon^2(n)} = \overline{y^2(n)} - \sum_{m=0}^{M} g_m^2 \overline{w_m^2(n)}. \tag{26}$$

It then follows that the MSE reduction given by the mth candidate function is given by

$$Q_m = g_m^2\, \overline{w_m^2(n)} = g_m^2 D(m,m), \tag{27}$$

The candidate with the greatest value for Q is selected as the model term, but optionally its addition to the model may be subject to its Q value exceeding a threshold level [5, 6, 11]. The residual MSE after the addition of each term can be computed by

$$\mathrm{MSE}_m = \mathrm{MSE}_{m-1} - Q_m. \tag{28}$$

The search algorithm may be stopped when an acceptably small residual MSE has been achieved (i.e., a ratio of the MSE over the mean-squared value of the input [12] or an acceptably small percentage of the variance of the time series being modeled [5]). The search may also stop when a certain number of terms have been fitted. Another stopping criterion is when none of the remaining candidates can yield a sufficient MSE reduction value (this criterion would be representative of not having any candidates that would yield an MSE reduction value greater than the addition of a white Gaussian noise series).

5. Experimental Results

The data were collected by a low-cost MEMS-based inertial measurement unit (IMU CC-300, Crossbow). These measurements were collected during a one-hour experiment to obtain stochastic error models of both gyroscopes and accelerometers. To illustrate the performance, two sensors were selected as an example (accelerometer-Y, Gyro-Y) while the other inertial sensors gave similar results. Figure 5 shows one hour of sampled accelerometer-Y and Gyro-Y acquired at 200 Hz.

FOS is applied directly on the raw inertial sensor 200 Hz data without any preprocessing or denoising. Traditional methods like Yule Walker, Covariance, and Burg perform poorly on the raw data, so we first applied wavelet denoising of up to 4 levels of decomposition that resulted in band limiting the spectrum of the raw inertial sensor data to 12.5 Hz.

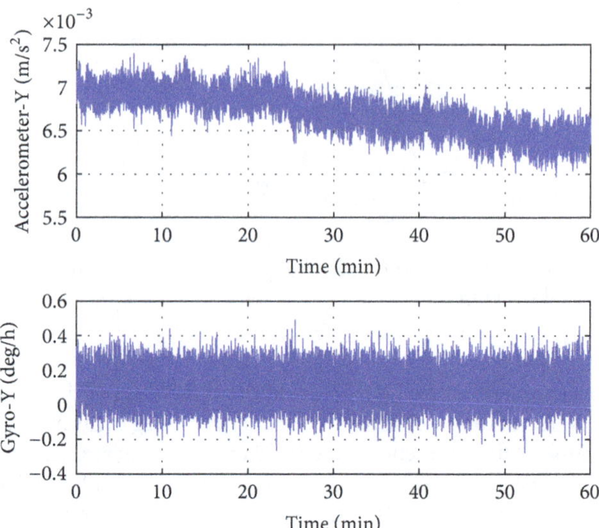

Figure 5: Accelerometer-Y and Gyro-Y specific force measurements.

Therefore, unlike FOS, the other 3 methods operate on the denoised version of the same data. After denoising, AR model parameters were then estimated as well as the corresponding prediction MSE for all sensors using Yule-Walker, Covariance, and Burg methods.

For FOS, the raw INS data were divided into three datasets for model training, evaluation, and prediction stages. The first 3 minutes of the INS raw data were utilized for model training, which uses the FOS algorithm to identify several possibly nonlinear AR equations. Different models can be obtained by changing the maximum delay L in the output and the degree of output cross-products (CP). The next 3 minutes of the data were used for the evaluation stage. Here, models are compared and the best one, fitting the real output with minimum MSE, is chosen. As an example, the FOS model (CP = 1) of the accelerometer-Y is shown as follows:

$$\begin{aligned} Y[n] = {} & 6.03 \times 10^{-6} + 2.25y[n-1] - 1.2y[n-2] \\ & + 1.02y[n-4] - 0.67y[n-3] - 0.26y[n-5] \\ & + 0.62y[n-7] - 0.51y[n-6] - 0.32y[n-8] \\ & + 0.08y[n-9]. \end{aligned} \tag{29}$$

In the prediction stage, the output and MSE of the chosen model are computed over the remaining (novel) raw INS data. Figure 6 shows prediction MSE of Accelerometer-Y samples by Yule-Walker, Covariance, Burg, and FOS methods. For FOS, an AR model of order 3 or 4 suffices, and MSE decreases when the degree of the cross-product terms is raised to 2 (nonlinear model). An AR model of order 7 or 8 is required for Burg or Covariance method and order 9 or 10 for Yule-Walker. Large AR model order complicates the estimation method (like KF) used for the INS/GPS integration.

Table 1 shows a summary of the performance of three conventional stochastic modeling methods (Yule-Walker,

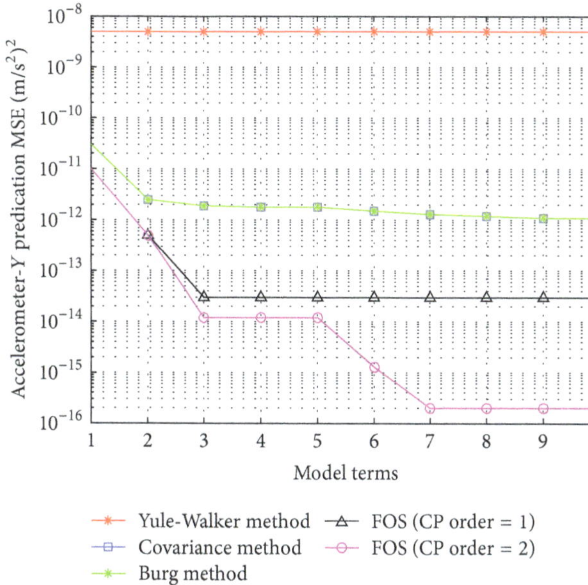

FIGURE 6: Accelerometer-Y prediction MSE using Yule-Walker, covariance, Burg, and FOS methods. For CP degree = 1, only linear candidate terms up to a maximum output lag $L = 10$ were allowed. For CP degree = 2, both linear and $y(n-l_1)y(n-l_2)$ candidates were allowed, up to maximum output lag $L = 10$.

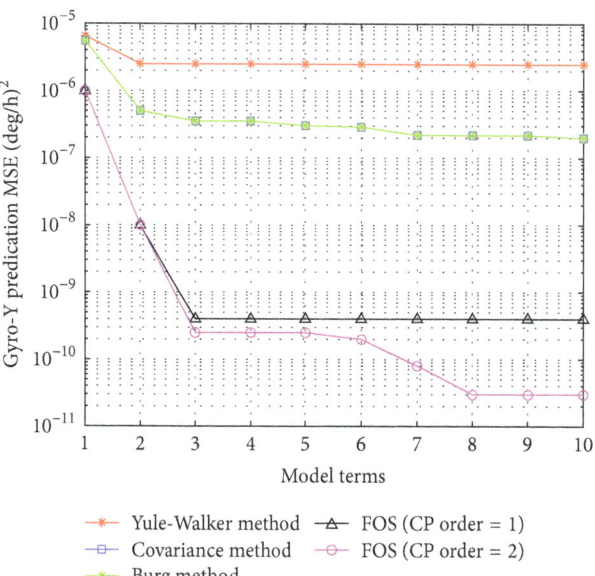

FIGURE 7: Gyro-Y prediction MSE using Yule-Walker, covariance, Burg, and FOS methods. For CP order = 1, only linear candidate terms up to a maximum output lag $L = 10$ were allowed. For CP order = 2, both linear and $y(n-l_1)y(n-l_2)$ candidates were allowed, up to a maximum output lag $L = 10$.

TABLE 1: Performance summary of AR models obtained by Yule-Walker, Burg, and FOS over one hour of accelerometer-Y measurements: first-order FOS: only linear model terms; second-order FOS: linear and cross-product model terms.

Modelling technique	Model MSE $(m/s^2)^2$	Corresponding position error (m)	Computational time (s)
Model order (maximum output lag L) 1			
Yule-Walker	5×10^{-9}	458	0.25
Covariance/Burg	3×10^{-11}	35	0.23
CP degree = 1 FOS	1×10^{-11}	20	0.13
CP degree = 2 FOS	1×10^{-11}	20	0.36
Model order (maximum output lag L) 10			
Yule-Walker	5×10^{-9}	458	0.69
Covariance/Burg	2×10^{-12}	9	0.45
CP degree = 1 FOS	3×10^{-14}	1	0.30
CP degree = 2 FOS	2×10^{-16}	0.09	0.60

TABLE 2: Performance summary of both AR Yule-Walker and Burg models and FOS model over one hour of Gyro-Y measurements: first-order FOS: only linear model terms; second-order FOS: linear and cross-product model terms.

Modeling technique	Model MSE $(deg/h)^2$	Corresponding position error (m)	Computational time (s)
model order (maximum output lag L) 1			
Yule-Walker	7×10^{-6}	978	0.40
Covariance/Burg	5×10^{-6}	826	0.23
CP degree = 1 FOS	1×10^{-6}	369	0.09
CP degree = 2 FOS	1×10^{-6}	369	0.35
Model order (maximum output lag L) 10			
Yule-Walker	2×10^{-6}	523	0.44
Covariance/Burg	2×10^{-7}	165	0.28
CP degree = 1 FOS	4×10^{-10}	7	0.18
CP degree = 2 FOS	3×10^{-11}	2	0.6

Covariance, and Burg) and the proposed FOS-based method with cross-product order set to 1 (i.e., linear model) and 2 (i.e., nonlinear model) for different model orders 1 and 10 over one hour of MEMS Accelerometer-Y measurements. The FOS model is capable of denoising the Accelerometer-Y measurements without appreciable degradation of the original signal. FOS achieves better performance in terms of lower MSE and less computational time than the traditional methods with no need for any denoising techniques. Increasing the cross-product degree to 2 for FOS improves model accuracy and lessens position error by an order of magnitude (for $L = 10$) but increases computation time.

A similar procedure was performed for the Gyro-Y sensor measurements. Figure 7 shows the prediction MSE of Gyro-Y samples by using Yule-Walker, the Covariance, Burg, and FOS methods. It is clear that FOS method achieves minimum MSE error with less model order than the other AR methods. Table 2 shows a summary of the performance of three conventional stochastic modeling methods (Yule-Walker, Covariance, and Burg) and the proposed FOS-based method with cross-product order set to 1 (i.e., linear model) and 2 (i.e., nonlinear model) for different model orders 1 and 10 over one hour of Gyro-Y measurements.

Similar to the accelerometer case, the stochastic model obtained for Gyro-Y using FOS surpasses the models obtained by the other methods in the MSE, the corresponding position error that would result from the residual errors and the computation time.

6. Conclusions

Inertial sensor errors are the most significant contributors to INS errors. Thus, techniques to model these sensor errors are of interest to researchers. The current state of the art in modeling inertial sensor signals includes low-pass filtering and using wavelet denoising techniques, which have had limited success in removing long-term inertial sensor errors.

This paper suggested using FOS to model the MEMS sensor errors in time domain. The FOS MSE and computational time are compared with those from Yule-Walker, the covariance, and Burg methods. FOS was applied directly to the one-hour raw inertial sensor 200 Hz data without any pre-processing or denoising. The other traditional 3 methods operated on the denoised version of the same data after we applied the wavelet denoising of up to 4 levels of decomposition.

For either gyroscope or accelerometer case, the FOS model surpasses those obtained by traditional methods. The results demonstrate the advantages of the proposed FOS-based method including the absence of a need for preprocessing or denoising, lower computation time and MSE, and achieving better performance with smaller model order. Increasing the cross-product degree for FOS improves model accuracy and lessens position error but increases computation time.

References

[1] A. Noureldin, T. Karamat, and J. Georgy, *Fundamentals of Inertial Navigation, Satellite-Based Positioning and Their Integration*, Springer, New York, NY, USA, 2012.

[2] M. Grewal, L. Weil, and A. Andrews, *Global Positioning Systems, Inertial Navigation and Integration*, John Wiley & Sons, New York, NY, USA, 2nd edition, 2007.

[3] S. Nassar, K. Schwarz, N. El-Sheimy, and A. Noureldin, "Modeling inertial sensor errors using autoregressive (AR) models," *Navigation, Journal of the Institute of Navigation*, vol. 51, no. 4, pp. 259–268, 2004.

[4] J. Georgy and A. Noureldin, "Vehicle navigator using a mixture particle filter for inertial sensors/odometer/map data/GPS integration," *IEEE Transactions on Consumer Electronics*, vol. 58, no. 2, pp. 544–522, 2012.

[5] M. J. Korenberg, "A robust orthogonal algorithm for system identification and time-series analysis," *Biological Cybernetics*, vol. 60, no. 4, pp. 267–276, 1989.

[6] M. J. Korenberg, "Fast orthogonal identification of nonlinear difference equation models," in *Proceedings of the 30th Midwest Symposium on Circuits and Systems*, vol. 1, August 1987.

[7] J. Armstrong, A. Noureldin, and D. McGaughey, "Application of fast orthogonal search techniques to accuracy enhancement of inertial sensors for land vehicle navigation," in *Proceedings of the Institute of Navigation, National Technical Meeting (NTM '06)*, pp. 604–614, Monterey, Calif, USA, January 2006.

[8] A. Noureldin, A. Osman, and N. El-Sheimy, "A neuro-wavelet method for multi-sensor system integration for vehicular navigation," *Measurement Science and Technology*, vol. 15, no. 2, pp. 404–412, 2004.

[9] S. Nassar, A. Noureldin, and N. El-Sheimy, "Improving positioning accuracy during kinematic DGPS outage periods using SINS/DGPS integration and SINS data de-noising," *Survey Review*, vol. 37, no. 292, pp. 426–438, 2004.

[10] K. M. Adeney and M. J. Korenberg, "Fast orthogonal search for array processing and spectrum estimation," *IEE Proceedings: Vision, Image and Signal Processing*, vol. 141, no. 1, pp. 13–18, 1994.

[11] M. J. Korenberg, "Fast orthogonal algorithms for nonlinear system identification and time-series analysis," in *Advanced Methods of Physiological System Modeling*, V. Z. Marmarelis, Ed., vol. 2, pp. 165–177, Plenum Press, New York, NY, USA, 1989.

[12] D. R. McGaughey, M. J. Korenberg, K. M. Adeney, S. D. Collins, and G. J. M. Aitken, "Using the fast orthogonal search with first term reselection to find subharmonic terms in spectral analysis," *Annals of Biomedical Engineering*, vol. 31, no. 6, pp. 741–751, 2003.

[13] Z. Shen, J. Georgy, M. J. Korenberg, and A. Noureldin, "FOS-based modelling of reduced inertial sensor system errors for 2D vehicular navigation," *Electronics Letters*, vol. 46, no. 4, pp. 298–299, 2010.

[14] S. Haykin, *Adaptive Filter Theory*, Prentice Hall, Upper Saddle River, NJ, USA, 2002.

[15] L. Jackson, *Digital Filters and Signal Processing*, Kluwer Academic, Norwell, Mass, USA, 1996.

[16] R. Kless and P. Broersen, *How to Handle Colored Noise in Large Least-Squares Problems*, Delft University of Technology, Delft, The Netherlands, 2002.

[17] M. Hayes, *Statistical Digital Signal Processing and Modeling*, John Wiley & Sons, New York, NY, USA, 1996.

[18] M. J. L. de Hoon, T. H. J. J. van der Hagen, H. Schoonewelle, and H. van Dam, "Why Yule-Walker should not be used for autoregressive modelling," *Annals of Nuclear Energy*, vol. 23, no. 15, pp. 1219–1228, 1996.

[19] J. Burg, *Maximum entropy spectral analysis [Ph.D. thesis]*, Department of Geophysics, Stanford University, Stanford Calif, USA, 1975.

[20] S. Orfanidis, *Optimum Signal Processing: An Introduction*, Macmillan, New York, NY, USA, 1988.

[21] J. M. Pimbley, "Recursive autoregressive spectral estimation by minimization of the free energy," *IEEE Transactions on Signal Processing*, vol. 40, no. 6, pp. 1518–1527, 1992.

[22] S. Marple, *Digital Spectral Analysis with Applications*, Prentice-Hall, Englewood Cliffs, NJ, USA, 1987.

[23] I. A. Rezek and S. J. Roberts, "Parametric model order estimation: a brief review," in *Proceedings of the IEE Colloquium on the Use of Model Based Digital Signal Processing Techniques in the Analysis of Biomedical Signals*, London, UK, April 1997.

[24] E. Ifeachor and B. Jervis, *Digital Signal Processing*, Prentice-Hall, Englewood Cliffs, NJ, USA, 4th edition, 2001.

[25] K. H. Chon, "Accurate identification of periodic oscillations buried in white or colored noise using fast orthogonal search," *IEEE Transactions on Biomedical Engineering*, vol. 48, no. 6, pp. 622–629, 2001.

2

Suppression of Instability on Sensing Signal of Optical Pulse Correlation Measurement in Remote Fiber Sensing

Hirokazu Kobayashi,[1] Toshimasa Tsuzuki,[1] Toshitake Onishi,[1] Yuhei Masaoka,[1] Xunjian Xu,[2] and Koji Nonaka[1]

[1] *Department of Electronic and Photonic Systems Engineering, Kochi University of Technology, Kochi 782-8502, Japan*
[2] *Key Laboratory, State Grid Corporation of China, Beijing 100031, China*

Correspondence should be addressed to Hirokazu Kobayashi, hirokazu.kobayashi@gmail.com

Academic Editor: Hao Zhang

Optical fiber sensing has the potential to overcome weak points of traditional electric sensors. Many types of optical fiber sensors have been proposed according to the modulation parameter of incident light. We have proposed an optical pulse correlation sensing system that focuses on the time drift values of the propagating optical pulses to monitor the temperature- or strain-induced extension along the optical fiber in the sensing region. In this study, we consider the instability in the optical pulse correlation sensing system applied to remote monitoring over a kilometer-long distance. We introduce a method to stabilize the instability of the pulse correlation signal resulting from the time drift fluctuation along a transmission line. By using this method, we can purify the response and improve the accuracy of signals at the focused sensing regions. We also experimentally demonstrate remote temperature monitoring over a 30 km-long distance using a remote reference technique, and we estimate the resolution and the measurable span of the temperature variation as $(1.1/L)^\circ$C and $(5.9 \times 10/L)^\circ$C, respectively, where L is the length of the fiber in the sensing region.

1. Introduction

In recent years, optical fiber sensors have attracted growing interest for a variety of reasons, including their corrosion resistance, electromagnetic immunity, long lifetime components, high sensitivity, and multiplexing capability [1–6]. This sensor impresses information onto the light beam in response to environmental parameters such as temperature and strain. The information could be impressed in terms of intensity, phase, wavelength, and time. In the course of their study, many types of optical fiber sensors have been proposed, such as in-fiber Bragg grating (FBG) sensors [7], low-coherence interferometric sensors [8], Fabry-Pérot sensors [9, 10], Brillouin scattering distributed sensors [11], and Raman scattering distributed sensors [12]. It has been expected that the optical fiber sensors would be used for the monitoring of complicated composite structures because they are capable of measuring strain and temperature over several tens of kilometers by accessing only one end of an optical fiber [13, 14]. Feasibility studies have been performed by using actual structures, for example, bridges [15–17], dams [18], marine vehicles [19], and aircrafts [20, 21].

In our laboratory, we have proposed and developed an optical pulse correlation sensing system in which the sensing signals are the time drift values of the optical pulses caused by the temperature- or strain-induced extension along the optical fiber in the sensing region [22–26]. This sensing system is not only simple and handy, but also highly accurate for temperature or strain measurement [24]. It is notable that our fiber sensor can work as a line sensor, that is, the time drift caused by the expansion of the optical fiber is integrated along the sensing region, and the sensing signal is proportional to the integrated time drift. Thus, we can select the sensitivity by changing the fiber length in the sensing region.

Herein, we consider the optical pulse correlation sensing system applied to remote monitoring over a kilometer-long distance. For this purpose, a kilometer-long optical fiber is used for the transmission line connecting the monitoring center and the sensing region. In this case, however, a minute temperature or strain fluctuation around the transmission line is integrated along the long optical fiber and induces some instability in the sensing signals, for example, power, polarization, and time drift fluctuations.

In this paper, we focus on the time drift fluctuations and estimate its effect on signal drift. Moreover, we introduce a method to suppress the fluctuation by using the region separation technique with a partial reflector [27]. We also experimentally demonstrate remote temperature sensing over a 30 km-long transmission line and estimate the resolution and the measurable span of the temperature variation.

The remainder of this paper is organized as follows. In Section 2, we introduce the principle of the optical pulse correlation sensing system. In Section 3, we experimentally confirm the time drift fluctuation of the sensing signal caused by the long transmission line. Then, we propose a modified experimental setup that can stabilize the sensing signal against the time drift fluctuation in Section 4. In this setup, the fluctuation is suppressed by use of a remote reference pulse. Moreover, we demonstrate remote temperature monitoring over a 30 km distance. A conclusion is presented in Section 5.

2. Principle of the Optical Pulse Correlation Sensing System

2.1. Conventional Sensing System Using Optical Pulse Correlation Measurement. A schematic of our conventional system is shown in Figure 1. First, we generate two optical pulses, a reference pulse (R) and a monitoring pulse (M), at the monitoring center. The monitoring pulse passes through a transmission line and then enters a sensing region. After that, the monitoring pulse is reflected back to the monitoring center and enters a pulse correlation measurement unit. In contrast, the reference pulse directly enters the pulse correlation measurement unit. If the temperature or strain in the sensing region is changed, the optical path in the measuring fiber produces a variation of relative time drift τ between the reference and monitoring pulses. From the variation of the time drift value measured by the pulse correlation measurement, we can estimate the temperature or strain variation.

2.2. Pulse Correlation Measurement Unit. A schematic of the pulse correlation measurement unit is shown in Figure 2. In this system, a reference pulse with 45° polarization is split into two orthogonally-polarized pulses, R_1 (horizontal polarization) and R_2 (vertical polarization), with a fixed timing separation τ_0 induced by a birefringence crystal (pulse doubler). The monitoring pulse with 45° polarization is combined with doubled reference pulses and separated into two channels by using a polarized beam splitter. Thus, one channel has a backward reference pulse R_1 and a monitoring pulse with a time drift $\tau_0 - \tau$, while the other channel has a forward reference pulse R_2 and a monitoring pulse with a time drift τ.

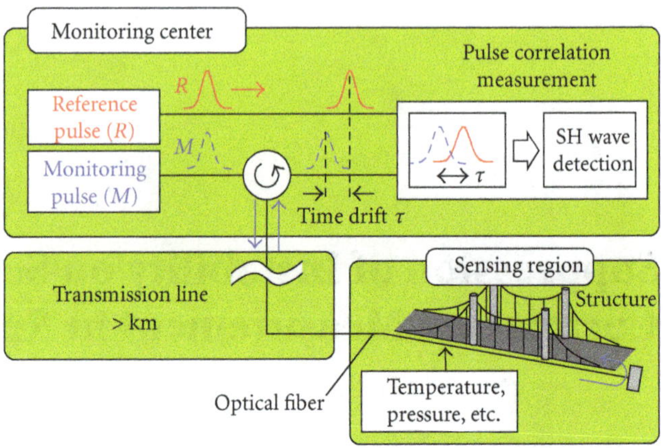

FIGURE 1: Schematic of remote fiber sensing using optical pulse correlation measurement.

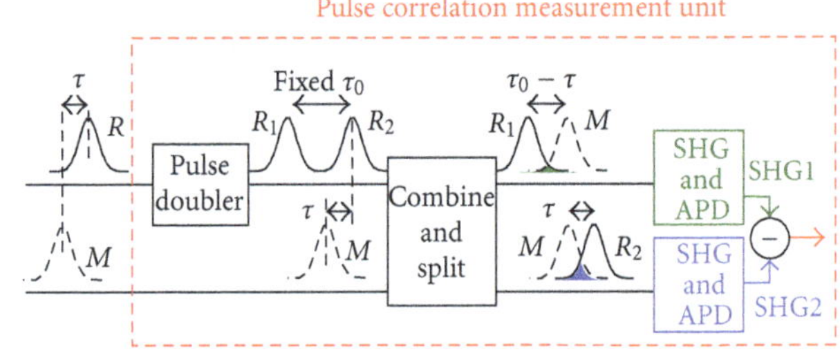

FIGURE 2: Schematic of pulse correlation measurement unit. R: reference pulse, M: monitoring pulse, SHG: second harmonic generation, and APD: avalanche photodiode.

In each channel, the second harmonic generation (SHG) signals are observed by two avalanche photodiodes (APDs). The SHG output is maximized when the monitoring pulse completely overlaps with the reference pulse and its full width at half maximum is related to the pulse width δt of the incident pulsed laser light. The green dashed line and the blue dotted line in Figure 3 show the theoretical plots of the typical relationship between the time drift τ and SHG outputs. To obtain a wide-range linear response with respect to the time drift value, we consider the differential signal between two SHG outputs (see the red solid line in Figure 3). The important part of the differential signal curve is highlighted inside a dashed oval in Figure 3. In this region, the output relationship shows good linearity with respect to the time drift value. In what follows, we estimate the time drift value from the differential signal by using this linear relationship.

2.3. The Variation of the Time Drift Value Owing to Temperature or Strain Changes. The variation of the time drift value $\Delta\tau$ within the fiber length L can be expressed in terms of the temperature change ΔT and the strain change ΔF:

$$\Delta\tau = L \cdot f, \tag{1}$$

$$f \equiv C_T \Delta T + C_F \Delta F, \tag{2}$$

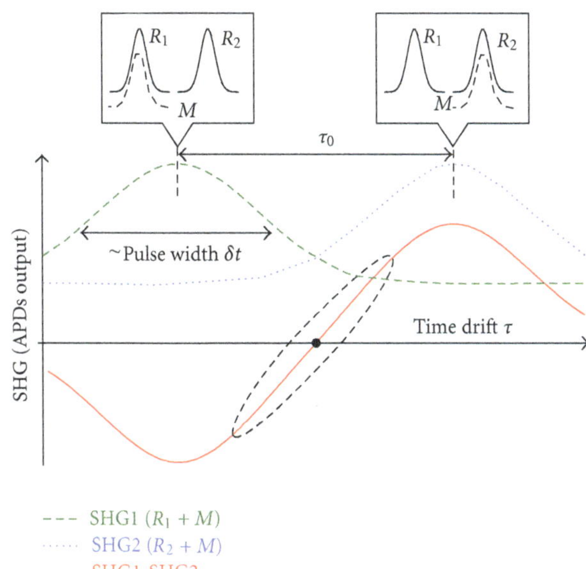

FIGURE 3: Theoretical plots of SHG signals (green dashed line and blue dotted line) and their differential signal (red solid line) with respect to the relative time drift τ.

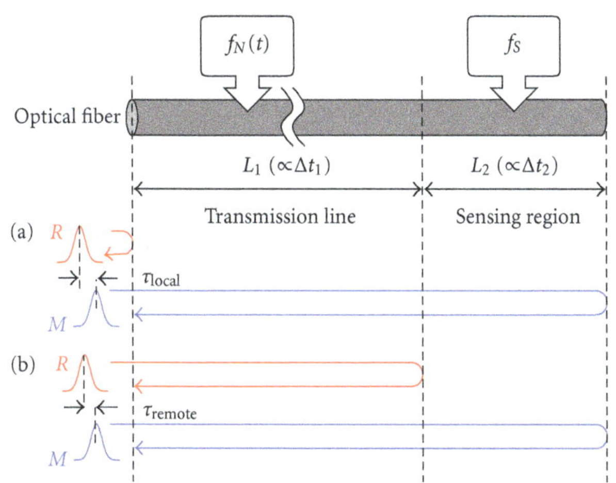

FIGURE 4: The time drift value in the optical pulse correlation sensing system when we use (a) a local reference pulse and (b) a remote reference pulse.

where f is the variation of the time drift value per unit length of the optical fiber, and C_T and C_F are the temperature sensitivity and the strain sensitivity, respectively. For standard commercial single mode fiber at wavelength 1550 nm with a diameter of 2 mm, the typical values are $C_T = 0.17$ ps/(m·°C) [26] and $C_F = 2.7$ ps/(m · N) [28].

Now, we consider the measurement of the temperature or strain changes around the sensing region with a length of L_2 over the transmission line with a length of L_1 (see Figure 4). f_S and f_N are variations of the time drift value per unit length along the sensing region and the transmission line, respectively. We assume that f_N is dependent on the time t, that is, there is the temporal fluctuation of temperature or strain around the transmission line.

The conventional situation of the optical pulse correlation sensing system is shown in Figure 4(a), where the monitoring pulse is sent to the sensing region through the transmission line while the reference pulse is maintained in the monitoring center (we call it the "local" reference pulse). We approximate the temporal variation of $f_N(t)$ caused by the propagation time in the transmission line (Δt_1) and in the sensing region (Δt_2) by a first-order Taylor expansion. The variation of the time drift value $\Delta\tau_{local}$ is calculated as

$$\begin{aligned}\Delta\tau_{local} &\simeq 2\left[L_2 f_S + L_1\left\{f_N(t) + \frac{df_N}{dt}(\Delta t_1 + \Delta t_2)\right\}\right] \\ &\simeq 2\{L_2 f_S + L_1 f_N(t)\},\end{aligned} \tag{3}$$

where we assume $f_N(t) \gg (df_N/dt)\Delta t_i$ ($i = 1,2$) and $\Delta t_i \equiv n_g L_i/c$ ($i = 1,2$) with the speed of light c and the refractive index for the group velocity n_g. The first term in (3) corresponds to the sensing signal, and the second term in (3) shows the effect of the time drift fluctuation caused by the transmission line. If the second term is sufficiently small compared to the first term, we can estimate the temperature or strain changes within the sensing region from the time drift value $\Delta\tau$. In case of the long transmission line, however, $L_1 f_N(t)$ could become large enough to induce the appreciable time drift fluctuation.

To suppress this fluctuation, we can utilize a "remote reference pulse," that is, an optical pulse reflected by a partial reflector just before the sensing region (see Figure 4(b)). This idea is inspired by region separation techniques used to realize multiple region sensing [27]. In this case, the variation of the time drift value $\Delta\tau$ is calculated as

$$\Delta\tau_{remote} \simeq 2\left[L_2 f_S + L_1 \frac{df_N}{dt}\Delta t_2\right]. \tag{4}$$

By using a remote reference pulse, the main factor $L_1 f_N(t)$ of the time drift fluctuation caused by the transmission line is canceled out because both the reference and monitoring pulses are passing through the transmission line. The second term in (4) shows the effect of temporal variation fluctuation caused by the difference of the propagation time between the reference and monitoring pulse. If the second term in (4) is sufficiently small compared to the first term, $\Delta\tau_{remote}$ can be stabilized against the time drift fluctuation by using remote reference pulses.

From (2), our sensing system has the cross sensitivity between temperature and strain changes. The simplest way of discriminating strain and temperature changes is to use two optical fibers where the first one is not fixed at the measuring object, that is, isolated from strain and experiences only temperature changes, and the second one is fixed at the measuring object, that is, affected by both strain and temperature. Assuming that the two sensing fibers are at the same temperature, the variation of the time drift value from the first sensor can be used to derive a temperature-corrected strain value from the second one. In what follows, the optical fiber is not fixed at the measuring object, and we monitor only temperature changes.

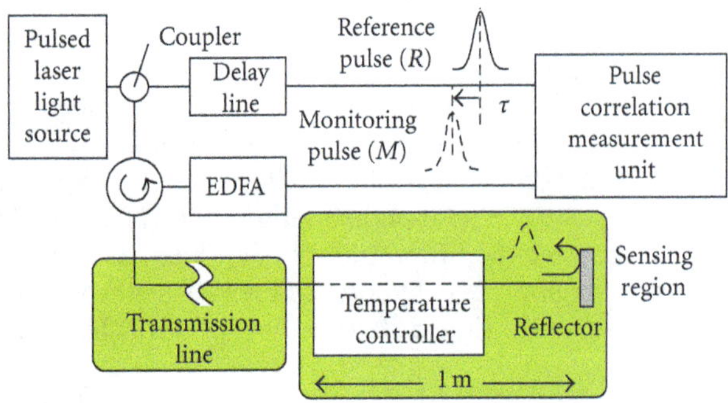

FIGURE 5: Experimental setup for remote fiber sensing using pulse correlation measurement.

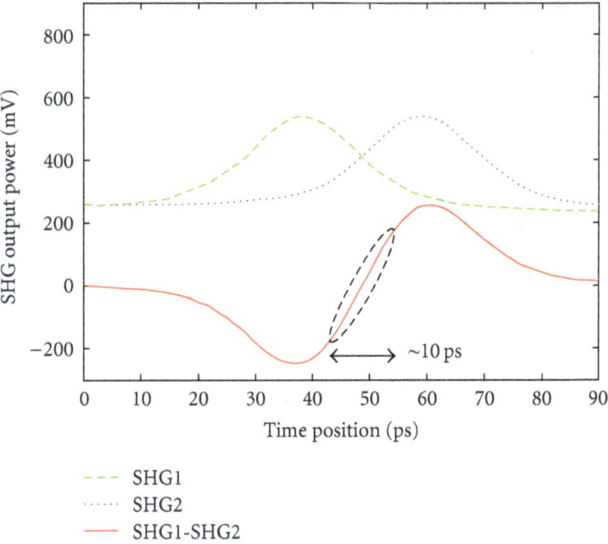

FIGURE 6: Experimentally obtained SHG outputs (green dashed line and blue dotted line) and their differential signal (red solid line).

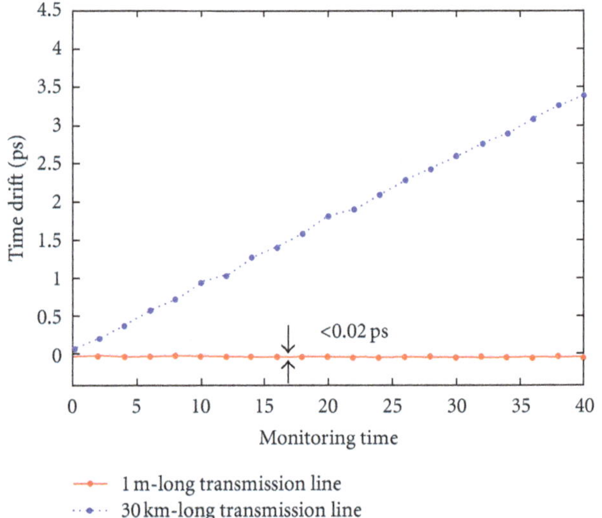

FIGURE 7: Time drift fluctuation in remote monitoring of the 1 m long sensing region at the fixed temperature 27.0 ± 0.1°C. The solid red line and the dotted blue line show the fluctuation of the time drift value with monitoring time through short (~1 m) and long (~30 km) transmission lines, respectively.

3. Experimental Observation of Time Drift Fluctuation in the Long Transmission Line

Figure 5 shows the experimental setup for remote fiber sensing using pulse correlation measurement. We use optical pulses with a repetition frequency of 1.4794 GHz, a center wavelength of 1550 nm, and a pulse width of around 10 ps as light source. First, the input pulses are split into two pulses, reference pulse and monitoring pulse, by a coupler. The monitoring pulse passes through an optical circulator and a transmission line and then enters a sensing region. After that, the monitoring pulse is reflected back to the circulator and then enters a pulse correlation measurement unit subsequent to amplification by an erbium-doped fiber amplifier (EDFA), which compensates for propagation loss in the transmission line. In contrast, the reference pulse directly enters the pulse correlation measurement unit through a tunable delay line that is fixed at a point to allow a partial overlap between the reference and monitoring pulses. The delay of the tunable delay line device (OZ Optics ODL300) can be adjusted with 0.005-ps increments between 0 and 350 ps. In the correlation measurement unit, the fixed timing separation τ_0 is 20 ps. The outputs of two APDs are connected to a 12-bit analog to digital converter (ADC) to obtain the SHG output power as the voltage value. Finally, the output differential signal between two SHG outputs is calculated and stored in a personal computer.

Figure 6 shows experimentally obtained SHG outputs and their differential signal with respect to the time position of the tunable delay line. The linear region with a width of about 10 ps can be used for the estimation of the time drift value. As described in Section 2.3, the temperature sensitivity is 0.17 ps/(m·°C). Thus, the measurable span of the temperature variation is calculated as $(5.9 \times 10/L_2)$°C, where L_2 is the length of the fiber in the sensing region. This indicates that the measurable span becomes smaller when we use a longer optical fiber.

Next, we stabilized the temperature of a 1 m long sensing region at room temperature (~27°C) by using a feedback loop of the temperature controller and continuously monitoring the time drift value to observe the time drift fluctuation at a fixed temperature in the sensing region. The experimental results are shown in Figure 7.

For the short transmission line (~1 m), the time drift value remains almost constant with respect to the monitoring time (see the solid red line in Figure 7). Actually, the time drift fluctuation is less than 0.02 ps, which is within the 12-bit resolution of the ADC. From this value and the sensitivity coefficient, the temperature resolution of this system is calculated as $(1.2 \times 10^{-1}/L_2)$°C. We can adjust the temperature resolution to satisfy the corresponding application requirements by changing the fiber length L_2.

For the long transmission line (~30 km), however, even in the constant-temperature sensing region, the measured time drift value shifts gradually with monitoring time at

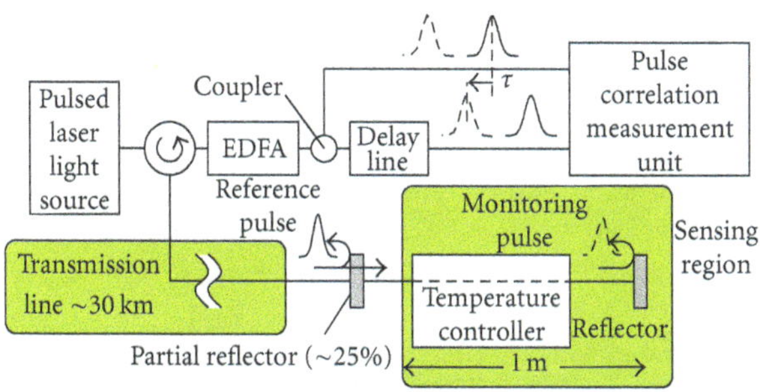

FIGURE 8: Experimental setup for remote optical fiber sensing using a remote reference pulse.

a speed of 0.09 ps per second (see the dotted blue line in Figure 7). Because the measurable span is about 10 ps, only a few minutes of the monitoring time is required to break the linear response with respect to the time drift value.

From (3), we can calculate the speed of the temperature variation around the transmission line as:

$$L_1 \frac{\mathrm{d}f_N}{\mathrm{d}t} = L_1 C_T \frac{\mathrm{d}\Delta T}{\mathrm{d}t} = 0.09\ \mathrm{ps/s}, \tag{5}$$

$$\therefore \frac{\mathrm{d}\Delta T}{\mathrm{d}t} = 1.8 \times 10^{-5}\ {}^{\circ}\mathrm{C/s}. \tag{6}$$

This minute variation is caused by the ambient temperature and the warming by optical pulses. It is integrated along the long transmission line and induces an appreciable time drift fluctuation. To realize remote temperature monitoring, we need to suppress the large time drift fluctuation in the long transmission line.

4. A Pulse Correlation Sensing System Using a Remote Reference Pulse

4.1. Experimental Setup. From (5), we can calculate the second term in (4) as 4.4×10^{-22} s. This result is sufficiently small compared to the first term in (4). Thus, we can suppress the time drift fluctuation by utilizing a remote reference pulse.

The experimental setup with the remote reference pulse is shown in Figure 8. In this setup, the reference pulse is reflected by a partial reflector with a reflectivity of about 25%, and the monitoring pulses are reflected at the end of the same optical fiber. After amplification by the EDFA, the reflected pulses including the reference and monitoring pulses are split into two pulses. The tunable delay line is set up to allow partial overlap between the reference pulse in one arm and the monitoring pulse in the other arm.

4.2. Experimental Results. The continuous monitoring of the time drift value is shown in Figure 9 when the temperature in the 1 m long sensing region is fixed at room temperature. We can easily confirm that the time drift value is sufficiently stabilized in monitoring time by using a remote reference pulse. The time drift fluctuation is below 0.2 ps even in an hour over the 30 km-long transmission line. This indicates

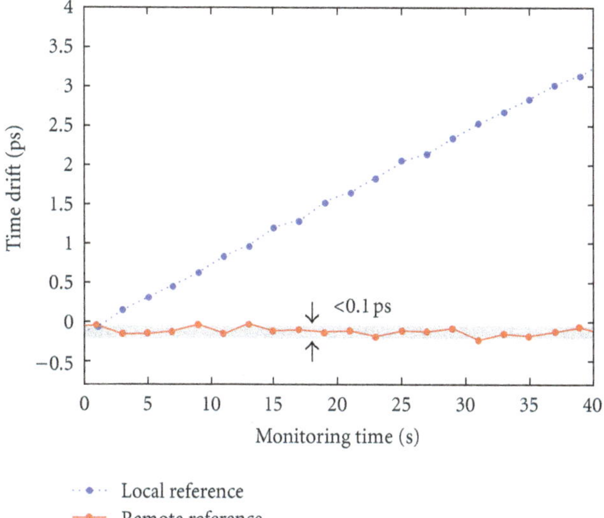

FIGURE 9: Time drift fluctuation in remote monitoring over a 30 km distance of the 1 m long sensing region at the fixed temperature 27.0 ± 0.1°C by using the local reference pulse (dotted blue line) and the remote reference pulse (red solid line).

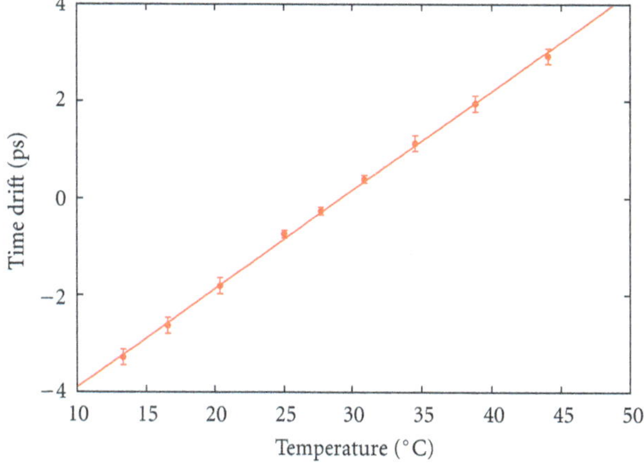

FIGURE 10: Remote temperature monitoring at a 1 m long sensing region over a 30 km-long transmission line by utilizing a remote reference.

that the temperature resolution is $(1.1/L_2)$°C and the measurable span of the temperature variation is $(5.9 \times 10/L_2)$°C.

Next, we demonstrate remote temperature monitoring. The time drift value is continuously monitored with the changing surrounding temperature of the optical fiber in the sensing region. Figure 10 shows the change of the measured time drift value using the remote reference with respect to the temperature around the 1 m long monitoring fiber. The good linearity between the measured time drift value and the temperature of the sensing region can be clearly observed by suppressing the effect of long-distance instability. From the result, the temperature sensitivity of the 1 m long monitoring fiber is found to be about 0.18 ps/°C, which is almost equal to the value in [26]. The temperature resolution estimated from the error bar in Figure 10 is less than 1.6°C.

5. Conclusions

In this paper, we introduced an optical fiber sensing system for long-distance remote monitoring using pulse correlation measurement. We estimated the effect of the time drift fluctuation from the long transmission lines and proposed a stabilizing method using a remote reference pulse against the time drift fluctuation caused by the long transmission line. Moreover, we demonstrated remote temperature monitoring over the 30 km-long transmission line by utilizing a remote reference pulse and obtained good linearity of the time drift value with respect to the temperature around the sensing region. The resolution and the measurable span of the temperature variation are $(1.1/L_2)^{\circ}$C and $(5.9 \times 10/L_2)^{\circ}$C, respectively. We can select the resolution and the measurable span by changing the length of the optical fiber in the sensing region. This system can be used for remote temperature monitoring of complicated structures, such as bridges, oil tanks, power transformers, and pipelines.

Acknowledgments

This work was partially supported by Y2009-2010 Practical Application Research no. 1513 of Japan Science and Technology Agency (JST) and the Japan Society for the Promotion of Science (JSPS) no. 18360180 and no. 22657062.

References

[1] E. Udd and W. B. Spillman Jr., Eds., *Fiber Optic Sensors: An Introduction for Engineers and Scientists*, John Wiley & Sons, New York, NY, USA, 1991.

[2] J. Dakin and B. Culshaw, *Optical Fiber Sensors: Principles and Components*, vol. 1, Artech House, Boston, Mass, USA, 1988.

[3] K. T. V. Grattan and B. T. Meggitt, Eds., *Optical Fiber Sensor Technology, Volume 2: Devices and Technology*, Chapman & Hall, London, 1997.

[4] T. G. Giallorenzi, J. A. Bucaro, A. Dandridge et al., "Optical fiber sensor technology," *IEEE Journal of Quantum Electronics*, vol. 18, no. 4, pp. 626–665, 1982.

[5] F. T. S. Yu and S. Yin, Eds., *Fiber Optic Sensor*, Marcel Dekker, New York, NY, USA, 2002.

[6] L. Thévenaz, "Review and progress in distributed fiber sensing," in *Proceedings of the Optical Fiber Sensors*, Cancuún, Mexico, 2006, Paper ThC1.

[7] Y. J. Rao, "In-fibre Bragg grating sensors," *Measurement Science and Technology*, vol. 8, no. 4, p. 355, 1997.

[8] Y. J. Rao and D. A. Jackson, "Recent progress in fibre optic low-coherence interferometry," *Measurement Science and Technology*, vol. 7, no. 7, p. 981, 1996.

[9] A. D. Kersey, D. A. Jackson, and M. Corke, "A simple fibre Fabry-Perot sensor," *Optics Communications*, vol. 45, no. 2, pp. 71–74, 1983.

[10] P. J. Henderson, Y. J. Rao, D. A. Jackson, L. Zhang, and I. Bennion, "Simultaneous multi-parameter monitoring using a serial fibre-Fabry-Perot array with low-coherence and wavelength-domain detection," *Measurement Science and Technology*, vol. 9, no. 11, p. 1837, 1998.

[11] X. Bao, J. Dhliwayo, N. Heron, D. J. Webb, and D. A. Jackson, "Experimental and theoretical studies on a distributed temperature sensor based on Brillouin scattering," *Journal of Lightwave Technology*, vol. 13, no. 7, pp. 1340–1348, 1995.

[12] M. A. Farahani and T. Gogolla, "Spontaneous Raman scattering in optical fibers with modulated probe light for distributed temperature Raman remote sensing," *Journal of Lightwave Technology*, vol. 17, no. 8, pp. 1379–1391, 1999.

[13] E. Udd, Ed., *Fiber Optic Smart Structures*, Wiley, New York, NY, USA, 1995.

[14] D. Balageas, C.-P. Fritzen, and A. Guemes, Eds., *Structural Health Monitoring*, Wiley-ISTE, London, UK, 2006.

[15] R. Maaskanta, T. Alaviea, R. Measuresa, G. Tadrosb, S. Rizkallac, and A. Guha-Thakurtad, "Fiber-optic Bragg grating sensors for bridge monitoring," *Cement Concrete Composites*, vol. 19, no. 1, pp. 21–33, 1997.

[16] D. Inaudi, S. Vurpillot, N. Casanova, and P. Kronenberg, "Structural monitoring by curvature analysis using interferometric fiber optic sensors," *Smart Materials and Structures*, vol. 7, no. 2, pp. 199–208, 1998.

[17] W. L. Schulz, E. Udd, J. M. Seim, and G. E. McGill, "Advanced fiber grating strain sensor systems for bridges, structures, and highways," in *Smart Structures and Materials 1998 Smart Systems for Bridges, Structures, and Highways*, vol. 3325 of *Proceedings of SPIE*, pp. 212–221, March 1998.

[18] L. Thévenaz, M. Facchini, A. Fellay, P. Robert, D. Inaudi, and B. Dardel, "Monitoring of large structure using distributed brillouin fibre sensing," in *Proceedings of the 13th International Conference on Optical Fiber Sensor*, vol. 3746, pp. 345–348, 1999.

[19] D. R. Hjelme, L. Bjerkan, S. Neegard, J. S. Rambech, and J. V. Aarsnes, "Application of Bragg grating sensors in the characterization of scaled marine vehicle models," *Applied Optics*, vol. 36, no. 1, pp. 328–336, 1997.

[20] P. D. Foote, "Fibre Bragg grating strain sensors for aerospace smart structures. ," in *Second European Conference on Smart Structures and Materials*, vol. 2361 of *Proceedings of SPIE*, pp. 290–293, 1994.

[21] W. Staszewski, C. Boller, and G. Tomlinson, Eds., *Health Monitoring of Aerospace Structures*, John Wiley & Sons, Chichester, UK, 2004.

[22] K. Uchiyama, K. Nonaka, and H. Takara, "Subpicosecond timing control using optical double-pulses correlation measurement," *IEEE Photonics Technology Letters*, vol. 16, no. 2, pp. 626–628, 2004.

[23] K. Nonaka, M. Sato, and T. Suzuki, "Optical pulse timing drift sensing for fiber delay monitoring using pulse correlation and differential detection," in *Proceedings of the 17th International Conference on Optical Fibre Sensors*, vol. 5855 of *Proceedings of SPIE*, p. 76, 2005.

[24] H. B. Song, T. Suzuki, M. Sato, and K. Nonaka, "High time resolution fibre optic sensing system based on correlation and differential technique," *Measurement Science and Technology*, vol. 17, no. 4, p. 631, 2006.

[25] H. B. Song, T. Suzuki, T. Fujimura, K. Nonaka, T. Shioda, and T. Kurokawa, "Polarization fluctuation suppression and sensitivity enhancement of an optical correlation sensing system," *Measurement Science and Technology*, vol. 18, no. 10, p. 3230, 2007.

[26] X. J. Xu and K. Nonaka, "High-sensitivity fiber-optic temperature sensing system based on optical pulse correlation and time-division multiplexer technique," *Japanese Journal of Applied Physics*, vol. 48, Article ID 102403, 5 pages, 2009.

[27] X. Xu, A. Bueno, K. Nonaka, and S. Sales, "Fiber strain measurement for wide region quasidistributed sensing by optical correlation sensor with region separation techniques," *Journal of Sensors*, vol. 2010, Article ID 839803, 10 pages, 2010.

[28] "We experimentally measured the strain sensitivity by stretching the optical fiber".

3

Flexible GMR Sensor Array for Magnetic Flux Leakage Testing of Steel Track Ropes

W. Sharatchandra Singh, B. P. C. Rao, S. Thirunavukkarasu, and T. Jayakumar

NDE Division, Indira Gandhi Centre for Atomic Research, Kalpakkam 603 102, India

Correspondence should be addressed to B. P. C. Rao, bpcrao@igcar.gov.in

Academic Editor: Gui Yun Tian

This paper presents design and development of a flexible GMR sensor array for nondestructive detection of service-induced defects on the outer surface of 64 mm diameter steel track rope. The number of GMR elements and their locations within saddle-type magnetizing coils are optimized using a three dimensional finite element model. The performance of the sensor array has been evaluated by measuring the axial component of leakage flux from localized flaw (LF) and loss of metallic cross-sectional area (LMA) type defects introduced on the track rope. Studies reveal that the GMR sensor array can reliably detect both LF and LMA type defects in the track rope. The sensor array has a fast detection speed along the length of the track rope and does not require circumferential scanning. It is also possible to image defects using the array sensor for obtaining their spatial information.

1. Introduction

Track ropes are a type of wire ropes used for transportation of coal in mining industries. One such rope system is operated for about 10 hours every day to transport 3000 tons of coal with the help of 256 numbers of buckets, each carrying nearly 1.6 tons of coal. The track rope is stationary and is rigidly supported by towers at periodic intervals. As shown in Figure 1, the track rope has 8 layers of stranded wires of different diameters. The 6 inner layers are round-type wires, while the outer two layers are Z-type wires. The round wires are locked by two Z wires to get the strength of the rope. The width of the outer surface of the first Z wire is 6.45 mm, and the gap width between two outer Z wires is 0.76 mm.

During the operation of the rope system, the carriage wheels of the bucket come in contact with the top surface of the outer Z wire as shown in Figure 2(a). Prolonged use of the rope system is expected to cause abrasion and wear, resulting in loss of metallic cross-sectional area (LMA) or localized flaw (LF) type defects (Figure 2(b)). Also wire breakage and formation of fatigue cracks, pitting corrosion, inter strand nicking or martensitic embrittlement, and so forth are likely to occur [1, 2]. When more than two Z wires of the outer layer are broken, they will be separated from the adjacent layers. Detection of damage in track rope is essential as part of the condition monitoring and life management programs. Nondestructive detection of damage in the track rope is challenging due to heterogeneous structure of the rope, multiplicity and uncertainty of broken wires, and hostile working environment.

Among various nondestructive evaluation (NDE) techniques, visual and magnetic flux leakage (MFL) techniques are widely used for monitoring the health of steel track ropes [3, 4]. Although visual inspection is simple and does not require special instrumentation, it is not suited for monitoring the internal deterioration in track ropes. On the contrary, the MFL technique is capable of detecting both LF and LMA type defects in wire ropes [4].

In MFL technique, wire ropes are locally magnetized using electromagnets or permanent magnets. If any defect is present in the rope, the magnetic flux produced in the rope takes a longer path around the defect and as a result, some amount of flux lines leak out of the surface. Measurement of this leakage flux using sensors forms an important step in MFL testing. The leakage flux has axial (along the rope length) and radial components which can be detected using

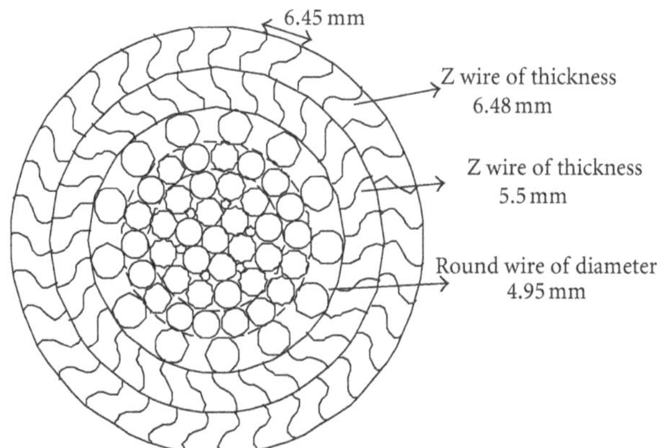

FIGURE 1: The cross-section of double locked track rope.

coils [5, 6] or sensors [7–9] and correlated to the size and location of the defects. The magnetization can be continuous [10] or pulsed [11, 12]. Recently, authors proposed GMR sensor [13, 14] -based technique for NDE of track ropes, and this requires the use of sophisticated scanning set-up for circumferential scanning over the ropes. Use of array sensors is attractive, and this enables fast and reliable inspection of track ropes [15]. This paper discusses the design and development of a flexible GMR sensor array for detection of damage on the outer surface of the track rope. The performance of the array sensor has been evaluated for fast inspection of track rope and imaging of LF and LMA type defects on the outer Z-wire of the track rope.

2. GMR Sensor

GMR sensor consists of a few nm thick multilayer structures (Co/Cu/Co) in which ferromagnetic layers are separated by nonmagnetic layers. The sensor works on GMR effect in which there is a large drop in electrical resistance of multilayer for an incident magnetic field, due to the spin dependent scattering of electrons [16]. The GMR sensors are characterized by high sensitivity at low magnetic field and high spatial resolution. They are inexpensive and consume less power [17]. In this study, the GMR sensors are connected in bridge configuration to measure the differential output with high stability. The sensor is encapsulated with standard SOIC-8 package of $5.9 \times 4.9 \times 1.4\,\text{mm}^3$ size. The maximum hysteresis of the GMR sensor is 2% unit.

3. Three Dimensional Finite Element Modeling

In order to identify the number of GMR elements required to cover the top surface of the track rope and to determine the sensor locations, 3D finite element modeling has been performed using COMSOL 3.4 Multiphysics software package. Figure 3(a) shows the mesh generated for the geometry which consists of a track rope (length 300.0 mm, outer diameter 64.0 mm) and magnetizing coils. Equation (1) has been solved in three dimensions using the finite element method:

$$\nabla \times \frac{1}{\mu_0 \mu_r} \nabla \times A = J, \qquad (1)$$

where A is the magnetic vector potential, μ_0 is the magnetic permeability of free space, μ_r is the relative permeability, and J is current density. Two saddle coils (length 120.0 mm, width 35.0 mm) each consisting of 90 turns with a cross-sectional area of $20.0 \times 10.0\,\text{mm}^2$ are used for magnetization of the track rope at a current of 5 A, as shown in Figure 3(a). The magnetizing current in the saddle coils is set in opposite directions to ensure axial magnetization of the rope region between the saddle coils.

For simplicity, in the model the track rope is assumed as a solid rod, and GMR sensor as well as velocity effects is not modeled. The relative magnetic permeability of the rope is assumed constant as 100. Magnetic insulation ($n \times A = 0$) boundary condition is applied at the outer boundaries constructed for the model. The computation time for solving (1) with 5673392 degrees of freedom is approximately 50 minutes in a dual-core 64 bit processor workstation with 8 GB primary memory.

The magnetic vector potential is computed in the solution region, and the axial component of the magnetic flux density (B_z) between the two saddle coils is predicted. As can be seen from Figure 3(b), the magnetic flux density is nearly uniform for an optimum circumferential intercoil distance of 80.0 mm that completely covers the expected damage region on the top surface of the Z-wire (dotted region in Figure 3(b)). This region can accommodate 12 GMR sensors. Hence, a flexible array of 12 GMR sensors has been fabricated and used for detection of damage on the track rope.

4. Design of GMR Sensor Array

The layout of the flexible GMR sensor array and its photograph are shown in Figures 4(a) and 4(b), respectively. Each sensor element in the array has a common power input of 5 V, and the array has 12 differential outputs. The overall size of the sensor array is $100 \times 12\,\text{mm}^2$ with a centre-to-centre distance (pitch) between two sensors of 6.6 mm. The sensor array is kept at the middle of the magnetizing coils. The GMR sensors measure the axial component (along the scan direction) of leakage flux from defects. The array sensor maintains a constant lift-off of 1 mm. The sensors' outputs are acquired and analysed using a LabVIEW-based data acquisition system incorporating averaging and low-pass filter to minimize noise.

5. Performance Evaluation

The performance of the sensor array has been evaluated by measuring the axial component of leakage fields from two LF and two LMA type defects in track rope. The two LFs are simulated by electro discharge machining (EDM) notches of size $5.5 \times 2.0 \times 2.0\,\text{mm}^3$ (length × width × depth) oriented

(a)

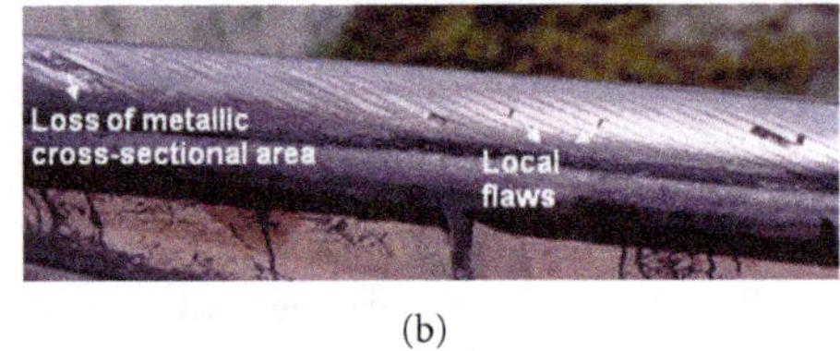

(b)

FIGURE 2: (a) photograph of the track rope system with bucket carrying coal and (b) local flaws and loss of metallic area at the outer surface of the track rope.

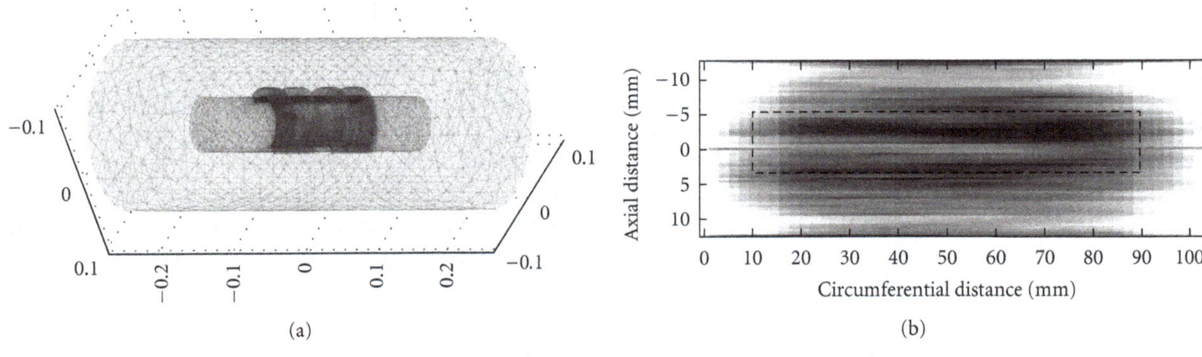

(a) (b)

FIGURE 3: (a) 3D finite element mesh and (b) predicted magnetic flux density between the saddle coils along half of circumferential distance.

along axial and circumferential directions in the track rope, as shown in Figures 5(a) and 5(b). The two LMAs of sizes $42.0 \times 9.0 \times 3.0\,\text{mm}^3$and $33.5 \times 14.2 \times 4.9\,\text{mm}^3$ are made along axial and circumferential directions, respectively (refer to Figures 5(c) and 5(d)).

The test set-up used for evaluation of performance of the GMR sensor array is shown in Figure 6. It consists of two saddle coils, variable DC power supply, track rope, flexible sensor array, GMR field meter, and a personal computer. Each saddle coil consists of 90 turns with a cross-sectional area of $20 \times 10\,\text{mm}^2$. The centre-to-centre distance between the two saddle coils is 80 mm. Measurements are made by moving the sensor array and the magnetization coils together as a single unit over the track rope. For this, a DC servo motor is used, as shown in Figure 6. In order to enhance the sensitivity, each GMR sensor output is amplified using low-noise differential amplifiers and notch rejection filter at 50 Hz, followed by 100 kHz low-pass filter. The sensor outputs are digitized using a 16-channel data acquisition system of 16-bit resolution.

The MFL signals of the sensor array for a circumferential LF of $5.5 \times 2.0 \times 2.0\,\text{mm}^3$ in the track rope are shown in Figure 7. As the length of the flaw is 5.5 mm, only two GMR sensors, namely, S6 and S7 have shown the output of the leakage flux.

The GMR array sensor output has been processed for removing background noise and formatted to obtain images. Typical MFL images of axial and circumferential LFs of size $5.5 \times 2.0 \times 2.0\,\text{mm}^3$ are shown in Figures 8(a) and 8(b), respectively. As compared to the MFL signals, it is possible to readily discern the spatial extent of the flaws from the MFL images produced by the sensor array. The MFL image of the axial LF is found to be extended as compared to that of the circumferential LF.

Typical MFL images of axial and circumferential LMAs are shown in Figures 9(a) and 9(b), respectively. As can be

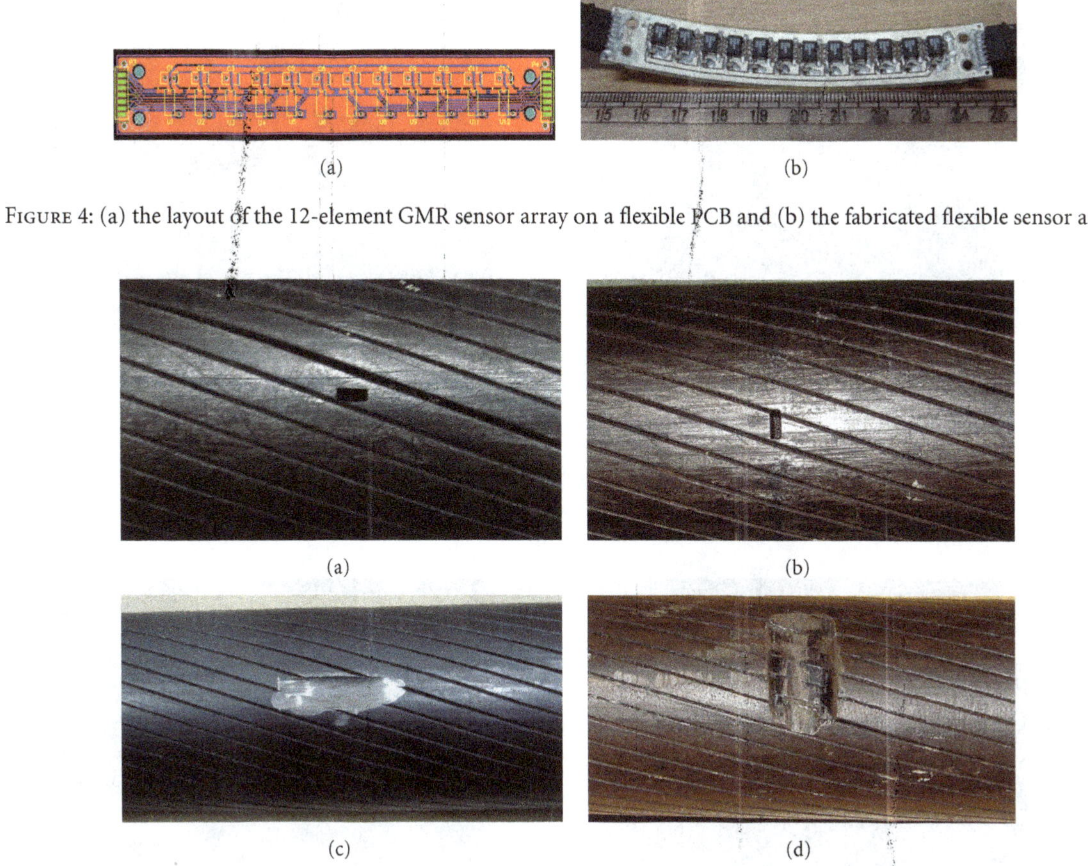

FIGURE 4: (a) the layout of the 12-element GMR sensor array on a flexible PCB and (b) the fabricated flexible sensor array.

FIGURE 5: photographs of (a) axial LF, (b) circumferential LF, (c) axial LMA, and (d) circumferential LMA type defects in the track rope.

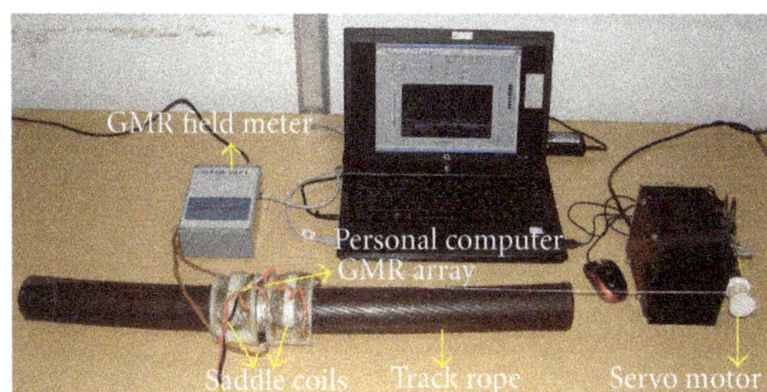

FIGURE 6: Test set-up used for performance evaluation of flexible GMR sensor array.

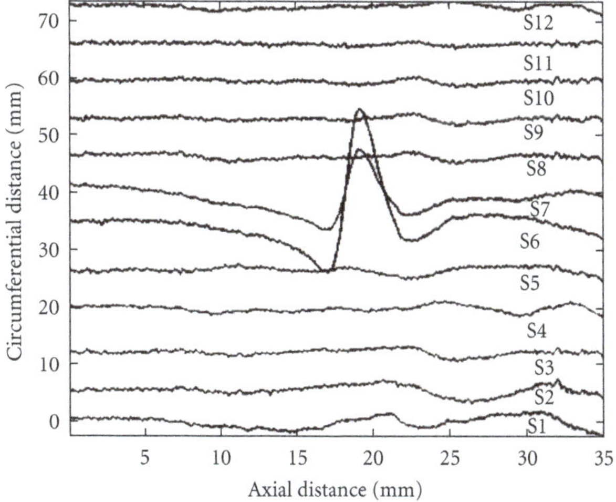

FIGURE 7: GMR sensor array response for a 5.5 mm long circumferential LF.

noted, the spatial extents of the LMAs could be readily felt from the images, despite some random noise. In the case of circumferential LMA, the output of three sensors, namely, S5, S6, and S7 that are exactly over the LMA defect have been found saturated due to high leakage field.

The flexible GMR sensor array designed has shown detection capability for both LF and LMA type defects oriented along the axial as well as circumferential directions. The sensor array has a fast detection speed along the length of the track rope and does not require circumferential scanning like in [13]. The images of circumferential notches have been found to be sharp and localized as compared to that of the axial notches. Thus, the flexible GMR array sensor proposed in this paper can be used for rapid nondestructive inspection of track ropes. Towards deployment of the sensor array for field use, studies are in progress to assess the probability of detection (POD) of the MFL technique. Studies are also

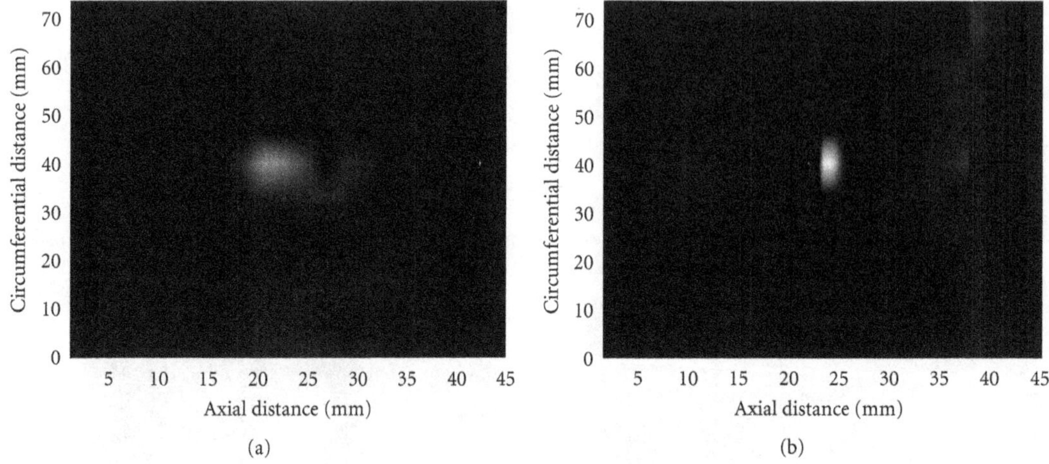

FIGURE 8: MFL images for (a) axial and (b) circumferential LFs.

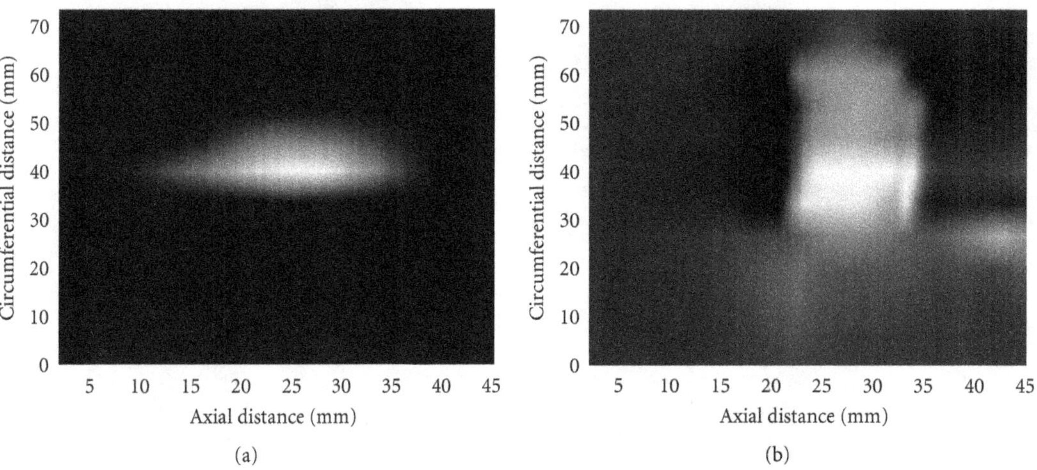

FIGURE 9: MFL images of (a) axial LMA ($42.0 \times 9.0 \times 3.0\ mm^3$) and (b) circumferential LMA ($33.5 \times 14.2 \times 4.9\ mm^3$).

in progress to enhance the detection capability for defects located in between sensor elements as well as the resolution of the sensor array, through the use of another tandem sensor array beside, with a slight angular shift.

6. Summary

A flexible GMR array sensor has been developed for fast magnetic flux leakage testing of 64 mm diameter steel track rope through detection of leakage magnetic fields from LF and LMA type defects on Z-wire rope. Three dimensional finite element modeling has been performed to identify the number of GMR sensor elements and their locations. The performance of the sensor array has been evaluated using machined LF and LMA type defects in the outer Z-wire. The flexible array sensor has shown detection of both types of defects in the track rope, with the possibility for imaging to obtain the spatial information of the defects.

Acknowledgments

The authors thank Mr. P. Krishnaiah, NDE Division, IGCAR, Kalpakkam for winding of the saddle coils and for his help during the experimental studies. They also thank Dr. C. K. Mukhopadhyay and Mr. S. Mahadevan for many useful discussions.

References

[1] D. Basak, "Performance comparison of drive ropes in a cable belt conveyor system using an NDT technique," *Insight*, vol. 51, no. 8, pp. 439–441, 2009.

[2] H. R. Weischedel and R. P. Ramsey, "Electromagnetic testing, a reliable method for the inspection of wire ropes in service," *NDT International*, vol. 22, no. 3, pp. 155–161, 1989.

[3] H. R. Weischedel, "The inspection of wire ropes in service: a critical review," *Materials Evaluation*, vol. 43, no. 13, pp. 1592–1605, 1985.

[4] N. Sumyong, A. Prateepasen, and P. Kaewtrakulpong, "Influence of scanning velocity and gap distance on magnetic flux leakage measurement," *ECTI Transactions on Electrical Engineering, Electronics and Communications*, vol. 5, no. 1, pp. 118–122, 2007.

[5] C. Jomdecha and A. Prateepasen, "Design of modified electromagnetic main-flux for steel wire rope inspection," *NDT and E International*, vol. 42, no. 1, pp. 77–83, 2009.

[6] E. Kalwa and K. Piekarski, "Design of inductive sensors for magnetic testing of steel ropes," *NDT International*, vol. 20, no. 6, pp. 347–353, 1987.

[7] E. Kalwa and K. Piekarski, "Design of Hall-effect sensors for magnetic testing of steel ropes," *NDT International*, vol. 20, no. 5, pp. 295–301, 1987.

[8] W. S. Singh, B. P. C. Rao, S. Vaidyanathan, T. Jayakumar, and B. Raj, "Detection of leakage magnetic flux from near-side and far-side defects in carbon steel plates using a giant magneto-resistive sensor," *Measurement Science and Technology*, vol. 19, no. 1, Article ID 015702, 2008.

[9] J. W. Wilson and G. Y. Tian, "3D magnetic field sensing for magnetic flux leakage defect characterisation," *Insight: Non-Destructive Testing and Condition Monitoring*, vol. 48, no. 6, pp. 357–359, 2006.

[10] W. S. Singh, B. P. C. Rao, T. Jayakumar, and B. Raj, "Simultaneous measurement of tangential and normal component of leakage magnetic flux using GMR sensors," *Journal of Non-Destructive Testing & Evaluation*, vol. 8, no. 2, pp. 23–28, 2009.

[11] A. Sophian, G. Y. Tian, and S. Zairi, "Pulsed magnetic flux leakage techniques for crack detection and characterisation," *Sensors and Actuators, A*, vol. 125, no. 2, pp. 186–191, 2006.

[12] J. W. Wilson and G. Y. Tian, "Pulsed electromagnetic methods for defect detection and characterisation," *NDT and E International*, vol. 40, no. 4, pp. 275–283, 2007.

[13] W. S. Singh, B. P. C. Rao, C. K. Mukhopadhyay, and T. Jayakumar, "GMR-based magnetic flux leakage technique for condition monitoring of steel track rope," *Insight*, vol. 53, no. 7, pp. 377–381, 2011.

[14] W. S. Singh, B. P. C. Rao, S. Mahadevan, and T. Jayakumar and B. Raj, "Giant magneto-resistive sensor based magnetic flux leakage technique for inspection of track ropes," *Studies in Applied Electromagnetics and Mechanics, Electromagnetic Nondestructive Evaluation (XIV)*, vol. 35, pp. 256–263, 2011.

[15] R. Grimberg, L. Udpa, A. Savin, R. Steigmann, V. Palihovici, and S. S. Udpa, "2D Eddy current sensor array," *NDT and E International*, vol. 39, no. 4, pp. 264–271, 2006.

[16] M. N. Baibich, J. M. Broto, A. Fert et al., "Giant magnetoresistance of (001)Fe/(001)Cr magnetic superlattices," *Physical Review Letters*, vol. 61, no. 21, pp. 2472–2475, 1988.

[17] B. P. C. Rao, T. Jayakumar, and B. Raj, "Electromagnetic NDE techniques for materials characterization," in *Ultrasonic and Advanced Methods for Non-Destructive Testing and Material Characterisation*, C. H. Chen, Ed., chapter 11, pp. 262–265, World Scientific Publishing, Singapore, 2007.

4

Decision Making in Reinforcement Learning Using a Modified Learning Space Based on the Importance of Sensors

Yasutaka Kishima, Kentarou Kurashige, and Toshihisa Kimura

Muroran Institute of Technology, 27-1 Mizumoto, Hokkaido, Muroran 0508585, Japan

Correspondence should be addressed to Kentarou Kurashige; kentarou@csse.muroran-it.ac.jp

Academic Editor: Guangming Song

Many studies have been conducted on the application of reinforcement learning (RL) to robots. A robot which is made for general purpose has redundant sensors or actuators because it is difficult to assume an environment that the robot will face and a task that the robot must execute. In this case, Q-space on RL contains redundancy so that the robot must take much time to learn a given task. In this study, we focus on the importance of sensors with regard to a robot's performance of a particular task. The sensors that are applicable to a task differ according to the task. By using the importance of the sensors, we try to adjust the state number of the sensors and to reduce the size of Q-space. In this paper, we define the measure of importance of a sensor for a task with the correlation between the value of each sensor and reward. A robot calculates the importance of the sensors and makes the size of Q-space smaller. We propose the method which reduces learning space and construct the learning system by putting it in RL. In this paper, we confirm the effectiveness of our proposed system with an experimental robot.

1. Introduction

In recent years, reinforcement learning (RL) [1] has been actively studied, and many studies on its application to robots have been conducted [2–4]. A matter of concern in RL is the learning time. In RL, information from sensors is projected onto a state space. A robot learns the correspondence between each state action in the state space and determines the best correspondence. When the state space expands according to the number of sensors, the number of correspondences learned by the robot is also increased. In addition, the robot needs considerable much experience in each state to perform a task. Therefore, learning the best correspondence becomes time-consuming.

To overcome this problem, many studies have investigated accelerated RL [5–15] for which there are two approaches: a multirobot system and autonomous construction of the state space. In the former approach, multiple robots exchange experience information [5–9], so that each robot augments its own knowledge. Therefore, in this system robots can find the best correspondence between each state and action faster than an individual robot in a single-robot system. In addition Nishi et al. [10] proposed a learning method in which a robot learns behavior through observations of the behavior of other robots, constructing its own relationships between state and behavior. However, in this approach, a robot needs other robots with whom to exchange experience information, and hence, if there are no additional robots in the system, this approach becomes irrelevant. We focus on the state construction of a single robot.

In contrast to the above approaches, in the approach that applies autonomous state space construction [11–16], a single robot is sufficient. The robot constructs a suitable state space based on its experience. Moreover, it can reduce the state space and learn correspondences faster. However, in the studies on this approach, all the sensors installed in the robot were considered to be equally important, and their number of states was the same. The installed sensors, which influence how well a robot executes a task, can be divided into important and unimportant sensors according to the task to be performed. However, in this approach, the robot has to learn using unnecessary inputs because all the sensors are considered equally important. For example, Takahashi et al. [16] proposed a state that is constructed autonomously from state space by incremental state segmentation. In this method, the division rule of state is applied to the consecutiveness of the sensor data. Ishiguro et al. [12] proposed a state construction method using empirically obtained perceivers (EOPs). These methods are not focused on the importance of each sensor for performing a task. In fact, although the sensors installed on a robot have varying levels of importance in terms of performing a task, few studies focused on this aspect. We focus on the importance of a sensor for a particular task and propose a novel efficient learning method.

In this paper, we propose a system in which the robot constructs a temporary Q-space for decision making based on which sensors are considered important for the execution of a particular task, which facilitates high-speed learning. Since very important sensors affect the performance of a task significantly, they should sense the environment circumstantially. Thus, the number of their states is increased. On the other hand, since less important sensors do not affect the performance of a task, they may sense their environment less exactly. Thus, the number of their states, which is determined based on the importance of the sensors, is decreased. In this study, the importance of the sensors is defined as the correlation between the sensor value and the reward.

A temporary Q-space is constructed from the Q-space of the robot based on the importance of the sensors. The number of states of the Q-space is the maximum number that the sensors can describe. A Q-space is reduced by merging Q-values according to the number of the states of the sensors. Using the reduced Q-space, the robot can efficiently select an action, which is based on more information because the Q-values of low importance are merged. Therefore, the robot can learn correspondences using fewer experiences.

This system is effective for a variety of tasks. When this method is implemented, the amount of information that the robot requires in order to learn correspondences is reduced. As a result, when our proposed system is applied, a robot can learn correspondences faster than when an ordinary RL is applied.

2. Concept of Importance of Sensors

To select the sensors that are important to a certain task, a robot needs to measure the importance of each sensor. We focus on the correlation of the sensor value and the reward, which is specific to each task, as the measure of the importance of each sensor for a task. For example, in a garbage collection task, a robot is expected to approach a garbage heap and lift it. For this task, the reward is expressed by the distance between the robot and garbage heap and increases as the robot moves closer to the garbage heap. This implies that there is a correlation between the reward and the distance between the robot and garbage heap.

We show in Figure 1 an outline of the determination of the importance of the sensors where two types of sensor are installed on the robot. Via its sensors, the robot recognizes its environment, which is expressed as a group of all the sensor values. The robot collects the sensor value of each sensor and the reward of the task to be performed and then determines the correlation between them. In Figure 1, the robot conducts this determination for sensors 1 and 2. The robot estimates the importance of the sensors according to the two types of correlation between the sensor value and reward: negative and positive.

Very important sensors affect the performance of a task. Therefore they should sense their environment circumstantially, and thus the number of their states is increased. On the other hand, less important sensors do not affect the performance of a task, and therefore they may sense the environment less exactly, and thus, the number of their states is decreased. The number of states is therefore determined based on the importance of the sensors.

3. Decision Making in Reinforcement Learning Using a Q-Space Based on the Importance of Sensors

3.1. Outline of the Proposed System. We show our proposed system in Figure 2. In the figure, the proposed system is divided into two stages. The first stage constitutes the proposed method whereby the robot determines the importance of its sensors. The next stage is RL.

In the first stage, the robot calculates the importance of the sensors for a task based on the correlations between each sensor and reward. The robot first estimates each sensor value and the reward and then calculates the coefficient of the correlation between them. Finally, it determines the important sensors based on this coefficient of correlation.

In RL, a robot learns the actions that are suitable for each state. This stage consists of an action evaluation element, wherein a temporary Q-space is constructed based on the determination of important sensors, and an action selection element. In the action evaluation element, each pair of state and action is evaluated and updated. A state contains the value of all the sensors. The robot constructs a temporary Q-space, by adding to it only those sensors that have been determined to be important. In the action selection element, the robot selects the action for a state recognized by the sensor based on the temporary Q-space.

We show the workflow of our proposed system in Figure 3. This workflow is executed by the robot for each action. We define this flow as one trial.

3.2. Determination of the Importance of Sensors. In the proposed method, the robot determines the importance of its sensors based on the correlation between each sensor value and reward, which it experiences. The robot stores each sensor value and the averaged reward as a list called a knowledge list, an example of which is shown in Figure 4. When a robot identifies a state that is not in the list, it adds it to the list. Then, the robot calculates the averaged reward and adds it to the list. On the other hand, when a recognized state is already in the list, it calculates the averaged reward and updates the list.

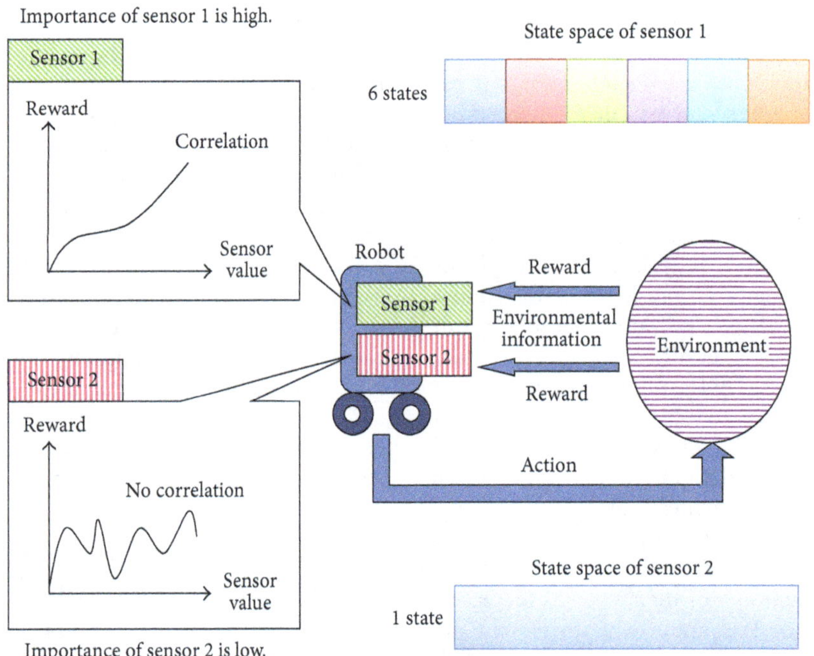

FIGURE 1: Outline of determination of importance.

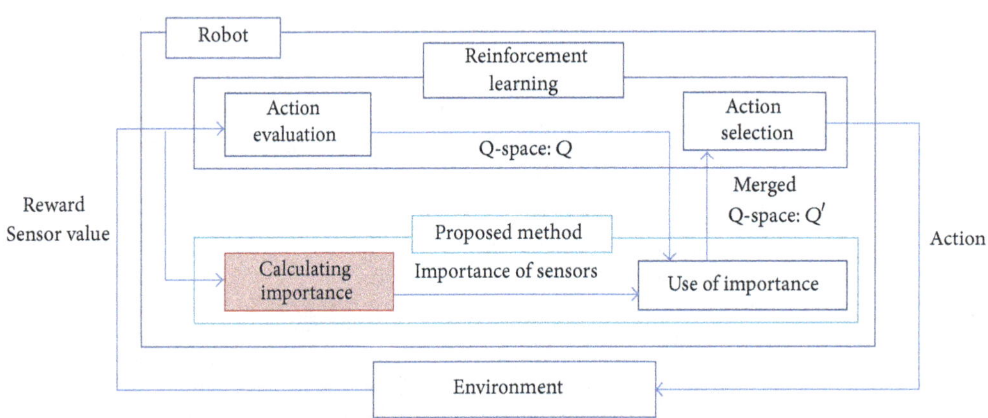

FIGURE 2: Outline of the proposed system.

A state s_i is defined as (1), where i is the state ID in Figure 4, $e_{i,j}$ is the value of sensor j, and E_j is a group of values that describe a sensor:

$$s_i = \{e_{i,1}, e_{i,2}, \ldots, e_{i,j}, e_{i,n} \mid e_{i,1} \in E_1, e_{i,2} \in E_2, \ldots, e_{i,j} \in E_j, e_{i,n} \in E_n\}. \quad (1)$$

In this study, the rewards for the state experienced by the robot are weight averaged. Weighted averaging gives a greater weight to more recently obtained rewards. The averaged reward in state s_i is denoted by $r'_t(s_i)$ and it is updated as (2), where $r_t(s_i)$ is the reward obtained by the robot at time t:

$$r'_t(s_i) \longleftarrow r_t(s_i) + \alpha_{\text{ave}} \left(r_t(s_i) - r'_{t-1}(s_i) \right). \quad (2)$$

The robot calculates the importance of each sensor based on the knowledge list. We use an equation for multiple coefficient correlation to calculate the sensor's level of importance. In multiple coefficients, each partial regression coefficient represents the importance of each sensor. A sensor with a higher regression coefficient has a higher importance level.

The multiple regression equation is defined by (3), where $a_{t,1}, a_{t,2}, \ldots, a_{t,j}$ are the regression coefficients for each sensor, $r'_t(t, s)$ is the averaged reward in state s, and a_0 is the constant term:

$$r'_t(s_i) = a_1 e_{i,1} + a_2 e_{i,2} + \cdots + a_i e_{i,n} + a_0. \quad (3)$$

A robot needs to calculate the regression coefficients of each sensor to calculate its importance. Each regression coefficient $a_1, a_2, \ldots, a_n$ is the solution resulting from the multiple simultaneous equation (4). Here, b_{ij}, calculated by (5), is the covariance of sensor i and sensor j, and b_i^2, calculated by (6), is the variance of sensor i. The average of sensor i values is calculated by (7), and the $\overline{e_i}$ and $\overline{e_j}$ in (5) are

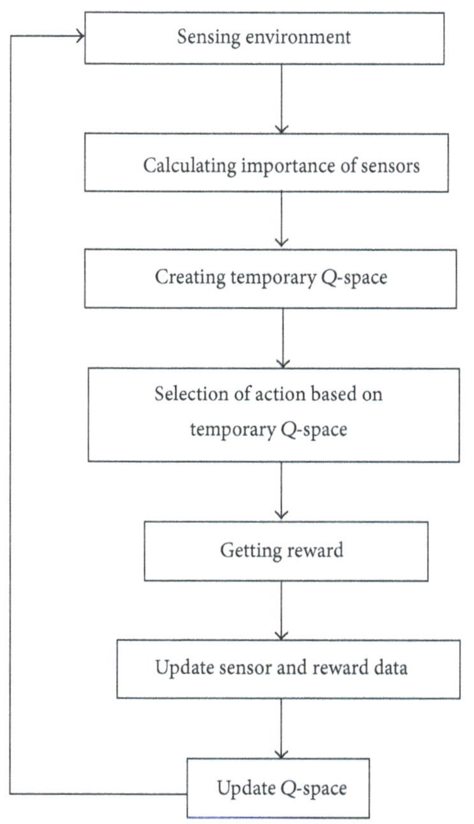

FIGURE 3: Flow of the proposed system.

State ID	Sensor value of sensor 1	Sensor value of sensor 2	⋯	Sensor value of sensor n	Averaged reward $r'(s_i)$
1	10	56	⋯	111	45
2	23	45	⋯	123	23
⋮					
m	90	10	⋯	15	132

FIGURE 4: List for calculating correlation.

calculated by (7). b_{ir}, calculated by (8), is the covariance of sensor i and reward r, and $\overline{r'}$, calculated by (9), is the average value of the averaged reward:

$$\begin{aligned} a_1 b_1^2 + a_2 b_{12} + \cdots + a_n b_{1n} &= b_{1r}, \\ a_1 b_{12} + a_2 b_2^2 + \cdots + a_n b_{2n} &= b_{2r}, \\ &\vdots \\ a_1 b_{1n} + a_2 b_{2n} + \cdots + a_n b_n^2 &= b_{nr}, \end{aligned} \tag{4}$$

$$b_{ij} = \frac{1}{m} \sum^{m} k = 1 \left(e_{i,k} - \overline{e_i}\right)\left(e_{j,k} - \overline{e_j}\right), \tag{5}$$

$$b_i^2 = \frac{1}{m} \sum^{m} k = 1 (e_i, k - \overline{e_i})^2, \tag{6}$$

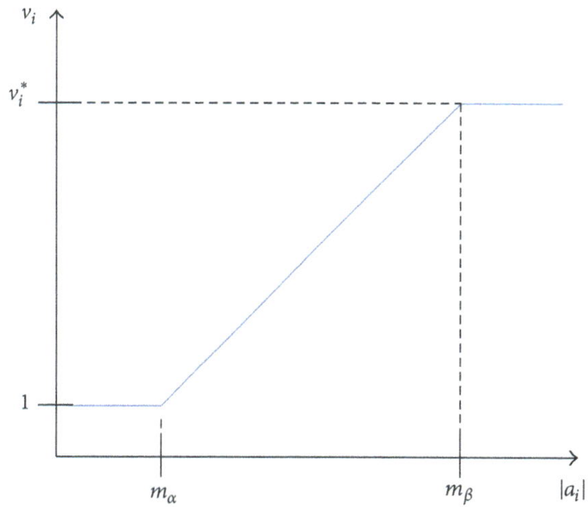

FIGURE 5: Variation property.

$$\overline{e_j} = \frac{1}{m} \sum_{k=1}^{m} e_{i,k}, \tag{7}$$

$$b_{ir} = \frac{1}{m} \sum^{m} k = 1 \left(e_{i,k} - \overline{e_i}\right)\left(r_k - \overline{r'}\right), \tag{8}$$

$$\overline{r'} = \frac{1}{m} \sum_{k=1}^{m} r'(s_k). \tag{9}$$

3.3. Determination of the Number of States of Each Sensor. The number of states is determined based on the number of states of each sensor, which is determined based on the regression coefficient of each sensor. When the absolute value of the regression coefficient of a sensor is higher, the number of states of the sensor is increased.

We use a variation property, shown in Figure 5, as the number of states based on the regression coefficient. When the regression coefficient of a sensor is less than m_β, the minimum number of states is 1; when it is greater than m_α, the maximum number of states is v_i^*; and when it is from m_β to m_α, the number of state increases gradually. The parameters m_β and m_α are determined by a human. The formulation of the property is determined by (10). v_i is the number of states of sensor i:

$$v_i = \begin{cases} 1 & (|a_i| < m_\alpha), \\ \left[\dfrac{v_i^* - 1}{m_\beta - m_\alpha} |a_i| + \dfrac{m_\beta - m_\alpha v_i^*}{m_\beta - m_\alpha} \right] & (m_\alpha \le |a_i| \le m_\beta), \\ v_i^* & (m_\beta < |a_i|). \end{cases} \tag{10}$$

Here, v_i^* is determined based on the performances of the sensor. We focus on the resolution and maximum range as the performance of the sensor. When the performance of the resolution and maximum range of a sensor are high, the robot can describe more states. v_i^* is the state number, which can be calculated based on the resolution and maximum range of the sensor, as in (11), where $g_{\max,i}$ is the maximum range of sensor

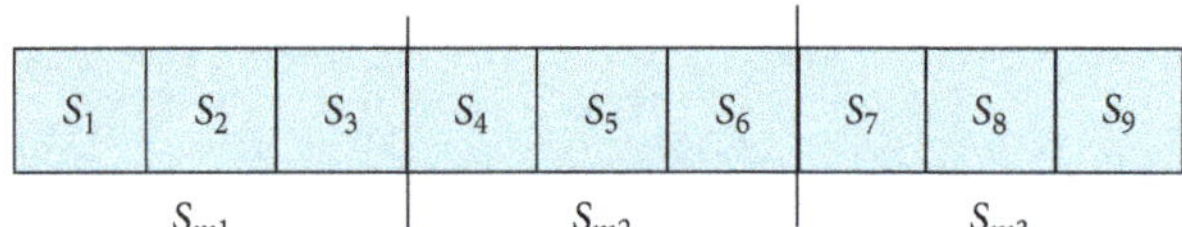

FIGURE 6: Example of merged state of a sensor.

i, $g_{\min,i}$ is the minimum range of sensor i, b_i is resolution of sensor i, and v_i^* is calculated for each sensor. We define a state of a sensor on v_i^* as a "state unit":

$$v_i^* = \frac{g_{\max,i} - g_{\min,i}}{b_i}. \tag{11}$$

3.4. Construction of a Temporary Q-Space for Action Selection. The robot constructs a temporary Q-space based on v_i. The temporary Q-space consists of a unit, which is several state units merged together. We show an example of a merged state space in Figure 6. In this example, v_i^* is 9 and the number of state units is 9. When v_i is 3, the temporary Q-space is constructed of three states obtained by merging three state units. This example is focused on one sensor. All the installed sensors are merged according to v_i.

Here, when the number of states is v_i, the number of state units in a merged state v_i' is calculated by the following equation:

$$v_i' = \left\lceil \frac{v_i^*}{v_i} \right\rceil. \tag{12}$$

The Q-value of each state unit in merged targets is averaged, as shown in Figure 7, where an example of the temporary Q-space when v_1^* and v_2^* are 6, v_1' is 3, and v_2' is 3 is depicted. For sensor S_1, each of the states is three merged state units. For sensor S_2, each of the states is two merged state units. Merged state units are averaged Q-values.

The robot selects an action based on the temporary Q-space for the current state. It recognizes the current state as state unit s. When the current state is s, the merged target group of the state unit is $C_i = \{v_{i,o}, v_{i,p}, \ldots, v_{i,q}\}$, where $v_{i,o}, v_{i,p}, \ldots, v_{i,q}$ in the group are the state units of a sensor. Similarly, the target group of sensor t is $C_t = \{v_{t,o}, v_{t,p}, \ldots, v_{t,q}\}$, and the target group of sensor u is $C_u = \{v_{u,o}, v_{u,p}, \ldots, v_{u,q}\}$. Here, the Q-value of the merged target in Q-space $Q(s_m, i)$ is defined by (13), where $R(s_m, a)$ is the total reward of merged state units, defined as (14), where $R(s_w, a)$ is the total reward at state $s_w = \{v_{i,w}, v_{t,w}, \ldots, v_{u,w}\}$ and action a and $E(s_m, a)$ is the total number of the experiences of the state action pair (s_m, a), defined as (15). Merging is performed for all the actions in state s_m:

$$Q(s_m, a) = \frac{R(s_m, a)}{E(s_m, a)}, \tag{13}$$

$$R(s_m, a) = \sum_{v_{i,w} \in C_i} \sum_{v_{t,w} \in C_t} \cdots \sum_{v_{u,w} \in C_i} (Q(s_w, a) \cdot E(s_w, a)), \tag{14}$$

$$E(s_m, a) = \sum_{v_{i,w} \in C_i} \sum_{v_{t,w} \in C_t} \cdots \sum_{v_{u,w} \in C_i} E(s_w, a). \tag{15}$$

3.5. Action Selection. The robot selects an action based on the temporary Q-space. We apply the ϵ-greedy method for action selection. This method selects the action that has the highest Q-value in the current state unit s. However, the method selects an action randomly with probability ϵ.

3.6. Action Evaluation. We apply the weighted averaging method as the action evaluation method. This method evaluates actions by assigning a weight to each reward recently obtained by the robot. When the current state unit of the robot is s and the selected action is a, Q-value $(Q(s,a))$ is updated by (16). α is a step size parameter $(0 \le \alpha \le 1)$:

$$Q(s,a) \longleftarrow Q(s,a) + \alpha [r - Q(s,a)]. \tag{16}$$

4. Experiment to Confirm the Effectiveness of the Proposed System

4.1. Outline of the Experiment. In this section, we describe our evaluation of the effectiveness of the proposed system via an experimental robot. The experimental environment is shown in Figure 8. This environment is surrounded by walls of length 1100 mm. We prepared an experimental robot, as shown in Figure 9. The robot has two distance sensors, which measure the distance between the current position of the robot and the walls. It can recognize the current state of a sensor as the state value u, as shown in Figure 10. Its sensors are divided into 11 states every 70 mm, and each state is given a state value. In this experiment, each sensor had 11 states and the total number of states was 121. The robot has omniwheels and can move in the forward, back, left, right, and each diagonal direction but cannot turn. It can move 70 mm in one action when it is not moving in a diagonal direction. When the robot moves in a diagonal direction, it moves in the cross direction and then in the lengthwise direction.

In the experiment, the task of the robot was to move close to wall A. The robot could obtain rewards according to its distance only from wall A. It was placed at the lower right corner of the environment and when it reached wall A, an episode was considered. The experiment was concluded at n_e episodes.

This task seems simple at first glance. However, it is difficult for an RL robot. In this task, when a robot strays into an unexperienced area, its action selection becomes random, because the Q-values of the states in the area are at their initial value. Therefore, the robot strays more by repeating a random action selection and takes time to estimate the Q-values of the states and to leave the area. In addition, in this experiment, the task was being performed by an experimental robot. Therefore, it is possible that the sensor noise affected its learning of, for example, gap state recognition.

In real world use scenes, a robot is rarely required to move in a maze-like environment with many obstacles. Most use scenes are open spaces, such as a warehouse or park. Therefore, the environment we adopted is appropriate for this experiment.

We confirmed the effectiveness of the proposed system by comparing it with a conventional RL. Therefore, we prepared two types of robot for comparison, to one of which the

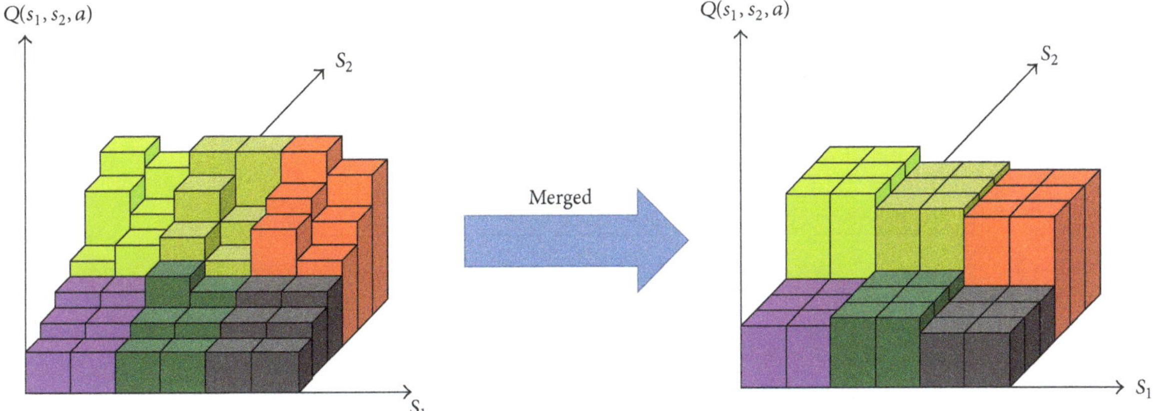

FIGURE 7: Example of projection for a temporary Q-space.

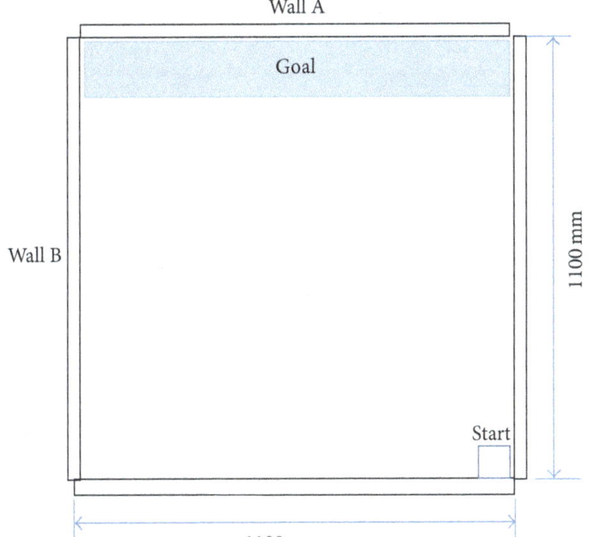

FIGURE 8: Experimental environment.

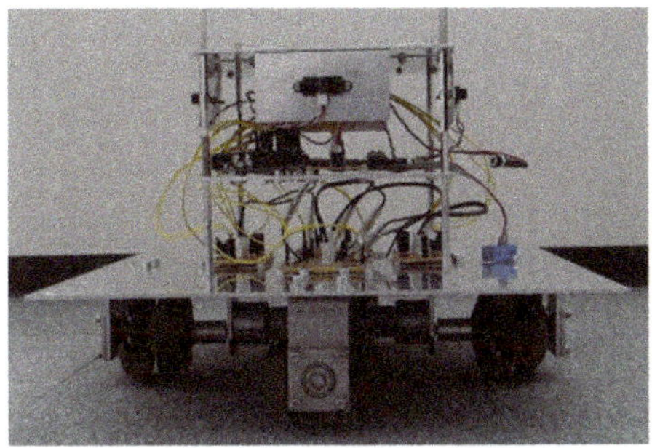

FIGURE 9: Experimental robot.

proposed system was applied, and to the other RL. The number of states of each sensor of the robot to which RL was applied was v^*. We compared these agents in terms of the total reward obtained after n_e episodes were completed.

TABLE 1: Experimental settings.

n_e	100
α_{ave}	0.1
ϵ	0.1
α	0.1
Initial value of Q-value	0
Maximum range of sensors $g_{\max}$	1100 mm
Minimum range of sensors $g_{\min}$	0
Resolution of sensors b	100 mm
m_α	0.2
m_β	0.8
The number of initial states	11
Initial importance	1.0

4.2. Experimental Setup. In this section, we explain the reward for each task and discuss the parameter settings. In task A, the robot can obtain a higher reward by increasing the distance between its current position and wall A, as defined in (17), where d_A is the state value shown in Figure 10. d_A is determined based on the actual measurement value of the distance between the current position and wall A:

$$r = 11 - d_A. \tag{17}$$

A list of the parameter settings of this experiment is given in Table 1. In this experiment, the settings of the maximum range, minimum range, and resolution of sensors remained the same. v^* was 11 according to (11). When the robot started a new episode, the Q-space and state knowledge list from previous episodes were adopted.

4.3. Experimental Results. The experimental results are shown in Figures 11–15. Figure 11 shows the importance of the sensors for the final action in each episode. The importance of the sensors is represented by the regression coefficients. In first episode, the regression coefficients of each sensor are converged in the early phase of learning. The regression of the coefficient of sensor A is greater than the threshold $m_\beta = 0.8$. On the other hand, the regression of the coefficient of sensor

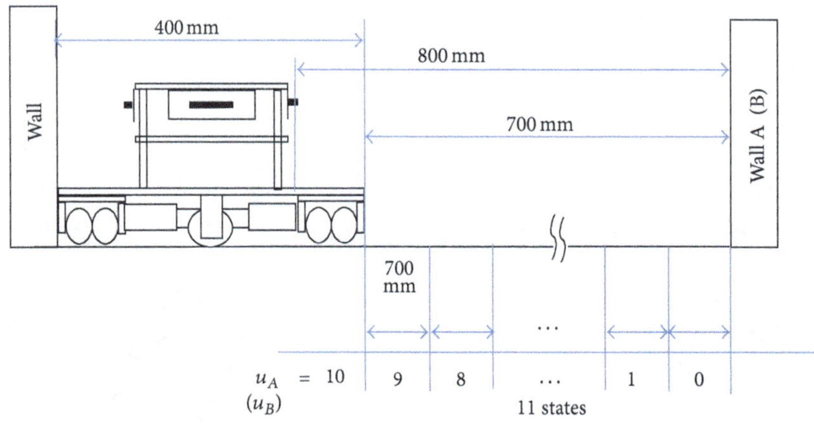

FIGURE 10: State recognition of the robot.

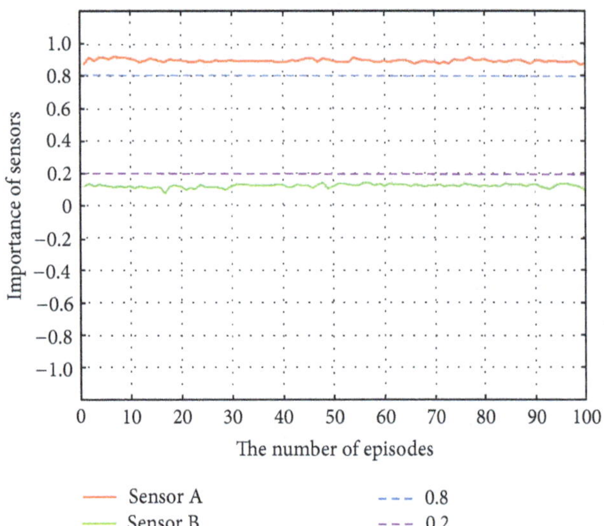

FIGURE 11: Importance of sensors in each episode.

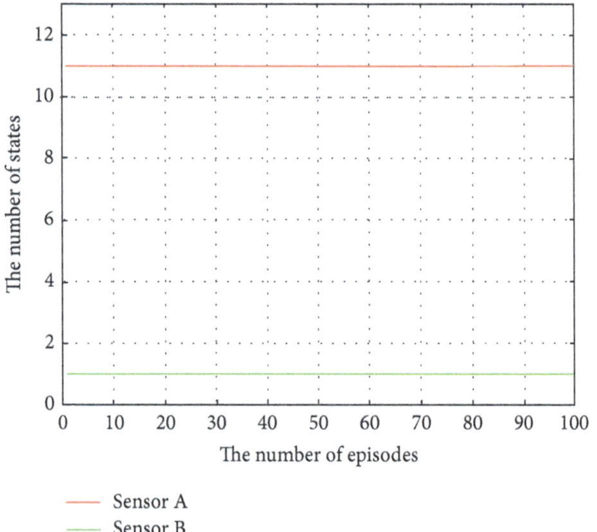

FIGURE 12: The number of states in each episode.

B is smaller than the threshold $m_\alpha = 0.2$. In this task, only wall A is related to reward and its importance is high. It is valid that the regression of the coefficient of the wall A sensor is high and that of wall B is low.

Figure 12 shows the number of the states of the sensors in the final action in each episode. The number of states of sensor A is the maximum, 11. The number of states of sensor B is the minimum, 1. Using these results, the robot can construct a correct temporary Q-space.

Figures 13 and 14 show the importance of the sensors and the number of states in each action in the first episode, respectively. Until the 30th action, the importance of the sensors is unstable. The reason is that the robot has insufficient knowledge to calculate the regression of the coefficient correctly. After the 31st action, the robot has sufficient knowledge and can therefore calculate the regression of the coefficient correctly. Until the 30th action, because of the instability in the number of the states of the sensors, the

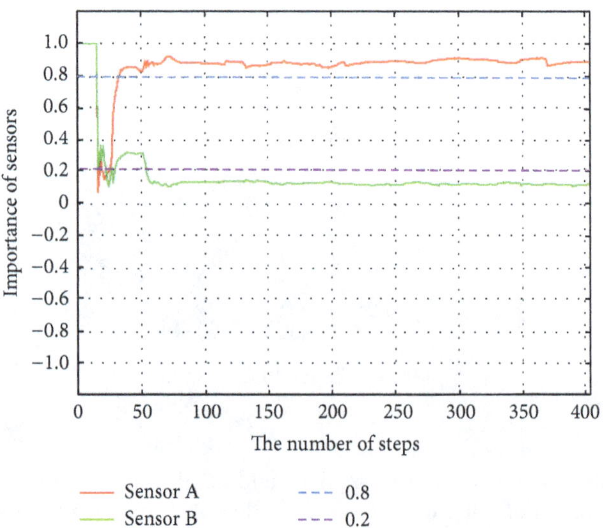

FIGURE 13: Importance of sensors in each action in episode 1.

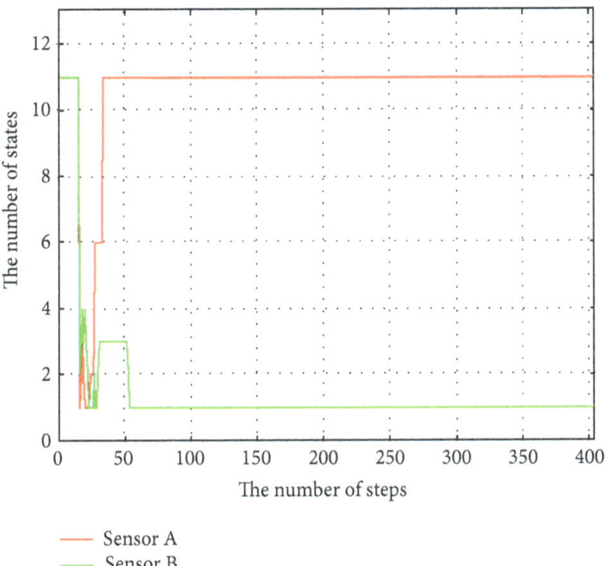

Figure 14: The number of states in each action in episode 1.

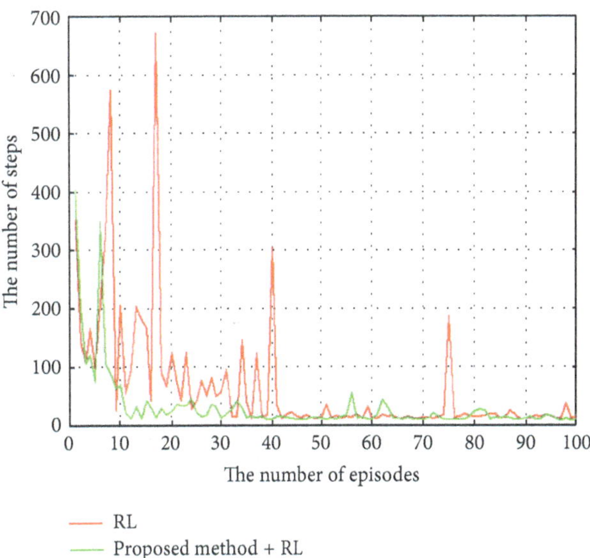

Figure 15: The number of total actions at each episode.

robot's ability to calculate the importance of the sensors is negatively affected by its insufficient knowledge.

Figure 15 shows the total number of actions in each episode. The robot using the proposed method achieves a high convergence of the number of actions as compared to that using RL. This is because the robot using RL strays into an area for which it has no experience and takes time to leave it. On the other hand, this does not occur when the proposed method is used. This is because, when the robot strays into an area for which it has no experience, it can use other Q-values learned in other states based on the constructed temporary Q-space, which consists of only important sensors. Therefore, the robot does not need to take time for learning in the unexperienced area. Thus, the robot focuses on only the important sensors and selects the suitable action by using this Q-space. These results show that the proposed method is effective for learning.

5. Conclusion

In this paper, we proposed a method in which a robot selects an action by using a temporary Q-space based on the importance of its sensors. This method assumes that there is a correlation between the sensor value and reward. The robot calculates the regression coefficient using a multiple regression equation of the sensor value and reward. The robot determines the importance of its sensors according to the regression coefficient. The higher the level of importance, the larger the number of states. To select an action, the robot constructs a temporary Q-space based on the importance of the sensors. It then selects actions based on the temporary Q-space. Thus, the robot is able to learn faster.

We examined the effectiveness of the proposed system using an experimental robot. We investigated a task for the execution of which only one sensor of a sensor pair was important. We compared the proposed system with conventional RL. The robot used sensors whose number of states is common in the case of conventional RL.

The results showed that using the proposed system the robot could calculate the importance of the sensors correctly. In addition, convergence speed was faster than that in conventional RL. Thus, we confirmed the effectiveness of the constructed system and the proposed method.

In future studies, first, we will examine the effectiveness of the proposed system by comparing it with other autonomous state construction systems. In this study, we examined only normal reinforcement learning. It is necessary to examine the proposed system by comparing it with those proposed in related studies. Then, we will modify the proposed method. Currently, the proposed method cannot be applied in a delayed reward task, because the regression coefficient is calculated using the immediate reward in each state. When a task is a delayed reward task, the robot cannot calculate the importance of its sensors. Therefore, we will modify the proposed method so that it can be applied in a delayed reward task by using information that does not include the reward.

References

[1] R. S. Sutton and A. G. Barto, *Reinforcement Learning: An Introduction*, MIT Press, 1998.

[2] T. Kondo and K. Ito, "A reinforcement learning with evolutionary state recruitment strategy for autonomous mobile robots control," *Robotics and Autonomous Systems*, vol. 46, no. 2, pp. 111–124, 2004.

[3] K. J. Person, E. Oztop, and J. Peters, "Reinforcement learning to adjust robot movements to new situations," in *Proceedings of the 22 nd International Joint Conference on Artificial Intelligence*, pp. 2650–2655, 2010.

[4] N. Navarro, C. Weber, and S. Wermter, "Real-world reinforcement learning for autonomous humanoid robot charging in a home environment," in *Towards Autonomous Robotic Systems*,

vol. 6856 of *Lecture Notes in Computer Science*, pp. 231–240, Springer, 2011.

[5] M. Tan, "Multi-agent reinforcement learning: independent vs. cooperative agents," in *Proceedings of the 10th International Conference on Machine Learning*, 1993.

[6] M. N. Ahmadabadi, M. Asadpur, S. H. Khodaabakhsh, and E. Nakano, "Expertness measuring in cooperative learning," in *Proceedings of the IEEE/RSJ International Conference on Intelligent Robots and Systems (IROS '00)*, vol. 3, pp. 2261–2267, November 2000.

[7] M. N. Ahmadabadi and M. Asadpour, "Expertness based cooperative Q-learning," *IEEE Transactions on Systems, Man, and Cybernetics, Part B*, vol. 32, no. 1, pp. 66–76, 2002.

[8] H. Iima and Y. Kuroe, "Swarm reinforcement learning algorithm based on exchanging information among agents," *Transactions of the Society of Instrument and Control Engineers*, vol. 42, no. 11, pp. 1244–1251, 2006.

[9] Y. Yongming, T. Yantao, and M. Hao, "Cooperative Q learning based on blackboard architecture," in *Proceedings of International Conference on Computational Intelligence and Security Workshops (CIS '07)*, pp. 224–227, December 2007.

[10] T. Nishi, Y. Takahashi, and M. Asada, "Incremental behavior acquisition based on reliability of observed behavior recognition," in *Proceedings of IEEE/RSJ International Conference on Intelligent Robots and Systems (IROS '07)*, pp. 70–75, November 2007.

[11] M. Asada, S. Noda, and K. Hosoda, "Action-based sensor space categorization for robot learning," in *Proceedings of IEEE/RSJ International Conference on Intelligent Robots and Systems (IROS '96)*, pp. 1502–1509, November 1996.

[12] H. Ishiguro, R. Sato, and T. Ishida, "Robot oriented state space construction," in *Proceedings of IEEE/RSJ International Conference on Intelligent Robots and Systems (IROS '96)*, pp. 1496–1501, November 1996.

[13] K. Samejima and T. Omori, "Adaptive internal state space construction method for reinforcement learning of a real-world agent," *Neural Networks*, vol. 12, no. 7-8, pp. 1143–1155, 1999.

[14] A. J. Smith, "Applications of the self-organising map to reinforcement learning," *Neural Networks*, vol. 15, no. 8-9, pp. 1107–1124, 2002.

[15] K. T. Aung and T. Fuchda, "A proposition of adaptive state space partition in reinforcement learning with voronoi tessellation," in *Proceedings of the 17th International Symposium on Artificial Life and Robotics*, pp. 638–641, 2012.

[16] Y. Takahashi, M. Asada, and K. Hosoda, "Reasonable performance in less learning time by real robot based on incremental state space segmentation," in *Proceedings of IEEE/RSJ International Conference on Intelligent Robots and Systems (IROS 96)*, pp. 1518–1524, November 1996.

5

Energy-Efficient Adaptive Geosource Multicast Routing for Wireless Sensor Networks

Daehee Kim,[1] Sejun Song,[2] and Baek-Young Choi[1]

[1] *Department of Computer Science Electrical Engineering, School of Computing and Engineering, University of Missouri, Kansas City, MO 64110, USA*
[2] *Department of Engineering Technology Industrial Distribution, Dwight Look College of Engineering, Texas A&M University, College Station, TX 77843, USA*

Correspondence should be addressed to Baek-Young Choi; choiby@umkc.edu

Academic Editor: Eugenio Martinelli

We propose an energy-efficient adaptive geosource multicast routing (EAGER) for WSNs. It addresses the energy and scalability issues of previous location based stateless multicast protocols in WSNs. EAGER is a novel stateless multicast protocol that optimizes location-based and source-based multicast approaches in various ways. First, it uses the receiver's geographic location information to save the cost of building a multicast tree. The information can be obtained during the receiver's membership establishment stage without flooding. Second, it reduces packet overhead, and in turn, energy usage by encoding with a small sized node ID instead of potentially large bytes of location information and by dynamically using branch geographic information for common source routing path segments. Third, it decreases computation overhead at each forwarding node by determining the multicast routing paths at a multicast node (or rendezvous point (RP)). Our extensive simulation results validate that EAGER outperforms existing stateless multicast protocols in computation time, packet overhead, and energy consumption while maintaining the advantages of stateless protocols.

1. Introduction

Large self-organizing wireless sensor networks (WSNs) consist of a great number of sensor nodes with wireless communication and sensing capabilities. The sensor nodes can be deployed randomly close to or inside of the terrain of interest to provide cooperative wireless ad hoc network services. The sensed data and control messages are exchanged between sensor nodes and the control (sink) nodes via a multihop routing protocol. Potential applications of WSNs are numerous, and include environmental monitoring, industrial control and monitoring, and military surveillance to name a few.

Many sensor nodes have been commercially developed for various purposes (e.g., [1–6]). However, the sensor nodes have limitations such as a low capacity processor, small memory, and tiny storage as shown in Table 1, in addition to battery constraints.

Meanwhile, many WSN applications such as mission assignments, configuration updates, and phenomenon reports require one-to-many communications in nature, either from a sensor node to sink nodes or a sink node to sensor nodes. Multicast routing is an important routing service for such applications, as it provides an efficient means of distributing data to multiple recipients compared to multiple unicasts, using in-network replication. Considering both limitations of sensor nodes and the significance of multicast routing, it is critical to deliver multicast packets with a low overhead of resources such as energy, processing, memory, and storage in sensor nodes.

Multicast protocols can be classified into three categories including tree-based, source-based, and location-based multicast protocols. The tree-based multicast protocols [7–12] deliver a multicast packet relying on forwarding states maintained at nodes in a path. Its major drawbacks are control information flooding and storage for forwarding table establishment and maintenance, which produce a lot of overhead in WSNs. The source-based multicast protocols [13, 14] make a path tree at a source, and a multicast

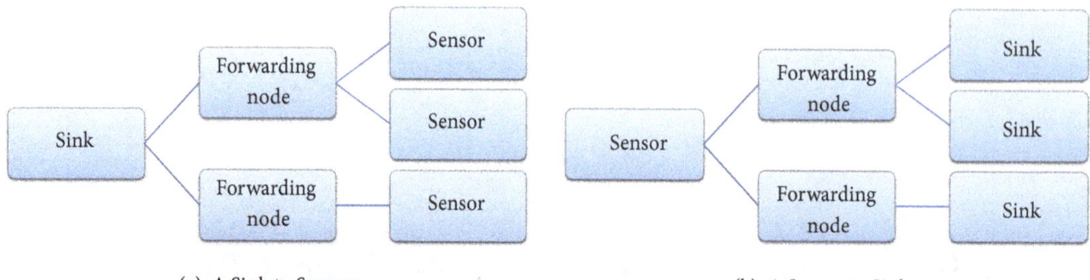

(a) A Sink to Sensors (b) A Sensor to Sinks

FIGURE 1: wireless sensor network multicast application types.

packet encoded with the path tree information is propagated, requiring no states in WSN nodes. However, as the network size expands, it accrues packet size elongation due to the increased path tree information, which in turn, causes a sharp increase in the overhead of CPU processing and energy consumption. In location-based multicast protocols [15, 16], a multicast packet contains the location information of the destination nodes. It is stateless, like source-based routing, but the packet header size is proportional to the number of destinations and does not increase with the network size. However, it requires computation at every forwarding node in a path while looking for the next forwarding node, resulting in excessive processing of CPU and energy consumption.

In this paper, we present an energy-efficient adaptive geosource multicast Routing (EAGER) protocol for WSNs. EAGER is a novel stateless multicast protocol to optimize the previous location-based and source-based multicast approaches in various ways. The unique contributions of the proposed protocol are as follows: (1) it builds a common path multicast tree during the group membership establishment period. This on-demand approach reduces the location flooding overhead of the network topology maintenance on each node; (2) it decreases the computational overhead of each forwarding node such as the forwarding decision and packet decoding/encoding, with simple serialized path information; (3) it reduces the packet encoding overhead by adaptively using geographic unicast and source multicast. Geographic unicast is more efficient for long nonbranching path segments, and source multicast is desirable with branching path segments; (4) it further decreases the packet header size by using the multicast packet with a small node ID instead of potentially large position information; (5) overall, the reduced computational overhead, encoding overhead, and packet header size enable EAGER to consume less energy than location-based or source-based multicast protocols.

The rest of the paper is organized as follows. Section 2 provides a survey of existing multicast protocols for WSNs. Section 3 describes the proposed EAGER scheme and its algorithms. Section 4 presents extensive evaluations of EAGER and its comparison with location-based or source-based multicast protocols in various scenarios. Section 5 offers the conclusions and future work.

2. Related Work

A large number of studies (e.g., [17–19]) have been conducted in the field of multicast applications in WSNs. Those applications operate in a multicast communication mode either from a sensor node to sink nodes or a sink node to sensor nodes, as depicted in Figure 1(a). Configuration updates [17] or mission assignments [18] are the examples of a sensor to sinks multicast. Sensor to multiple sinks [19] scenarios are common for monitoring applications that require reliability.

TABLE 1: Capacities of sensor nodes.

Categories	Name	CPU (MHz)	Memory	Storage
Sensor nodes	Egs	96	52 KB	256 KB
	Micaz	8	4 KB	128 KB
	TelosB	8	10 KB	48 KB
	Tmote sky	8	10 KB	48 KB
	IMote2	400	32 MB	32 MB
	SunSPOT	400	1 MB	8 MB

Multicast protocols in wireless networks can be classified into three categories including tree-based, source-based, and location-based multicast protocols. The examples of tree-based multicast algorithms include adaptive demand-driven multicast routing protocol (ADMR) [8], on-demand multicast routing protocol (ODMRP) [9], multicast ad hoc on demand distance vector routing (MAODV) [10], progressively adapted sub-tree in dynamic meshs (PAST-DM) [7], ad hoc multicast routing protocol utilizing Increasing id-numberS (AMRIS) [11], and ad hoc multicast routing protocol (AMRoute) [12]. They have been developed for traditional wireless ad hoc networks, and have evolved in support of WSNs. However, those traditional multicast routing techniques are designed as control centric approaches, and focus on solving the mobility issues under the assumption of enough processing and local storage capacity on each node. They maintain a forwarding table on each node through a multicast routing tree for each group. The distributed group forwarding states should be updated via periodic control flooding messages, consuming significant energy. Due to the resource limitations on sensor nodes, they cannot be directly used for WSNs.

A Source-based multicast protocol, such as dynamic source multicast (DSM), has been proposed to perform centralized membership management on a multicast root instead of distributed state maintenance. A root or a multicast source builds a multicast tree using the locally maintained network topology information and encodes the tree information into the packet header. Forwarding nodes relay the packet according to the tree path information carried in the packet

TABLE 2: Classification of WSN multicast protocols.

	Tree-based routing	Location-based routing	Source-based routing
Strengths	Small data packet overhead	No distributed routing control overhead, less path encoding and decoding overhead than source-based	No distributed routing control overhead, relatively smaller forwarding computation than location-based
Weaknesses	Stateful, large control overhead (flooding), large memory/storage	Large packet size (location information of destinations), large forwarding computation	Large packet size (path), path encoding and decoding overhead
Examples	MAODV, ADMR, ODMRP, AM-Route, AMRIS	PBM, LGT, GMR, DDM, RDG	DSM

header. Although the distributed stateless multicast protocols are typically considered to be better for resource constrained WSNs than stateful distributed tree-based protocols, for large-scale networks, those stateless multicast protocols suffer substantial energy consumption due to the packet encoding and decoding operations at a source node and forwarding nodes, respectively.

Several location-based multicast protocols such as Location Guided Tree construction algorithms (LGT) [15], differential destination route driven gossip (RDG) [20], differential destination multicast (DDM) [21], geographic multicast routing (GMR) [16], and position-based multicast routing protocol (PBM) [14] have been proposed. These protocols compute the next forwarding node at each node on the path, based on location information of destinations rather than path information. Therefore, these have less path encoding and decoding overhead than source-based multicast protocols. However, these protocols still require large packet sizes for the destinations' location information and large forwarding computation at all the path nodes. In order to address the issue of scalability for a large number of destinations, hierarchical rendezvous point multicast protocol (HRPM) [22], hierarchical geographic multicast routing (HGMR) [23], hierarchical differential destination multicast (HDDM) [24], and scalable position-based multicast (SPBM) [25] have been proposed.

Our work, EAGER, is unique. It adaptively uses location-based unicast and source-based multicast approaches in order to reduce the computational overhead of forwarding. It also minimizes packet header overhead using enhanced state encoding capability, as well as tree construction overhead using on-demand path information-based tree construction. Table 2 summarizes strengths, weaknesses, and examples of classified multicast protocols.

3. Energy-Efficient Adaptive Geosource Multicast Routing

In this section, we first give a brief background on GMR and DSM that are representative examples of location-based multicast routing and source-based multicast routing, respectively, as EAGER optimizes their advantages adaptively. We then describe the detailed EAGER protocol for the following three main operations: (1) multicast tree construction, (2) routing path encoding, and (3) packet forwarding method.

We have chosen GMR and DSM as the best representative schemes for large size multicasting. Inherently, LGT [15] and DDM [21] have been designed for small group multicast and are not scalable to large sized networks. GMR has shown that PBM [14] needs larger computation time and number of data transmissions than GMR. With the same packet size per transmission, the larger number of data transmissions result, in larger total packet sizes.

3.1. Background. GMR assumes to have the entire multicast membership information at the multicast root node like other stateless protocols. However, instead of building a multicast tree for all destinations, the multicast root node selects forwarding nodes among its neighbors according to the cost and progress ratio to the destinations. Hence, the packet header only carries selected forwarding neighbor IDs and a list of the destinations for each forwarding node. Figure 2(a) illustrates how GMR routing works. A source (A) broadcasts a packet that has a neighbor id (B) and $x - y$ coordinates of destinations $\{Ex, Ey, Dx, Dy\}$. Upon receiving the message packet, each selected forwarding node calculates the next forwarding nodes for the given destinations among its neighbors. That is, node B calculates a neighbor id for each path. In this example, a node C is selected as a neighbor for a destination E, and a node D is chosen for a destination D, respectively. Subsequently, the node B broadcasts the packet with chosen neighbors and the coordinates of the destinations. The multicast packet is eventually propagated to destinations using the next forwarding neighbor calculation on each forwarding node. Each forwarding node performs approximately $O(min(neighbors, destinations)^3)$ calculations to select forwarding neighbors.

DSM assumes that each node has the entire network topology using periodic location flooding information. The root node locally computes a Steiner tree for the multicast group. For example, the tree at Figure 2(b) is a Steiner tree that the root node A creates. The packet header carries encoded multicast tree path information (node IDs) using the Prüfer sequence [26]. Node IDs in the Prüfer sequence [26] represent interior nodes in paths (not leaf-nodes). Upon receiving the message packet, each child node, which is inside the Prüfer sequence, decodes the sequence, creates a Steiner subtree, and encodes a new Prüfer sequence to the packet header. For example, a node B receives $\{B, B, C\}$ sequence and knows it is a forwarding node because B is in the sequence.

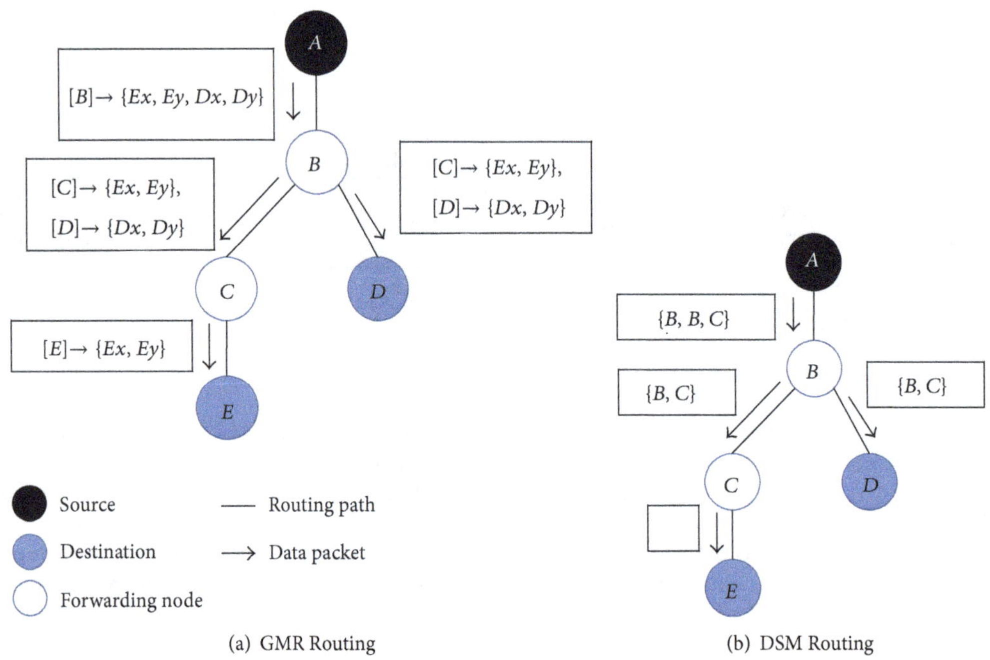

Figure 2: GMR versus DSM routing.

The node B decodes the sequence and creates a Steiner subtree without a node A. After that, node B creates and broadcasts the $\{B, C\}$ sequence. The packet is relayed until it reaches leaf nodes. When node D receives the packet, D is not in a sequence. Therefore, D knows that D is a leaf node and stops forwarding the packet. The complexity of tree encoding is $O((pathLength)^2)$ on each child node.

3.2. EAGER Algorithms and Operations. EAGER protocol consists of algorithms for multicast tree construction, routing path encoding, and a packet forwarding method. We next describe each of them in details.

3.2.1. Multicast Tree Construction. The existing source multicast routing protocols assume that every node maintains the entire network topology information using location flooding. Each multicast root node constructs a multicast tree for the given destinations using the network topology. However, the periodic location flooding is expensive as it consumes much of the energy, especially for large networks. To save the cost of building a multicast tree, in EAGER, each multicast root node (or rendezvous point (RP)) obtains the path information to the destination during the multicast membership establishment stage instead of the location flooding. Each join request message carries its path information toward the multicast root node along with its location information. For example, when a member node joins by using geographic unicast, each intermediate node in the path adds its location information to the join packet. The multicast root eventually receives reverse geographic shortest path information to the destination. A multicast tree is created on-demand using the path information in $O(no.paths)$. According to the path information obtained by each join message, the multicast root further optimizes the multicast tree identifying the common path segments among the destinations.

The scheme works as follows. First a destination node, dst, sends a join message, joinMsg, toward the source, src. The join message is relayed to the next forwarding node among the neighbor nodes that is geographically close to src. As illustrated in Algorithm 1, an intermediate node maintains a temporary multicast state table named MState with the information of [dstSeg, hopCnt]. dstSeg is a list of destination node IDs, and hopCnt is the longest hop count from the intermediate node to the destinations in the list. When a joinMsg arrives at an intermediate node, if any dstSeg in the MState table contains the same dst ID and the existing hopCnt is larger than the new join message, the join message will be dropped. For example, in Figure 3, if the destination node j has already joined the multicast group, the intermediate node d will maintain the destination nodes both i and j in the dstSeg.

When the join message from the destination node i arrives in the intermediate node d, d will stop sending the join message as the MState table contains i, and the hopCnt is greater than the hop count from the join message. If the hopCnt is less than the new join message's hop count, the dstSeg and hopCnt will be updated with the new path information and hop count, respectively. If the dst ID is new, a new entry will be added in the MState table. The intermediate node adds the location information to the join message and forwards the join message to the next forwarding node toward the src node. The MState table will be maintained temporarily during the membership establishment stage. The MState table on an intermediate node helps the src node to make compressed path information to the destinations, but is not used for data packet forwarding.

```
// MState: multicast state table, contains dst and hopCnt pairs
if joinMsg(dst) ∈ MState(dstSeg) then
  if hopCnt < MState(dstSeg).hopCnt then
    MState(dstSeg) = dst
    MState(dstSeg).hopCnt = hopCnt
    Add the location information to joinMsg
    Send the joinMsg to the next hop toward src
  end if
else
  Add [dstSeg, hopCnt] to Mstate
  Add the location information to joinMsg
  Send the joinMsg to the next hop toward src
end if
```

ALGORITHM 1: Group management at intermediate nodes.

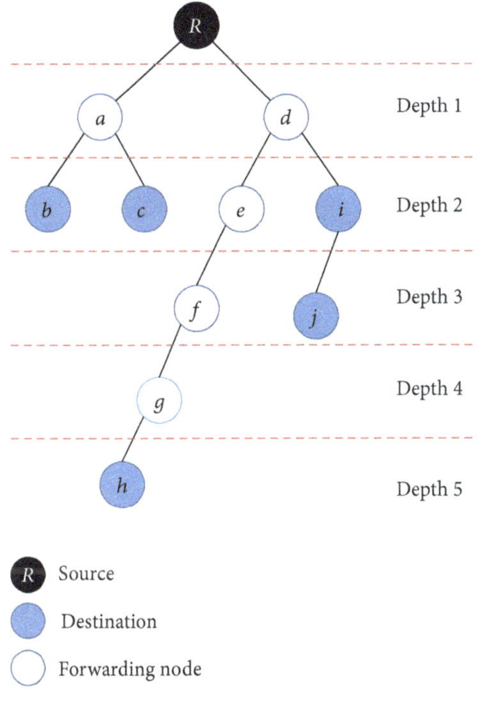

FIGURE 3: Construction of multicast tree.

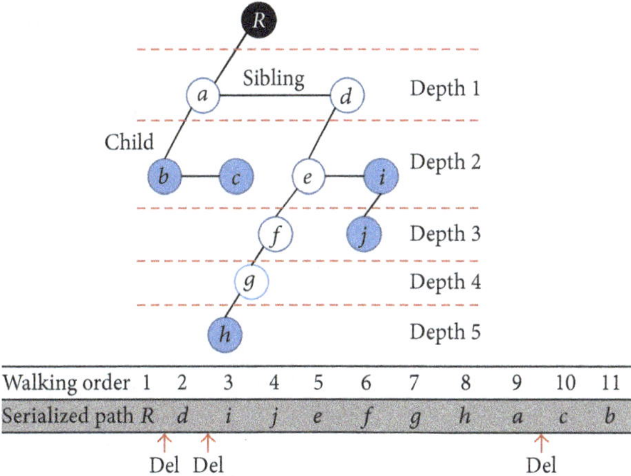

FIGURE 4: LCRS binary tree and a serialized path sequence.

Once a joinMsg is received by the src node, it constructs a multicast tree starting with a path segment that is used by the highest number of destinations, until it includes all the destinations. For example, in Figure 3, when the src node R has received join messages from h and j in sequence, the src node can identify a common path segment of $\{R, d\}$ from the paths $\{R, d, e, f, g, h\}$ and $\{R, d, i, j\}$.

3.2.2. Routing Path Encoding. In stateless source-based multicast routing protocols, the multicast root node encodes the multicast tree into the packet header using tree structure algorithms such as the Prüfer sequence algorithm. The encoded multicast tree information will be decoded on each intermediate node and re-encoded for the subtree entries before sending the packet. For example, the Prüfer sequence algorithm takes $O(pathLength)$ packet header size. However, the complexity of multicast tree encoding and decoding is $O((pathLength)^2)$ on each intermediate node. To avoid the expensive encoding and decoding overhead on each intermediate node, EAGER serializes the subtree path information using an LCRS (left child right sibling) binary tree [27]. As the tree serialization requires additional delimiters, it may result in a slightly bigger packet size than the other source multicast routing protocols. For example, compared with the pure source multicast scenarios of the spanning tree with n number of nodes, the encoding ratio $(n-2)$ of the Prüfer sequence algorithm used in DSM can be slightly better than our LCRS-based serialized path encoding algorithm $(n-1+\text{number of branch delimiters})$. However, EAGER is designed to have less computation overhead on each intermediate node. The computation complexity of EAGER is $O(1)$ while the Prüfer sequence algorithm has $O(n^2)$.

Algorithms 2, 3, and 4 show how serialization algorithms work. First, the encoding algorithm translates the original *n-ary* multicast tree to an LCRS (left child and right sibling) binary tree. Starting from the multicast root node, the left

```
Input: source node
Output: path
  path = ∅
  PathSerializing(a source node, path)
```

ALGORITHM 2: Routing path encoding.

```
Input: node
Output: path
  (1) if node == Null then
  (2)     return
  (3) end if
  (4) if node → right then
  (5)     PathSerializing(node → right)
  (6) end if
  (7) if IsBranch(node) then
  (8)     // if a node has more than two children
  (9)     if ID(node) ≠ ID(source node) then
  (10)        add(path, ID(node))
  (11)    end if
  (12)    add(path, Delimiter)
  (13)    PathSerializing(node → left)
  (14) else
  (15)    if IsLongPath(node) then
  (16)        // if there are more than three subsequent children
  (17)        // that is, child - grandchild - grandgrandchild ...
  (18)        TmpNode = PathSerializingForUnicast(node)
  (19)        add(path, XCoordinate(TmpNode))
  (20)        ‘add(path, YCoordinate(TmpNode))
  (21)        node = TmpNode
  (22)    else
  (23)        add(path, ID(node))
  (24)    end if
  (25)    PathSerializing(node → left)
  (26) end if
```

ALGORITHM 3: Path serializing.

most child of a node becomes the left child of the new binary tree, and the right most sibling becomes the right child of the new binary tree. For example, the original tree in Figure 3 becomes the new LCRS binary tree in Figure 4. Second, serialized path information is created by walking along the LCRS binary tree in the order of "sibling first, then child node." As illustrated in Figure 4, the serialized path $\{d, i, j, e, f, g, h, a, c, b\}$ is created by walking through the LCRS binary tree with a few additional delimiters. EAGER uses a fixed size information block to encode the state information. The serialized path is presented as consecutive information blocks. The information block can be used as a node ID, location coordinates, or a delimiter. Figure 7 shows a 2-byte delimiter format example. A delimiter can be distinguished from other information blocks by setting 1 in the most significant bit of an information block. Each delimiter has two 7-bit offsets. The node ID block can be identified by setting 00 in the left 2 bits. With a 2 byte information block, the maximum number of node IDs can be about 16 K (use only 14 bits). Branch delimiters are inserted next to the original tree branches' serialized path to indicate the original tree's sibling relationship. That is, the subtree information for each sibling node is separated by the branch delimiters.

Furthermore, EAGER optimizes the encoded packet size adaptively using branch geographic information for common source routing path segments as can be seen at line 15 to 21 of Algorithm 3. It identifies long nonbranchingrouting path segments and uses the branch locations for the source routing information instead of many node IDs along the path. A long nonbranchingpath segment is identified during the serialized path creation if a node has more than three subsequent children; that is, child-grandchild-great grandchild. The serialized path is minimized using the location information (i.e., x and y coordinates) instead of putting the entire node IDs in the path. As we see in Algorithm 4, if a forwarding node finds the location information (delimiter value is 01) in the serialized path, it uses a geographic unicast toward the next branching node. Although it requires each forwarding node to run a geographic unicast algorithm,

```
Input: node
Output: a branch or a destination node
if IsBranch(node) then
    return node
end if
// a leaf node
if node → left == Null then
    return node
end if
PathSerializingForUnicast(node → left)
```

ALGORITHM 4: Path serializing for unicast.

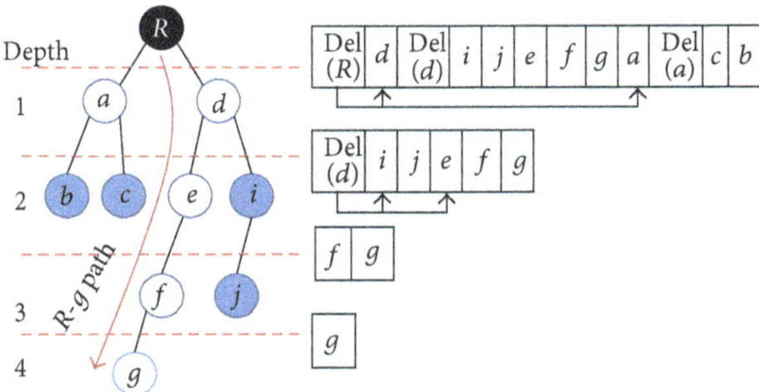

FIGURE 5: Packet decoding/forwarding on the short path nodes.

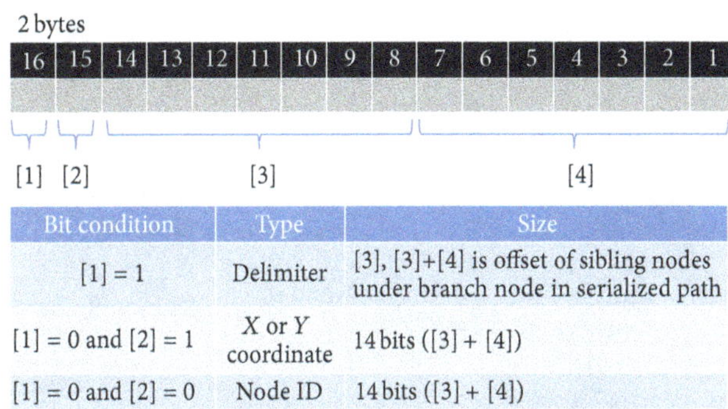

FIGURE 7: 2 Bytes information block example.

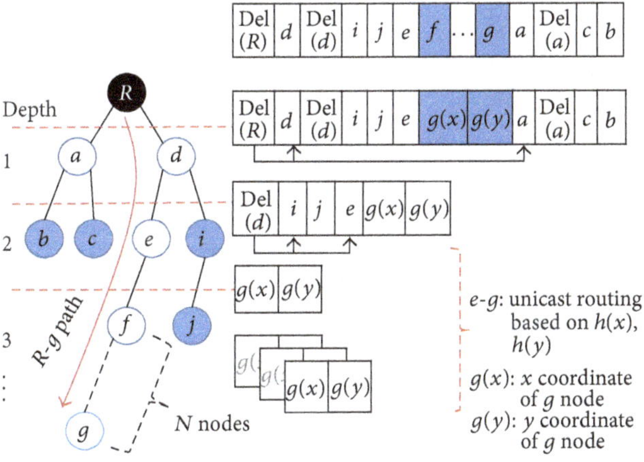

FIGURE 6: Packet decoding/forwarding with long path optimization.

the computation complexity is minimal. We name *long-path optimization* as the technique to reduce packet size by using x and y coordinates for a long nonbranchingpath.

3.2.3. Packet Forwarding. When a multicast packet is received by a forwarding node, the node selects the serialized path for its own subtree according to the branch delimiter information. Figure 5 illustrates how to utilize the serialized path information along the R to g forwarding track on the short path nodes. For example, node d can make a new serialized path for its subtree by checking the offset from the first delimiter (R). *Del(R)* points node d and node a, which are forwarding nodes. Node d recognizes that it is a forwarding node with *Del(R)*. By using line 3 to 7 of Algorithm 5, node d extracts the path of the next subtree, $\{Del(d), i, j, e, f, g\}$, and broadcasts with the path. In this case, there is no additional calculation overhead except the simple packet truncation for the next subtree, and computation overhead at each forwarding node.

However, in case of a long non-branch path, all node IDs in the long path should be contained into serialized paths. In order to reduce the size overhead of node IDs due to the long path, we use *long-path optimization*. As shown in Figure 6, there is a long (n nodes) nonbranchingpath segment between node f and g. In this case, a multicast packet is delivered to node e by using the same way as Figure 5, but node e sends the packet to node g by the geographic unicast routing because of the existence of the long and non-branch path (f to g). The path segment can be represented by only 2 information blocks; thus, the information is reduced by $2(n-2)$ bytes. It also results in a great saving of packet size along the path. The algorithms for the forwarding operation are shown in Algorithms 5 and 6.

4. Performance Evaluation

In this section, we evaluate the performance of EAGER and compare it with the performance of GMR and DSM. We implemented EAGER using an NS2 (v2.35) simulator. We used a grid network topology. Most evaluations were

```
Input: serialized path, node
if FirstBit(path[0]) == 1 then
    // the first byte in path is a delimiter
    if ID(node) ∈ {IDs pointed by the delimiter} then
        pos1 = indexInPath(node) + 1
        pos2 = indexInPath(next node pointed by delimiter after me) – 1
        path = {path[pos1], . . ., path[pos2]}
        Forward(path)
    end if
else
    if FirstTwoBits(path[0]) == 01 then
        // the first byte in path is x coordinate
        // path [0], path [1] are X and y coordinates
            of a branch or a destination
        if path[2] ≠ Null then
            path = {path[2], . . ., path[length(path) – 1]}
        end if
        Forward(path)
    else if FirstTwoBits(path[0]) == 00 then
        // the first byte in path is an ID
        path = {path[1], . . ., path[length(path) – 1]}
        Forward(path)
    end if
end if
```

Algorithm 5: Packet forwarding at each node.

```
Input: path
if FirstTwoBits(path[0]) == 01 then
    nextHop = GetNextHop(path[0], path[1])
    Unicast packet(with path) to nextHop
else
    Broadcast packet(with path)
end if
```

Algorithm 6: Forward(path): Decision of forwarding mechanism.

performed in the network with 2025 nodes unless the network size is not mentioned in this section. The number of neighbor nodes in the communication range is set up to 12 unless specified differently. We assume that there is no packet loss, and the size of the location coordinates of a node is 2 times bigger than the size of the node identifier. The evaluation metrics used were total packet overhead, average computation time, and consumed energy. *Total packet overhead* is the sum of all the multicast packets delivered from a multicast root node to all the destination nodes along a multicast path. *Average computation time* is the average time taken by each forwarding node on the multicast path for neighbor selection and packet re-encoding. *Consumed energy* is the total energy used by the nodes in the multicast path to perform transmission, reception, and computation. Consumed energy is computed by multiplying duration for transmission, reception, and computation by the power consumption (Watt). The power consumption (Watt) ratio of computation, transmission, and reception is shown in Table 3 that corresponds to the cc2420 [28] and ATMega128L [29] specifications.

Table 3: Power consumption ratio.

Operations	Consumed power (Watt)
Computation	0.0000459
Transmission	0.0001
Reception	0.000132

As for the placement of destinations, we used both random and clustered destinations. Many studies [30–34] have shown that clustered destinations are common for the group communication applications. Clustered WSNs were used to achieve efficient energy usage, a long network lifetime, and high network coverage. To evaluate clustered destinations, we used various configurable parameters, including the number of clusters, the number of nodes in a cluster, the distance between a source and a cluster, and the radius of a cluster.

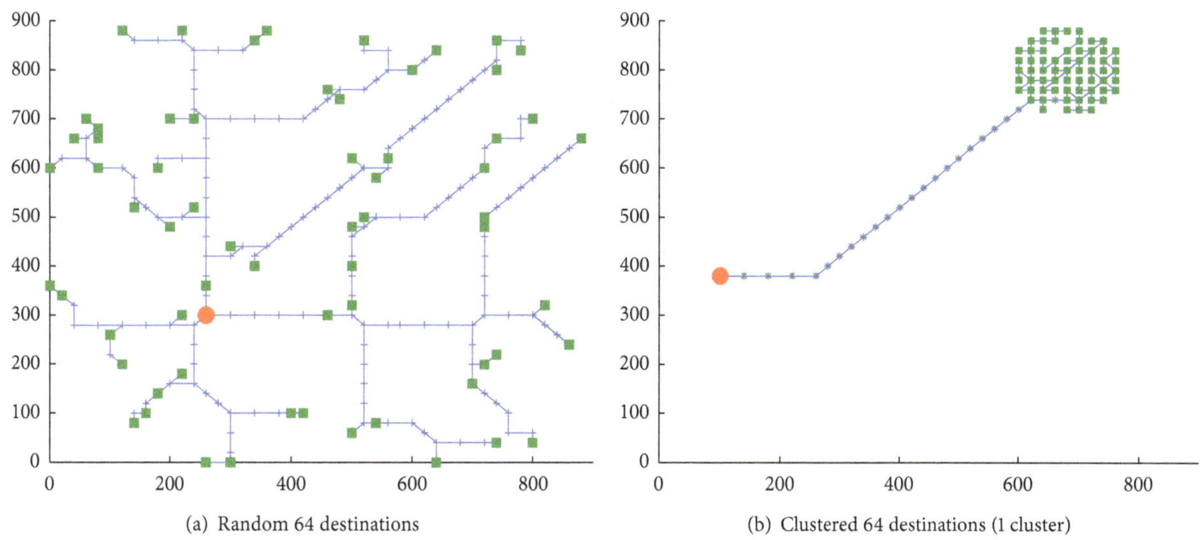

(a) Random 64 destinations

(b) Clustered 64 destinations (1 cluster)

FIGURE 8: Example of random and clustered destinations.

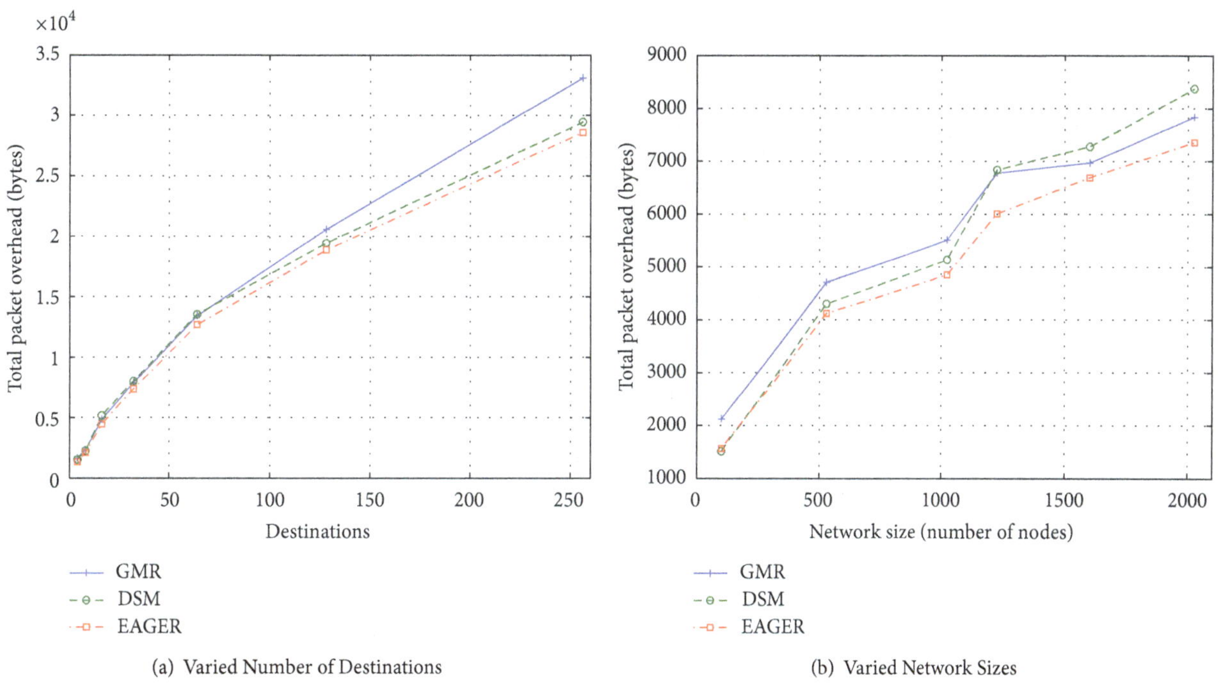

(a) Varied Number of Destinations

(b) Varied Network Sizes

FIGURE 9: Total packet overhead comparison with random destinations.

4.1. Random Destinations. We evaluated packet processing overhead including total packet overhead and average computation time and then quantified the consumed energy with random destinations for EAGER, GMR, and DSM. We randomly selected destinations as well as a multicast root node. Figure 8 shows examples of random and clustered destinations with 64 destinations. A solid circle, solid squares, crosses, and lines represent a source, destinations, forwarding nodes, and routing paths, respectively.

The *total packet overhead* for the different number of destinations is shown in Figure 9(a). The numbers of destinations are varied with 4, 8, 16, 32, 64, 128, and 256 nodes. While the number of destinations increases, EAGER and DSM use less packets than GMR. Since GMR encodes the packet header with the destination locations, the packet header size becomes larger as the number of destinations increases. For example, if the number of destinations is less than 64, the total packet overhead of GMR is smaller than

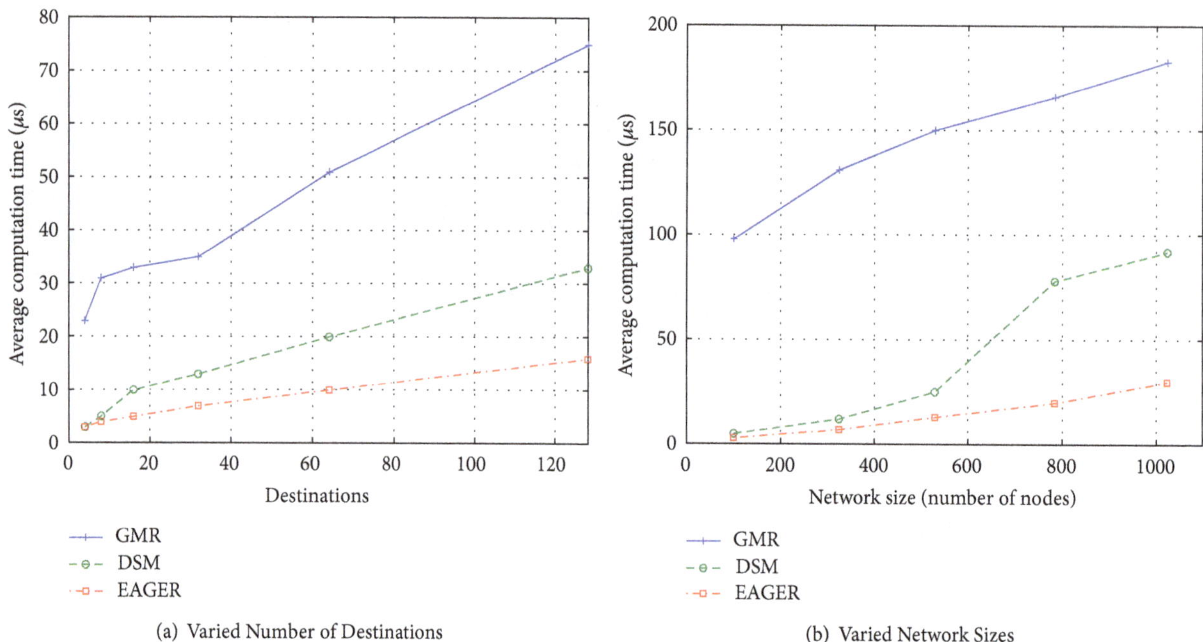

(a) Varied Number of Destinations

(b) Varied Network Sizes

FIGURE 10: Average computation time comparison with random destinations.

that of DSM. However, if the number of destinations becomes greater than 64, GMR has a bigger packet overhead than other protocols. The result also shows that EAGER uses slightly smaller packet sizes than DSM, as EAGER can reduce the packet sizes adaptively using branch geographic information for common source routing path segments.

Next, we examined the total packet overhead for the different network sizes in Figure 9(b). The network size ranges from 100 to 2025 nodes while the number of destinations is fixed with 30 nodes. The result also shows that the total packet overhead of EAGER is always smaller than that of GMR and DSM. Because DSM encodes the packet with the multicast tree path, DSM's packet size becomes bigger as the larger network size increases. It also illustrates that DSM is more sensitive to the network size than EAGER as well as GMR. EAGER has a smaller total packet overhead and is less sensitive to the network size than other protocols, as it has a larger chance of having longer and nonbranchingpaths for the large network.

As network size increases to more than 2025, we expect that EAGER has the lowest total packet overhead and DSM has the highest one. However, the network size at which DSM has more total packet overhead than GMR changes depending on the number of destinations. Specifically, the more numbers of destinations are used, the larger network size is needed so that DSM has more total packet overhead than GMR.

The average computation time is compared for the different number of destinations in Figure 10(a). The numbers of destinations are from 4 to 128 in the network with 2025 nodes. GMR requires the most computation time compared to other protocols, resulting in a high CPU overhead. This is because GMR calculates the next forwarding neighbors on each forwarding node and the algorithm complexity increases according to the number of destinations. Meanwhile, in DSM and EAGER, the multicast path information is calculated and encoded by the multicast root node, leading to a lower average computation time. However, for the large number of destinations, the computation time of DSM is higher than that of EAGER due to the encoding and decoding overhead of the forwarding nodes.

We show the average computation time for the varied network sizes in Figure 10(b). The network size is varied from 100 to 1024, while the number of destinations is set with 30% of the network size. The results display that GMR spends a much higher computation time than other protocols, but the time difference is bounded and not proportional to the increment of the network size. Both DSM and EAGER spend minimal computation time and have little dependency on the network size. However, for the larger network size, the computation time of DSM becomes much higher than that of EAGER because the encoding and decoding overhead increase proportionally to the number of nodes on the multicast routing path.

Figure 11 shows the energy consumption for the varied number of destinations. The number of destinations increases from 32 to 256 for the network of 2025 nodes. The result shows that EAGER consumes the least energy. It is because it has a smaller total packet overhead and a lower computation time than the other two protocols. The result also shows that EAGER becomes more energy-efficient than other protocols as the number of destinations increases. DSM shows worse energy efficiency than GMR for the small number of destinations. However, DSM has better energy consumption than GMR when the number of destinations increases. It is also observed that the energy consumption conforms more to

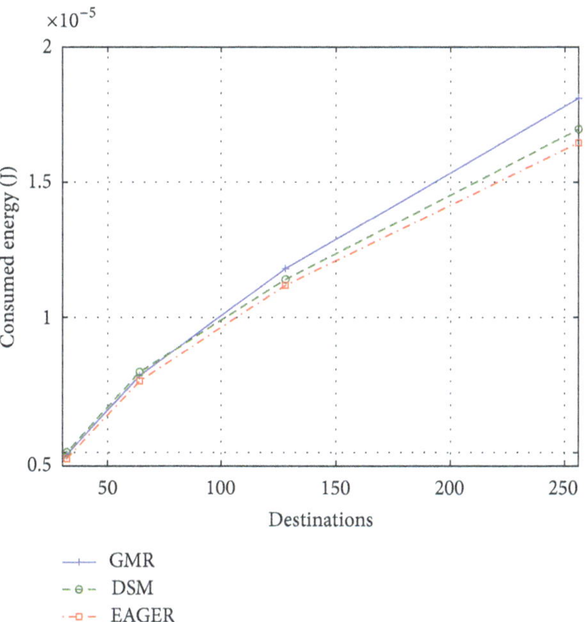

Figure 11: Consumed energy comparison with random destinations.

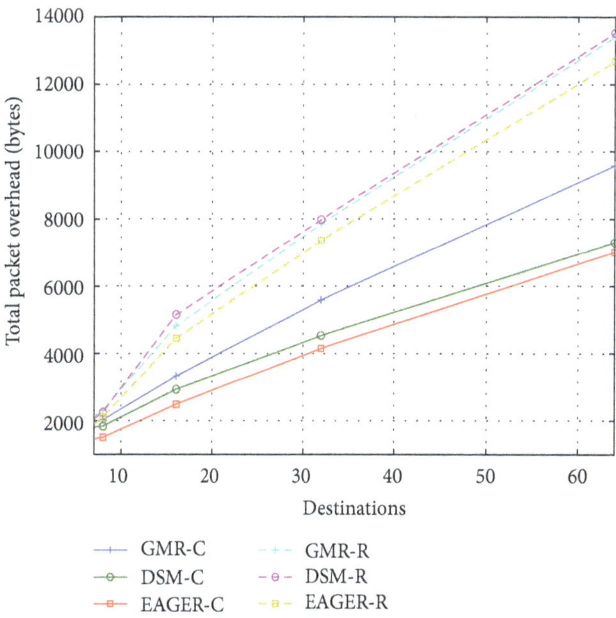

(a) Total Packet Overhead Comparison

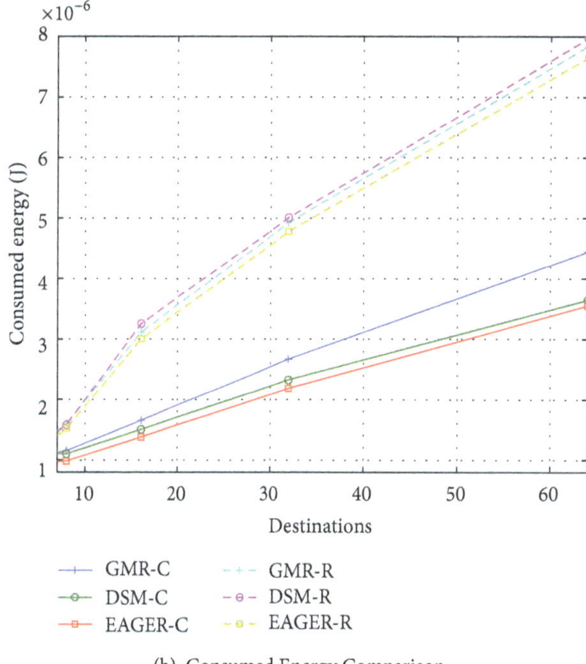

(b) Consumed Energy Comparison

Figure 12: Total packet overhead and consumed energy comparisons with varied number of destinations, (-R: random destinations, -C: clustered destinations).

the total packet overhead than the computation time, as the relative energy consumption for communication is set much higher than the one for computation as in Table 3.

4.2. Clustered Destinations. Here we used the clustered destinations for the comparative evaluations of total packet overhead and consumed energy.

Figures 12(a) and 12(b) illustrate total packet overhead and consumed energy, respectively, with the different number of destinations from 4 to 64. *Protocol*-C means the protocol measured with clustered destinations, and *protocol*-R means the protocol measured with random destinations. We found that all protocols with clustered destinations have smaller total packet overhead and energy consumption than those with random destinations. It was observed that the gap between GMR and DSM becomes larger in clustered destinations than in random ones while the number of destinations increases. This is because the total path lengths from a source and destinations have been reduced much faster in clustered destinations than in random destinations. As EAGER enjoys the benefit of GMR, it outperforms DSM in clustered destination scenarios. It also always has a smaller total packet overhead than GMR as well, due to compact packet encoding.

We next varied the different distances between a source and a cluster head from 600 to 850 meters while fixing the number of destinations as 8 and measured total packet overhead and consumed energy in Figures 13(a) and 13(b). EAGER exhibits the least energy consumption and total packet overhead than other protocols. DSM shows lower energy consumption and less total packet overhead than GMR in the short path lengths. However, as the path lengths become longer, DSM shows higher energy consumption and larger total packet overhead than GMR, since DSM's packet header has to include all of the path information.

Finally, we varied the number of clusters from 1 to 8 while the total number of destinations is fixed, in Figures 14(a) and 14(b). They demonstrate that the total packet overhead and energy consumption with EAGER shows less energy consumption and total packet overhead than other protocols as the number of clusters increases. DSM shows lower energy

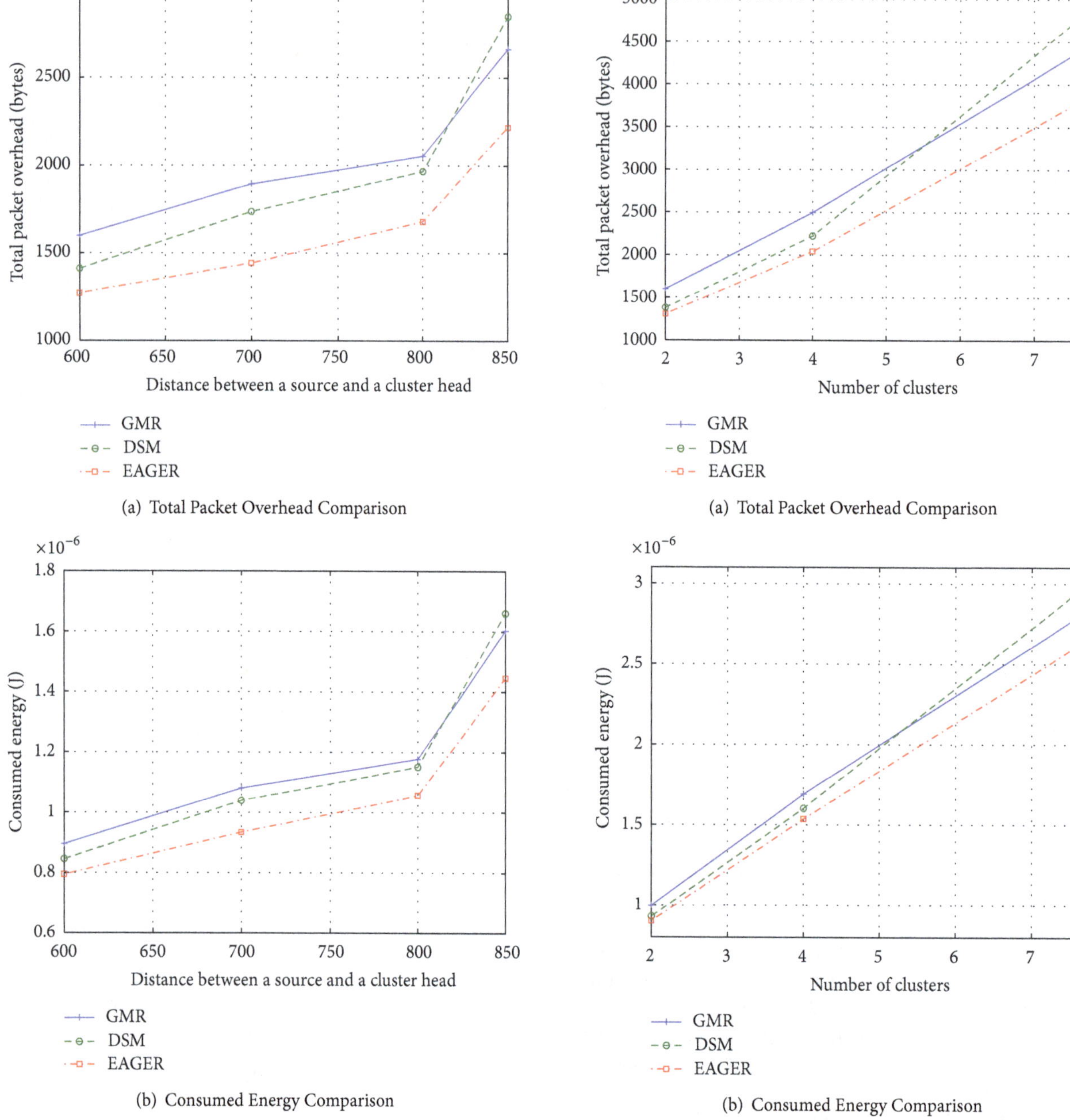

(a) Total Packet Overhead Comparison

(b) Consumed Energy Comparison

FIGURE 13: Total packet overhead and consumed energy comparisons with varied cluster path length.

(a) Total Packet Overhead Comparison

(b) Consumed Energy Comparison

FIGURE 14: Total packet overhead and consumed energy comparisons with varied number of clusters of destinations.

consumption and less total packet overhead than GMR in the small number of clusters. However, as the number of clusters increases, DSM shows higher energy consumption and larger total packet overhead than GMR. This is because the path length increases as the number of clusters increases.

In all the various evaluation scenarios, EAGER outperformed both source-based and location-based multicast protocols, not only taking the advantage of each effectively, but also enhancing each of them with efficient encoding and forwarding operations.

5. Conclusion

We have presented a novel stateless path information-based multicast protocol, named EAGER (energy-efficient adaptive geo-source multicast routing) for WSNs. EAGER optimizes the previous location-based multicast and source multicast approaches by adaptive usage of geographic unicast and source multicast routing. It is also equipped with unique features including on-demand tree construction using path information, light-weight forwarding, and enhanced state encoding capability. Our extensive simulation results exhibit that EAGER outperforms GMR and DSM in computation time, packet overhead, and energy consumption while maintaining the advantages of stateless protocols.

References

[1] "Imote2: High-Performance Wireless Sensor Network Node," http://docs.tinyos.net/tinywiki/index.php/IMote2.

[2] "TMote Sky," http://www.eecs.harvard.edu/~konrad/projects/shimmer/references/tmote-sky-datasheet.pdf.

[3] J. G. Ko, Q. Wang, T. Schmid, W. Hofer, P. Dutta, and A. Terzis, "Egs: a Cortex M3-based mote platform," in *Proceedings of the 7th Annual IEEE Communications Society Conference on Sensor, Mesh and Ad Hoc Communications and Networks (SECON '10)*, pp. 1–3, IEEE, June 2010.

[4] "Micaz," http://www.openautomation.net/uploadsproductos/micaz_datasheet.pdf.

[5] "SunSPOT," http://www.sunspotworld.com/products/index.html.

[6] "TelosB," http://www.willow.co.uk/TelosB_Datasheet.pdf .

[7] C. Gui and P. Mohapatra, "Efficient overlay multicast for mobile Ad Hoc networks," in *Proceedings of the IEEE Wireless Communications and Networking (WCNC '03)*, vol. 2, pp. 1118–1123, March 2003.

[8] J. G. Jetcheva and D. B. Johnson, "Adaptive demand-driven multicast routing in multi-hop wireless ad hoc networks," in *Proceedings of the 2001 ACM International Symposium on Mobile Ad Hoc Networking and Computing (MobiHoc '01)*, pp. 33–44, October 2001.

[9] S.-J. Lee, M. Gerla, and C. C. Chiang, "On-demand multicast routing protocol," in *Proceedings of the IEEE Wireless Communications and Networking (WCNC '99)*, September 1999.

[10] E. M. Royer and C. E. Perkins, "Multicast operation of the Ad-Hoc on-demand distance vector routing protocol," in *Proceedings of the 5th Annual ACM/IEEE International Conference on Mobile Computing and Networking (MobiCom '99)*, pp. 207–218, 1999.

[11] C. W. Wu and Y. C. Tay, "AMRIS: a multicast protocol for ad hoc wireless networks," in *Proceedings of the IEEE Military Communications Conference (MILCOM '99)*, pp. 25–29, November 1999.

[12] J. Xie, R. R. Talpade, A. McAuley, and M. Liu, "AMRoute: Ad Hoc multicast routing protocol," *Mobile Networks and Applications*, vol. 7, no. 6, pp. 429–439, 2002.

[13] S. Basagni, I. Chlamtac, and V. R. Syrotiuk, "Location aware, dependable multicast for mobile ad hoc networks," *Computer Networks*, vol. 36, no. 5-6, pp. 659–670, 2001.

[14] M. Mauve, H. Fuessler, J. Widmer, and T. Lang, "Positionbased multicast routing for mobile Ad-Hoc networks," Tech. Rep. CS TR-03-004, University of Mannheim, Baden-Württemberg, Germany, 2003.

[15] K. Chen and K. Nahrstedt, "Effective location-guided tree construction algorithms for small group multicast in MANET," in *Proceedings of the 21th Annual Joint Conference of the IEEE Computer and Communications Societies (INFOCOM '02)*, pp. 1180–1189, June 2002.

[16] X. Liu, J. A. Sanchez, P. M. Ruiz, and I. Stojmenovic, "GMR: geographic multicast routing for wireless sensor networks," in *Proceedings of the 3rd Annual IEEE Communications Society on Sensor and Ad hoc Communications and Networks (Secon '06)*, pp. 20–29, September 2006.

[17] S. Pennington, A. Waller, and T. Baugé, "RECOUP: efficient reconfiguration for wireless sensor networks," in *Proceedings of the 5th Annual International Conference on Mobile and Ubiquitous Systems: Computing, Networking, and Services (Mobiquitous)*, pp. 33:1–33:2, ICST (Institute for Computer Sciences, Social-Informatics and Telecommunications Engineering), 2008.

[18] H. Rowaihy, M. P. Johnson, O. Liu, A. Bar-Noy, T. Brown, and T. La Porta, "Sensor-mission assignment in wireless sensor networks," *ACM Transactions on Sensor Networks*, vol. 6, no. 4, 2010.

[19] R. Szewczyk, E. Osterweil, J. Polastre, M. Hamilton, A. Mainwaring, and D. Estrin, "Habitat monitoring with sensor networks," *Communications of the ACM*, vol. 47, no. 6, pp. 34–40, 2004.

[20] P. T. Eugster, J. Luo, and J. P. Hubaux, "Route driven gossip: probabilistic reliable multicast in ad hoc networks," in *Proceedings of the 22nd Annual Joint Conference on the IEEE Computer and Communications Societies (IEEE INFOCOM '03)*, pp. 2229–2239, San Francisco, Calif, USA, April 2003.

[21] L. Ji and M. S. Corson, "Differential destination multicast-a MANET multicast routing protocol for small groups," in *Proceedings of the 20th Annual Joint Conference of the IEEE Computer and Communications Societies (IEEE INFOCOM '01)*, pp. 1192–1201, April 2001.

[22] S. M. Das, H. Pucha, and Y. C. Hu, "Distributed hashing for scalablemulticast in wireless ad hoc networks," *IEEE Transactions on Parallel and Distributed Systems*, vol. 19, no. 3, pp. 347–361, 2008.

[23] Y. C. Hu, I. Stojmenovic, D. Koutsonikolas, and S. Das, "Hierarchical geographic multicast routing for wireless sensor networks," in *Proceedings of the International Conference on Sensor Technologies and Applications (SENSORCOMM '07)*, pp. 347–354, October 2007.

[24] M. Transie, H. Fuler, J. Widmer, M. Mauve, and W. Effelsberg, "Scalable multicasting in mobile ad hoc networks," in *Proceedings of the 23th Annual Joint Conference of the IEEE Computer and Communications Societies (IEEE INFOCOM '04)*, vol. 3, pp. 2119–2129, March 2004.

[25] M. Transie, H. Fuler, J. Widmer, M. Mauve, and W. Effelsberg, "Scalable position-based multicast for mobile Ad-Hoc networks," in *Proceedings of the International Workshop on Broad-band Wireless Multimedia: Algorithms, Architectures and Applications (Broad- Wim)*, October 2004.

[26] J. Gottlieb, B. A. Julstrom, G. R. Raidl, and F. Rothlauf, "Prüfer numbers: a poor representation of spanning trees for evolutionary search," in *Proceedings of the Genetic and Evolutionary Computation Conference (GECCO '01)*, pp. 343–350, 2001.

[27] NIST Dictionary of Algorithms and Data Structures, "Left child right sibling tree".

[28] "CC2420," http://focus.ti.com/lit/ds/symlink/cc2420.pdf.

[29] "ATMega128L," http://www.atmel.com/dyn/resources/prod_documents/2467s.pdf.

[30] W. Robertson, S. Sivakumar, N. Aslam, and W. Phillips, "A multi-criterion optimization technique for energy efficient cluster formation in wireless sensor networks," *Information Fusion*, vol. 12, no. 3, pp. 202–212, 2011.

[31] N. Vlajic and D. Xia, "Wireless sensor networks: to cluster or not to cluster?" in *Proceedings of the 2006 International Symposium on a World of Wireless, Mobile and Multimedia Networks (WoWMoM '06)*, pp. 258–266, June 2006.

[32] W. B. Heinzelman, A. P. Chandrakasan, and H. Balakrishnan, "An application-specific protocol architecture for wireless microsensor networks," *IEEE Transactions on Wireless Communications*, vol. 1, no. 4, pp. 660–670, 2002.

[33] N. Xu, A. Huang, T. W. Hou, and H. H. Chen, "Coverage and connectivity guaranteed topology control algorithm for cluster-based wireless sensor networks," *Wireless Communications and Mobile Computing*, vol. 12, no. 1, pp. 23–32, 2012.

[34] O. Younis and S. Fahmy, "HEED: a hybrid, energy-efficient, distributed clustering approach for Ad-Hoc sensor networks," *IEEE Transactions on Mobile Computing*, vol. 3, no. 4, pp. 366–379, 2004.

6

Remote-Time Division Multiplexing of Bending Sensors Using a Broadband Light Source

Mikel Bravo and Manuel López-Amo

Departamento de Ingeniería Eléctrica y Electrónica, Universidad Pública de Navarra, Campus Arrosadia S/N, Navarra, 31006 Pamplona, Spain

Correspondence should be addressed to Mikel Bravo, mikel.bravo@unavarra.es

Academic Editor: Weiqi Jin

This work experimentally demonstrates a remote sensing network which interrogates bending sensors using time-division multiplexing techniques and a broadband light source. The bending sensors are located 50 km away from the monitoring station. They are based on a simple tie displacement sensor and offer high-resolution measurements of displacement.

1. Introduction

Optical fiber intensity sensors based on bend mechanisms have been extensively used because they are simple, cheap, reliable, and can be multiplexed and used simultaneously in multiple locations inside a network. Furthermore, the technique of bending losses has been successfully applied to the measurement of many magnitudes like displacement, pressure, strain, vibration, or temperature [1]. Due to their properties of electromagnetic immunity, the utilization of these sensors has been growing, avoiding serious accidents and guaranteeing a higher level of safety.

Bends in optical fibers cause power propagating in guided modes to be lost by coupling to radiation modes. This type of loss is experienced in fibers wrapped around mandrels or deployed in flexible cables. Attenuations over tens of dB are possible to be achieved depending on the bending radius in single-mode fibers [2] or fiber tapers [3]. The main objective of the present paper is to report the multiplexing of a macrobending single mode fiber structure to be used as an inexpensive simple sensor, for displacement or high strain measurements.

The sensors used in this work consist of a continuous piece of optical fiber "tied" into the shape of a figure of eight. The sensor was initially demonstrated in [4] using multimode fiber; a singlemode fiber (SMF) version of the sensor was used in [5]. In that paper, 4 sensors were multiplexed using an optical time domain reflectometer (OTDR). Recently, we have used the singlemode version of the sensor to monitoring a concrete beam bending test [6] using also an OTDR. The sensors performed better than fiber Bragg grating (FBG) based sensors when high strain was applied to the concrete beam.

The utilization of an OTDR as the interrogation system of bending sensors, in combination with optical amplifiers, would allow also the remote utilization of them [7].

Remote sensing is related with the continuous monitoring of structures from a central station located tens or hundreds of kilometers away from the field through the critical location of sensors which send information to the central station. This remote capability allows immediate damage detection and consequently necessary actions can be taken.

However, when using an OTDR for remote monitoring of these sensors, this raises two issues:

A low dynamical range of the measurement (2-3 dB) [5, 7] and the cladding mode coupling effects, which causes a nonlinear response in the loss versus the bending radius variations [8]. This occurs when using a single wavelength laser source, as the incorporated one into an OTDR. Total loss of a bent fiber includes the pure bend loss in the bent section and the transition loss caused by the mismatch

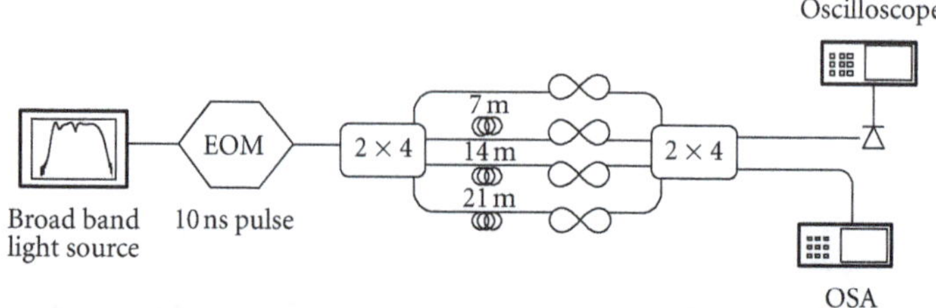

Figure 1: TDM sensing system schematic diagram used for the initial characterization of the sensors.

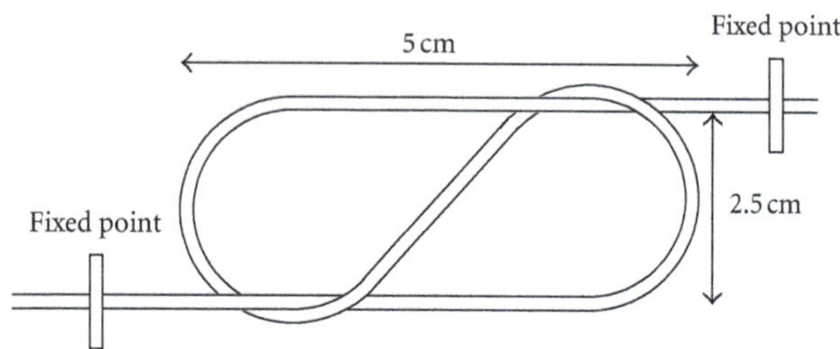

Figure 2: Bend tie sensors diagram, including initial dimensions for a nonstressed sensor.

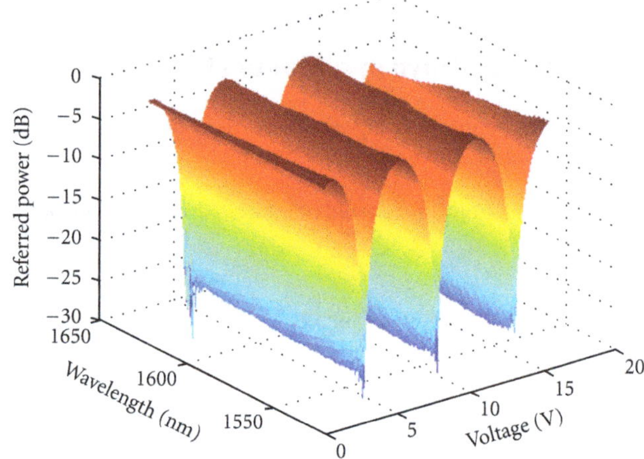

Figure 3: EOM response to the biasing point variation.

of propagation mode between the bent and the straight sections. For a single mode bent fiber of length L, the pure bend loss can be calculated by

$$L_s = 10 \log_{10}(\exp(2\alpha L)) = 8.686\alpha L, \quad (1)$$

where α is the so-called bend loss coefficient, which is determined by the fiber structure, bending radius, and wavelength of the light. The existence of the coating layer will produce a so-called whispering gallery mode for a bent fiber, due to the reflection of the radiated field at the interface between the cladding layer and the coating layer. In order to consider the effect of this reflection on the bend loss, more complicated formulas for the bend loss coefficient α have been developed [9] which justifies the power oscillation.

Because these oscillating peaks change with wavelength, a straightforward way of soften their effect is to use a broadband light source instead the used in conventional OTDRs.

In this work, we demonstrate the advantages of the utilization of a broadband light source for time division multiplexing (TDM) of remote bend loss sensors, increasing the resolution of the measurements and avoiding the dispersion problems caused by the light source.

2. Methodology

The initial developed system for TDM multiplexing of bend loss sensors is depicted in Figure 1.

The sensing section consists of four intensity sensors arranged in a parallel (star) configuration. The sensors are based on the bend tie sensor diagram firstly proposed by Sienkiewicz and Shukla (Figure 2) [4]. They are capable of measuring large displacements and consequently high deformations when their ends are attached to a concrete beam, for example [6]. In order to characterize these sensors, their input and output fiber sections (described as "fixed point" in Figure 2) are placed on two high precision motorized stages which makes that the tie shape changes with the displacement. Therefore, there is a curvature change which increases or decreases the losses. These stages provide a minimum displacement of 1.7 nm. For this work, we have developed a LabVIEW program which integrates all the instruments to make measurements of the displacement of the sensors each 3.4 mm. Also the fiber of the tie sensor is smeared of lithium grease for achieving a soft sensor glide and a good repeatability.

This characterization setup allows us to make a high-resolution characterization and to see the cladding mode effects that are present in the most part of the bending sensors systems.

In order to multiplex the sensors by using TDM techniques in a star configuration, two 2 × 4 couplers are used to divide the launched signal and to collect the intensity modulated outputs. To discriminate the sensors, different fiber delays are included in each sensing branch.

The characterization system includes a broad band light source whose spectrum is depicted in Figure 5, before and after the modulator, an EOM (electro-optic modulator), an optical spectrum analyzer (OSA), and an oscilloscope.

The operation mode is simple: a broad band light pulse of 10 ns enters into the first 2 × 4 coupler which divides this pulse in 4 almost identical pulses travelling through different

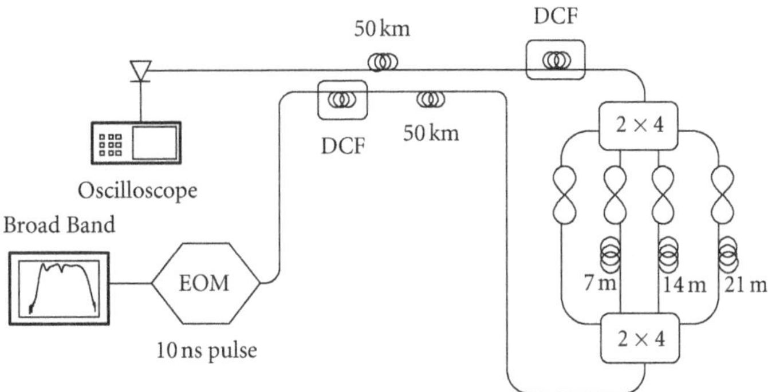

FIGURE 4: Remote (50 km) sensor system.

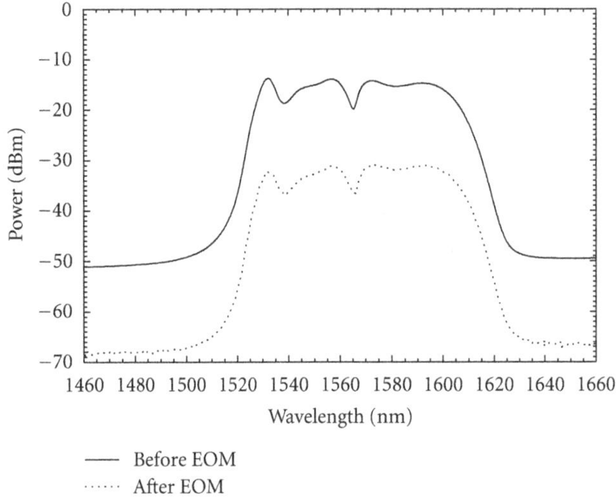

FIGURE 5: Broadband light source spectrum measured before and after the EOM.

fiber delay lines before being attenuated by the sensors, these sensors change their behavior while the motorized stages modify the distance between the fixed points. Afterwards, the pulses are recoupled in the second 2 × 4 coupler. Finally, four different pulses corresponding to the different sensors are detected by a New Focus 1811, 125 MHz, and 900–1700 nm high sensitivity detector. This detector shows a high trans impedance gain (4 × 104 V/A) and a low noise equivalent power (NEP), having a 3 ns rise time. Finally, the electrical signal offered by the detector is monitored by an oscilloscope. The high sensitivity characteristic of the detector is due to the low optical power present at the end of the remote network. The OSA was used to characterize the sensors spectral behavior using the spare output port of the 2 × 4 coupler.

We use a broadband light source from Nettest, model fiberwhite SPL, which covers a wavelength spectrum from 1525.5 to 1611.7 nm, with a maximum output power of 18.8 dBm and power stability of 0.005 dB. This broadband source avoids the oscillations measured in the curvature loss function because of the cladding mode coupling process when a narrow band laser is used [7]. The cladding mode excitation has a wavelength displacement according to the bending radius. Therefore, the losses caused by curvatures are very dependent on the wavelength. By using the broad band source, we achieve a loss averaging of this dependence, avoiding the unwanted effect of the cladding mode recoupling. Because of the nonuniform spectrum of the broadband source, we have referenced the measurements (Figures 3 and 7) regarding the first spectrum obtained with a nonstressed sensor. On the other hand, the utilization of short bending radius loops as mode scramblers, as we did in previous works [5, 6], contributes to attenuate the excited cladding modes improving also the accuracy of the measurements. These fiber loops were optimized for each sensor. The loops configuration was developed after several experimental tests, showing the optimum cladding mode suppression behavior. In our experimental set-up, we used only one loop before two of the sensors (the second and the fourth sensor).

A $LiNbO_3$ electro-optical "UTP APETM 2 × 2 interferometric switch" acts as an electro-optic amplitude modulator and it is used to generate short pulses from the light source. The EOM transfer function and its biasing point are two crucial parameters to be controlled. Figure 3 shows the characterization of the modulator response to the changes of the biasing point. These measurements were made by modulating the broad band spectrum, which is registered

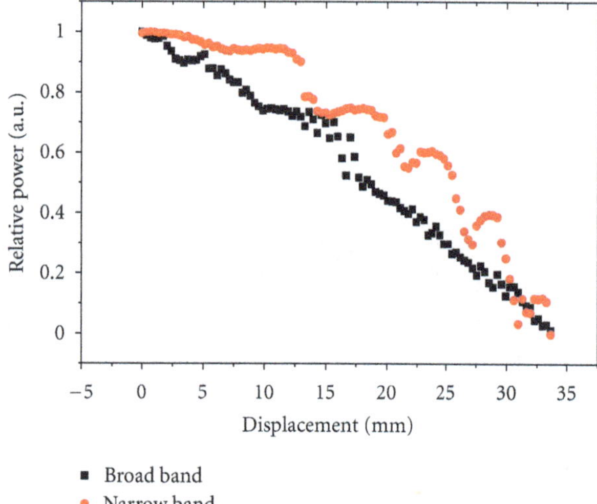

Figure 6: Measured transfer function of a displacement sensor when a laser diode (red) and a broadband light source (black) are used.

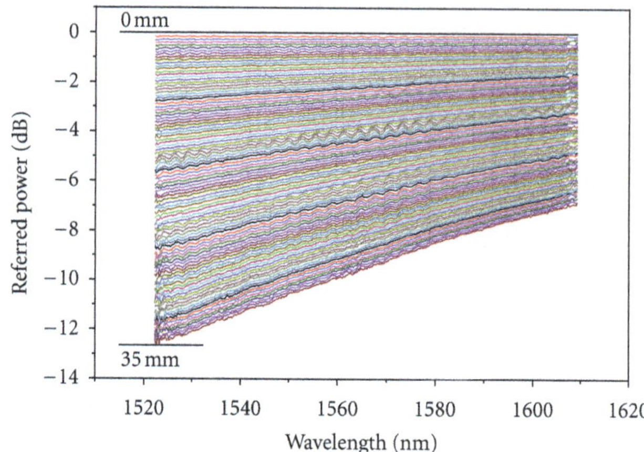

Figure 7: Spectrum evolution according the applied displacement after the sensor (the displacement applied is marked inside).

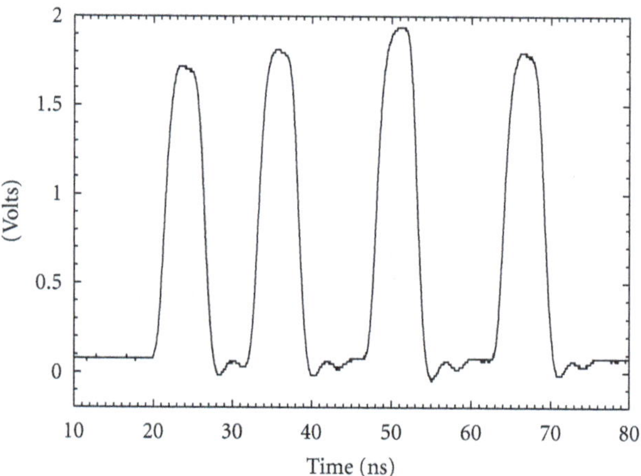

Figure 8: Oscilloscope view of the four sensors' received pulses.

by an OSA. A computer controlled DC source increased 0.01 V the biasing point in each measurement. This 3D graph shows the extinction ratio versus the wavelength and the polarization voltage applied. This polarization point is a crucial parameter to be controlled and to be stabilized. The transfer function of the EOM modifies the spectrum of the source while the biasing point controls the amplitude of the pulses.

In order to check the application of this structure for remote sensing, we introduced two 50 km fiber reels between the light source and the detector, as shown in Figure 4.

Dispersion is a crucial aspect to be taken into account when the optical path is enlarged and a modulated broadband light source is used. Because of this reason, two long dispersion compensation fiber (DCF) sections are used to compensate the dispersion caused by the 100 km of standard single mode fiber (SMF28). From our measurements we have obtained that the DCF reduces correctly the dispersion, although we have used a broadband source, allowing the sensors detection. However, the price to pay is an additional 7.8 dB loss.

3. Results

This section presents the experimental results for the proposed system. Firstly, we have compared the behavior of one of our displacement sensors using a laser diode from an OTDR (EXFO FTB-7423B-B) and the light from our modulated broadband light source. Figure 5 shows the broad band source spectrum both before and after the EOM. As it is explained before, the biasing point of the EOM is a crucial parameter to be controlled and the biasing point in this case is placed at the minimum point level of the EOM response for achieving the best dynamic range of the modulated pulses.

Figure 6 shows the characterization of one of our intensity sensors when displacement is applied and the two different light sources are used. The red trace corresponds to the measurement of the sensor when the OTDR laser is used in a transmissive configuration, as we did in [6]. A nonlinear behavior as in [8] is observed. This behavior limits the resolution and applicability of the sensor. The nonlinear behavior corresponds to the excitation of new cladding modes when the bending radius is reduced with the applied displacement, and this excitation shifts with wavelength [8]. Thus, when a broadband light source is used and all its spectral components are simultaneously introduced into the bending sensor, an average effect is obtained and the linearity of the transfer function is improved.

Figure 7 shows the evolution of the attenuation spectra as the bending radius changes. The first trace (corresponding to the shown in Figure 5, "After EOM") is taken as the reference one. Thus, to calculate the attenuation, the trace taken in each bending state is subtracted to the reference one. The results confirm the dependence of macrobendings with the wavelength.

Figure 8 shows the received signal from the 4 bend tie multiplexed sensors using a high sensitivity detector. For the

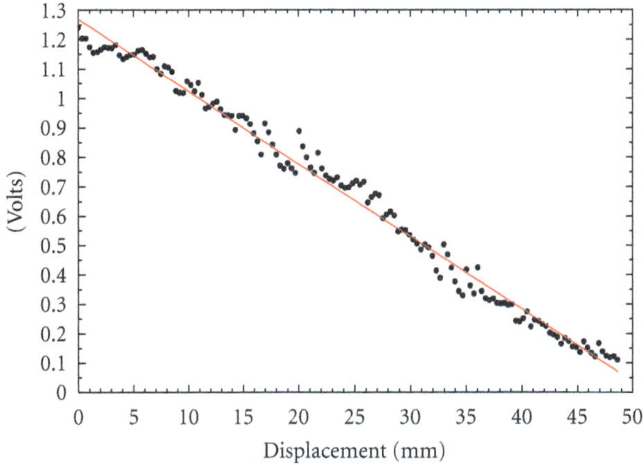

Figure 9: First sensor received signal evolution when displacement is applied.

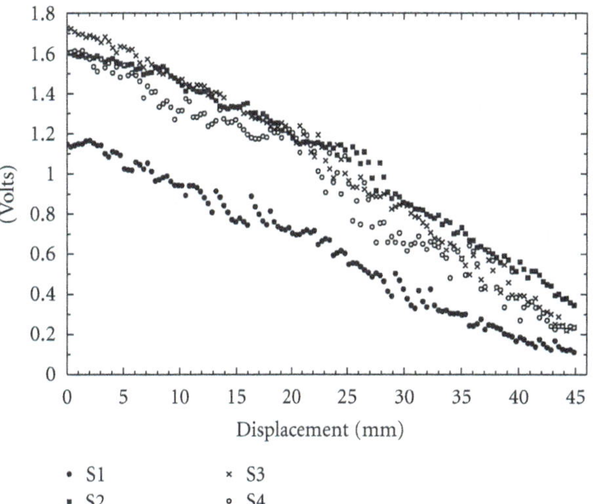

Figure 10: All sensor received signals evolution when displacement is applied.

experiment, we have tested each sensor separately and we have checked the crosstalk for the other sensors. A delay of approximately 20 ns (corresponding to a 7 m long fiber coil) has been applied to the pulse corresponding to this sensor. The number of sensors that can be multiplexed using this scheme is limited by the received optical power from each sensor. Theoretically, in a nonremote setup is possible to multiplex up to 32 sensors using our topology and equipments. However, for remote applications, this number should be reduced if a good received SNR is desired maintaining the same resolution. Figure 9 shows the evolution of the first sensor measured signal at the detector as it is displaced. In this result, we have a linear response and the cladding mode coupling effect does not appear. A linear fit was performed in the graph to show the linear response of the sensor.

Figure 10 shows the measured displacement behavior of the four sensors. The received signal levels are different because of the different mode scrambler loops calibration parameter explained before.

An important parameter to study is the crosstalk between sensors. In this work, the crosstalk was checked for each sensor. Figure 11 shows the second sensor behavior and the simultaneous signal variations obtained from the others. This figure demonstrates the crosstalk free behavior of this system because the received pulses corresponding to each sensor has its own time slot without suffering overlapping. Therefore, the size of the pulse and the delay of each branch must be well controlled to prevent pulse overlapping.

We have also made an identical characterization of the sensors using the remote sensor system configuration which has the sensors 50 km away from the monitoring equipments. On this occasion (Figure 12), the evolution of the sensor voltage, as the sensor is displaced, is as linear as the previous one but the dynamic range in this case is much lower. However, it is still sufficient to perform the desired measurements.

Finally, the system resolution, accuracy, and repeatability have been tested for both systems, local and remote systems.

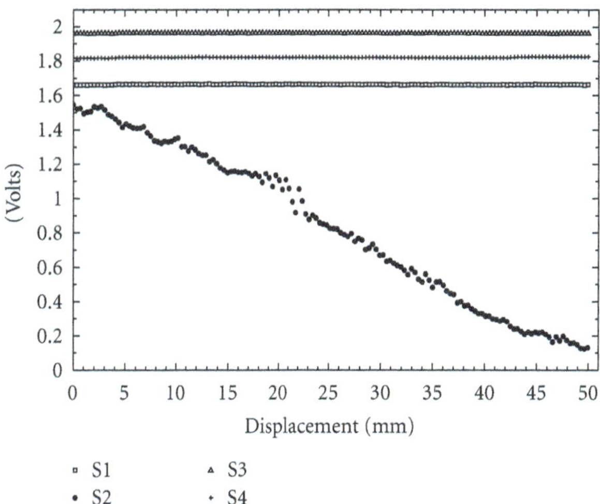

Figure 11: Sensors "S1, S3 and S4" crosstalk behavior when displacement is applied to "S2".

Figure 13 shows different measurements along time. With this aim, 30 measurements per minute have been taken without any sensor displacement. The aim of these measurements is to check the system noise and the measurement error versus time. The light source and the utilized photodetector will introduce the corresponding intensity noise and quantum and thermal noise, In order to reduce this noise, averaging of the received signals was used. Other possible measurement errors, coming from instabilities from the EOM and sensor network instabilities may be reduced using an intensity referenced scheme [10]. Using these data, we can estimate the system resolution and accuracy. The measurement frequency can be as fast as 1 Hz, being the real limitation the oscilloscope acquisition speed and data processing time.

The nonremote sensor system resolution is ±0.5 mm and the remote sensor configuration resolution is ±1 mm in the measurement range of 35 mm. These measurements were

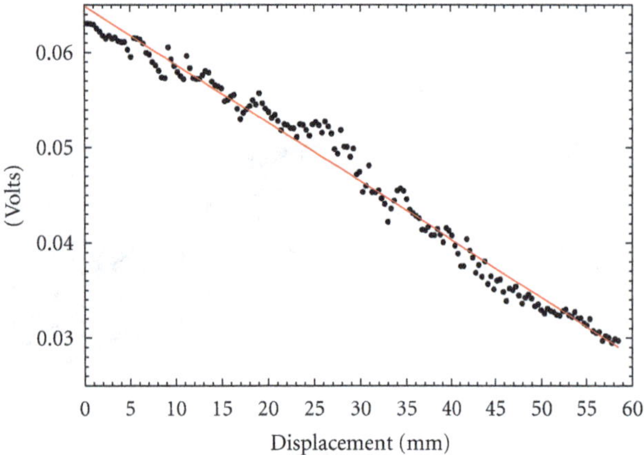

FIGURE 12: Remote sensor detected signal evolution versus displacement.

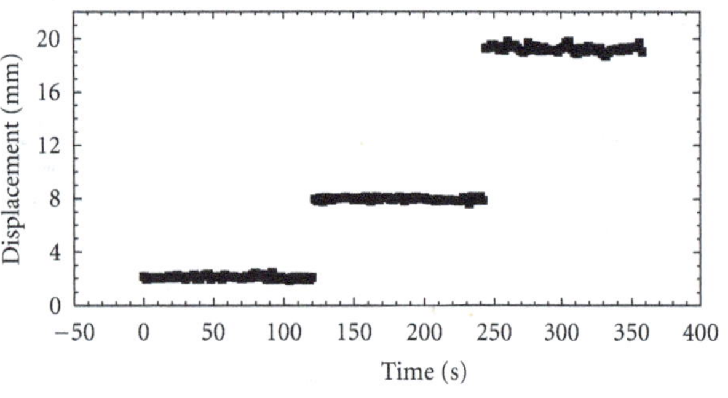

(a)

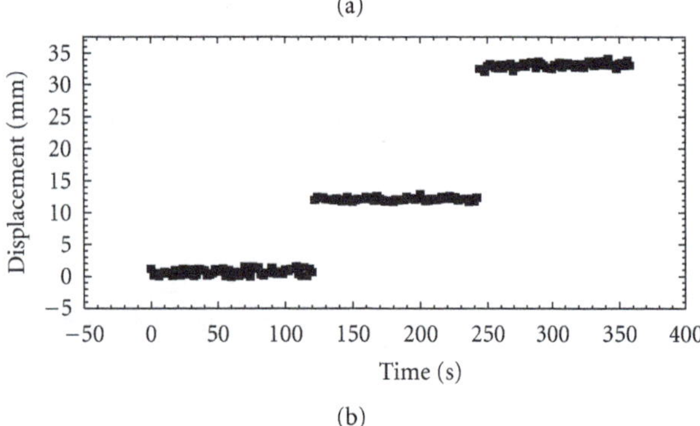

(b)

FIGURE 13: Local (a) and remote (b) measurements for resolution calculations.

obtained making an average of 64 oscilloscope traces in order to reduce noise. The measured repeatability, defined as the maximum difference between the consecutive readings of the sensors [11], was ±1.14%.

4. Conclusion

The utilization of broadband light sources for time-division multiplexing of bend sensors has been demonstrated. An in-depth study of the cladding-mode coupling effects has been carried out to validate the proposed solutions and to improve the sensing characteristics. The adopted light source improves the resolution and the linear behavior of the measurements. We have also demonstrated that these sources can also be used in remote sensing, multiplexing 4 bend loss sensors for displacement measurements 50 km away from the header of the system, without the utilization of optical amplifiers.

Acknowledgments

Thanks are due to Montserrat Fernández-Vallejo for her help in the paper preparation and useful discussions, and the authors are grateful to the Spanish Government Project TEC2010-20224-C02-01.

References

[1] J. M. López-Higuera, "Introduction to fibre optic sensing technology," in *Handbook of Optical Fibre Sensing Technology*, pp. 1–23, 2002.

[2] W. A. Gambling, H. Matsumura, C. M. Ragdale, and R. A. Sammut, "Measurement of radiation loss in curved single-mode fibres," *Microwaves Optics and Acoustics*, vol. 2, no. 4, pp. 134–140, 1978.

[3] F. J. Arregui, I. R. Matías, C. Bariain, and M. López-Amo, "Experimental design rules for implementing biconically tapered single mode optical fibre displacement sensors," in *European Workshop on Optical Fibre Sensors*, vol. 3483 of *Proceedings of SPIE*, pp. 164–168, July 1998.

[4] F. Sienkiewicz and A. Shukla, "A simple fiber-optic sensor for use over a large displacement range," *Optics and Lasers in Engineering*, vol. 28, no. 4, pp. 293–304, 1997.

[5] N. M. P. Pinto, O. Frazão, J. M. Baptista, and J. L. Santos, "Quasi-distributed displacement sensor for structural monitoring using a commercial OTDR," *Optics and Lasers in Engineering*, vol. 44, no. 8, pp. 771–778, 2006.

[6] M. Bravo, J. Sáenz, M. Bravo-Navas, and M. Lopez-Amo, "Fiber optic sensors for monitoring a concrete beam high strain bending test," *Journal of Lightwave Technology*, vol. 30, no. 8, pp. 1085–1089, 2012.

[7] M. Bravo, M. Fernandez-Vallejo, and M. Lopez-Amo, "Hybrid OTDR-fiber laser system for remote sensor multiplexing," *IEEE Sensors Journal*, vol. 12, no. 1, pp. 174–178, 2012.

[8] H. Renner, "Bending losses of coated single-mode fibers: a simple approach," *Journal of Lightwave Technology*, vol. 10, no. 5, pp. 544–551, 1992.

[9] Q. Wang, G. Farrell, and T. Freir, "Theoretical and experimental investigations of macro-bend losses for standard single mode fibers," *Optics Express*, vol. 13, no. 12, pp. 4476–4484, 2005.

[10] M. Bravo, A. M. R. Pinto, M. López-Amo, J. Kobelke, and K. Schuster, "High precision micro-displacement fiber sensor through a suspended-core Sagnac interferometer," *Optics Letters*, vol. 37, no. 2, pp. 202–204, 2012.

[11] D. S. Nyce, *Linear Position Sensor. Theory and Application*, John Wiley & Sons, 2004.

7

Investigation of Wireless Sensor Networks for Structural Health Monitoring

Ping Wang,[1] Yan Yan,[1] Gui Yun Tian,[2] Omar Bouzid,[2] and Zhiguo Ding[2]

[1] *College of Automation Engineering, Nanjing University of Aeronautics and Astronautics, Yudaojie Road 29, Jiangsu 210016, China*
[2] *School of Electrical and Electronic Engineering, Newcastle University, Newcastle upon Tyne NE1 7RU, UK*

Correspondence should be addressed to Ping Wang, zeit@263.net

Academic Editor: Raimond Grimberg

Wireless sensor networks (WSNs) are one of the most able technologies in the structural health monitoring (SHM) field. Through intelligent, self-organising means, the contents of this paper will test a variety of different objects and different working principles of sensor nodes connected into a network and integrated with data processing functions. In this paper the key issues of WSN applied in SHM are discussed, including the integration of different types of sensors with different operational modalities, sampling frequencies, issues of transmission bandwidth, real-time ability, and wireless transmitter frequency. Furthermore, the topology, data fusion, integration, energy saving, and self-powering nature of different systems will be investigated. In the FP7 project "Health Monitoring of Offshore Wind Farms," the above issues are explored.

1. Background

Wireless sensor networks (WSNs) are considered one of the most important technologies in the 21st century, which will have a profound impact on the future way of life for humankind. A typical wireless sensor network consists of wireless communication, data acquisition, processing, and fusion stages. The sensor nodes are self-organised through a specific protocol and are able to obtain information about the surrounding environment, working together to accomplish specific tasks. The technology spans many fields, such as wireless communication, network technology, integrated circuits, sensor technology, microelectromechanical systems (MEMS), and embedded systems, to name just a few. Figure 1(a) illustrates the typical basic architecture of a sensor network, and Figure 1(b) shows the typical hardware of a sensor network node. In February 2003, "Technology Review," a magazine published in the United States, selected ten far-reaching impacts of emerging technologies, where sensor networks were ranked in the first place. The US Department of Defense gave high priority to wireless sensor networks and put it as an important research area, establishing a series of military research projects. Intel, Microsoft, and other companies also have carried out some research in the field.

Not only industry and defense organisations have shown strong interest in wireless sensor networks, some world class universities such as UCLA, MIT, Cornell University, and the University of California, Berkeley [1] also have carried out research on WSN and achieved some results. The University of California, Berkeley, presented network connectivity reconstruction method for sensor location, based on the correlation of the data coding modes, to determine the location of sensor network nodes; they also developed a sensor operating system, TinyOS. The Massachusetts Institute of Technology studied sensor network data stream management systems integration frameworks, query optimisation, network congestion control for energy saving purposes, along with middleware technologies such as positioning, tracking, networking, and scalable algorithms for large-scaled sensor networks. These studies received NSF, DARPA, and the Air Force Space Laboratory support. The ZigBee Alliance and IEEE also developed standards of Chinese version IEEE 802.15.4c [2].

After developing further in the past several years, some applications of wireless sensor networks have been turned

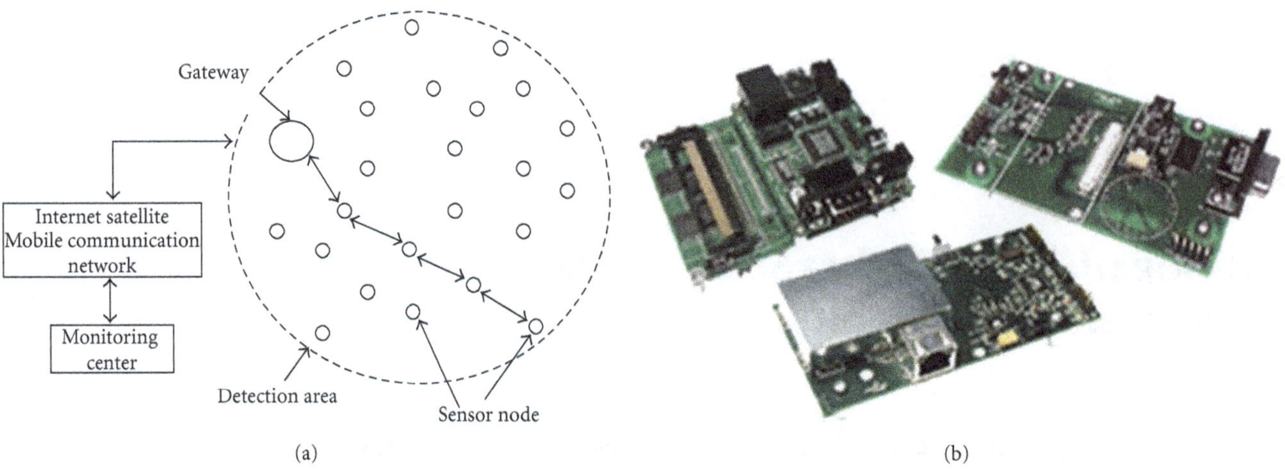

FIGURE 1: (a) a sensor network architecture. (b) hardware of sensor network nodes.

into products. Nowadays a number of WSN research platforms have been developed successfully, such as the University of California, Berkeley TinyOS/Mica, Smart Dust, and PicoRadio platforms. The ZigBee Alliance was codeveloped by Invensys, Mitsubishi Electric, TI, Motorola, Philips, and more than 20 other semiconductor and IT companies. Meanwhile, the IEEE has also developed the IEEE802.15.4 standards and has established a number of demonstration systems. Crossbow has developed a series of modules and products based on the ZigBee protocol and TinyOS, which can be used to form a WSN. TI has also developed some solutions based on ICs such as CC2431, which embed the ZigBee protocol of MAC and PHY on them and offer developments that contain the software and firmware of the complete ZigBee protocol stack [3].

Structural health monitoring (SHM) is a typical area amongst the many possible applications of wireless sensor networks. SHM is an emerging research area and is focussed on the field of infrastructure [4, 5], mainly on the integration and application of sensors, signal processing, and communication technologies. It also focuses on complex engineering systems and infrastructure to prevent structural failure and disaster, such as with the monitoring of bridges, large buildings, and dams. One of the possible SHM applications is that of a wind power monitoring system illustrated by the EU science and technology key project FP7 "Health Monitoring of Offshore Wind Farms." By 2020, China wind power capacity will reach 150 million kilowatts [6], and nondestructive testing (NDT) and SHM are useful means with which to reduce the maintenance costs of wind power, and to extend the lifetime of infrastructure, and to ensure the safety of power supply. As shown in Figure 2, the project aims to analyse wind power generation systems and to develop intelligent WSN and SHM technologies for the wind turbine blade, gearbox, generator, power electronics, and other structural components. In this project, a complete wind power system health analysis, life cycle assessment, fault diagnosis, maintenance management programs, planning and scheduling system of a complete windpower system is provided, including the design, production, installation, maintenance, and supply chain feedback.

In Section 2, the key problems acting against the implementation of WSNs for SHM on wind power systems will be identified and discussed.

2. The Key Problems of Wireless Sensor Networks Combined with Structural Health Monitoring

As briefly discussed in Section 1, the key problems to developing such an SHM system in conjunction with WSN are summarised as follows: the problem of compatibility between different sensors, their sampling frequencies, and operational modes, the problem of transmission bandwidth and real-time ability variance, the selection of a wireless transmission frequency, topology choice, data fusion method, and the contrast between the energy consumption requirements of different applications to that of each different device. In the following subsections, each individual problem will be discussed in detail.

2.1. Sampling Frequency and Operation Mode. In the field of structural health monitoring, various types of sensors are used, including, but not restricted to, resistance strain, piezoelectric vibration, optical fiber strain, dip angle, acoustic emission, and stress measurement sensors [7]. Each of these sensors has different physical mechanisms and should thus be operated in different ways; some examples are given as follows.

(1) The signals of strain, deformations, and dip angle are static or of low frequency, and they usually work at low sampling frequencies. For example, some of the signals in a strain test are sampled by a frequency lower than 1 Hz. The likelihood of a request for data processing and real-time transmission is low.

(2) Vibration sensors usually measure objects where the level of vibration can range from dozens of Hz to

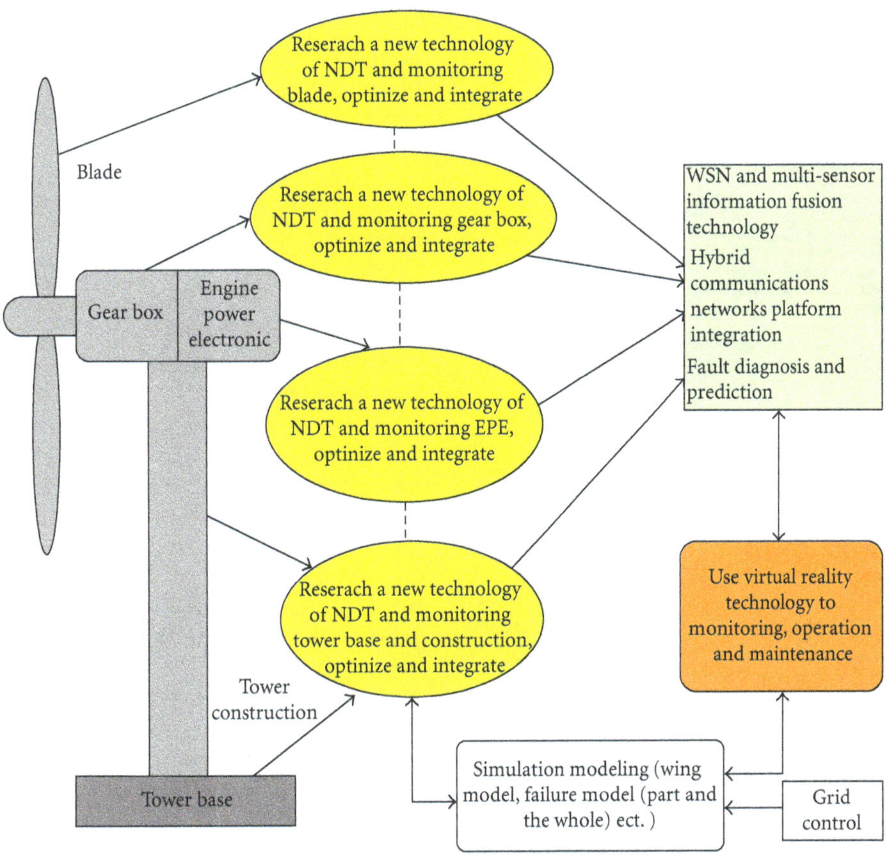

Figure 2: The system architecture of FP7 project "Health Monitoring of Offshore Wind Farms".

Table 1: The carrier frequency and transmission bandwidth of ZigBee Alliance and the IEEE 802.15.4 standard.

Carrier frequency	Band nature	Maximum bandwidth	Frequency point
2.4 GHz	ISM Worldwide	250 kbps	16
868 MHz	Europe	20 kbps	1
915 MHz	ISM Americas	40 kbps	10
780 MHz	802.15.4c (Chinese)	250 kbps	8

hundreds of KHz. Thus, there is a much greater requirement from the system in terms of sampling frequency, data processing, and transmission. The added need for synchronisation from the system adds an extra burden.

(3) Eddy current, pulsed magnetic flux leakage, and other nondestructive testing sensors are used for the monitoring of key components, where signals are sampled with a frequency of more than a few hundred Hz. In addition to this the requirement of data processing and analytical abilities from the WSN nodes is very high.

(4) The imaging sensors require the WSN nodes to have the ability to carry out high-speed data transmission. They also require WSN nodes to have certain decoding and image processing abilities.

As the task of structural health monitoring systems is a complicated one, consisting of the monitoring of many physical and electrical failures in different components, they need various sensors working together. The choices of the sensor network sampling frequency, from several Hz to several hundreds of kHz, working mode, and compatibility must be considered when choosing each node.

2.2. Transmission Bandwidth and Real-Time Monitoring. In general, with the requirements of low cost and low power, the design of WSN has mainly been for low bandwidth and non/nonurgent real-time applications, as shown in Table 1. For applications such as SHM, designs based on low-bandwidth and non-real-time systems can only be applied to strain deformation or dip monitoring, which require only slowly varying signal sensing and transmission. The data from vibration measurements as well as those resulting from image acquisition require a higher transmission bandwidth.

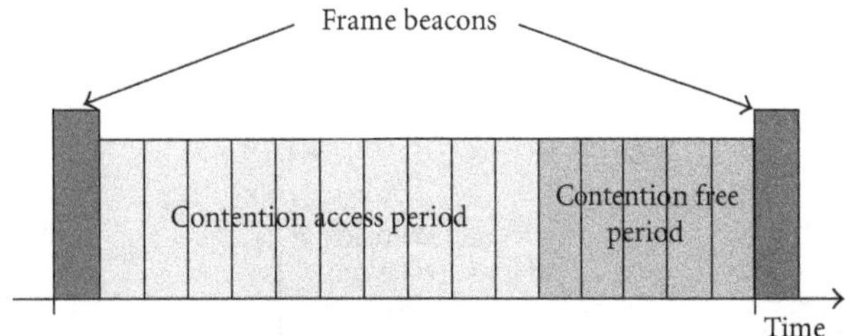

Figure 3: ZigBee MAC time slot with competition and sub-slot competition.

In addition to these, for some alarm transmission with high real-time requirements, such as those from pulsed eddy current measurement node (which can receive harbinger information of the sudden occurrence of crack), special considerations with regard to real-time network transmission need to be undertaken.

Except for the physical communication and processing capacity of a system, the protocols, especially MAC protocols, influence the real-time ability and transmission bandwidth of network communication [8, 9]. MAC technology usually consists of two different mechanisms, competitive and noncompetitive mechanisms. The competitive MAC mechanism, carrier sense multiple access with collision avoidance (CSMA/CA), is most commonly used. Distributed layer MAC protocols, which are based on the CSMA/CA protocol, use a sleep cycle mechanism to limit energy consumption. But a larger network time delay is produced as the system needs to be awoken from a sleep cycle before the data can first be sent to it and then transmitted onwards. In addition, the T-MAC (MAC) protocol transmits all of the data through a variable-length sudden transmission and sleeps during the two sudden transmissions to reduce spare detection.

Noncompetitive MAC protocols are generally used for cluster-based networks. The cluster head is responsible for allocating time slots to all of the sensor nodes within a cluster, collecting and processing data which is sent by the sensor nodes within a cluster, and forwarding data to the sink node.

In this project, we use a MAC mechanism, which is a combination of both competitive and noncompetitive methodologies, which is common practice in WSN technologies. One example of this is within ZigBee, with its purpose illustrated in Figure 3, allowing accommodation of the varied application requirements of a different network number of nodes and different real-time requirements.

Although the most common WSNs are for low-bandwidth applications, the development of technology and the demands of more diversified sensing technologies and more various information acquisition—especially for the application of acquisition of large amounts of data such as the vibration and image data—and the bandwidth requirements for sensor networks have enlarged and the management models required have become more strict.

2.3. Synchronization. In structural health monitoring, there are many requirements regarding synchronous and real-time data acquisition of the vibration information, which are distributed over different parts of an installation. It is especially important for the vibration model analysis of bridge structures, structural stability, and life assessments, which contain a large number of sensor nodes that are distributed over different positions, with different topology structures. The signals must be sampled synchronously by the nodes; otherwise there will be incorrect information (due to samples grouped together coming from different times) of the vibration phase, resulting in an incorrect vibration model judgement. In applications in which sampling frequencies usually exceed 1 KHz, the delay of the sensor nodes synchronisation is usually required to be less than 1 μs. This results in a higher requirement of synchronisation in an SHM WSN.

2.4. Operation Frequency. The wireless carrier frequency has a direct impact on the physical layer transmission accessibility of a network, which plays a significant role in the application of structural health monitoring. Carriers of 2.4 GHz and 868 MHz and 915 MHz are used according to IEEE 802.15.4 and ZigBee Alliance standards listed in Table 1. The penetration ability through buildings at these frequencies is acceptable, but the diffraction characteristics are relatively poor. When the interior structure of a building is under test, the performance using ZigBee is poor due to poor penetration. In contrast, the 433 MHz ISM band, which is used widely in fields such as automated meter reading (AMR), has achieved relatively good results. In addition, the frequency band of 470 MHz has been selected to be the instrumentation and sensor network special frequency band in China.

Frequencies used in applications such as AMR can be a reference for comparison, classification, and standardisation of the carrier frequency of WSN in SHM.

2.5. Topology and Data Fusion. WSNs need different topologies to meet the needs of different application characteristics in SHM. Typical topologies include star, cluster tree, and mesh networks [9]. Taking the ZigBee network for example, shown in Figure 4, the node is composed of both a reduced function device (RFD) and a full function device (FFD). The ZigBee standard supports the three kinds of topologies that were stated above [10, 11].

In the FP7 project mentioned previously, "Health Monitoring of Offshore Wind Farms," the project team compared the characteristics of three different topologies, as we can see in Table 2.

In our practical application, the star network is used for deformation monitoring of low-speed, nontimely applications. In addition to the above three forms, a chain structure (a simplification of mesh) and a combination of the chain structure and the three previous mentioned methodologies has become a commonly used topology.

2.6. Energy Issues. A lot of WSN nodes in SHM applications work in environments where a direct power supply is not available, such as with nodes located upon spinning wind power blades. Therefore, any enhancement of a single node lifetime through the use of low-power technology is a major

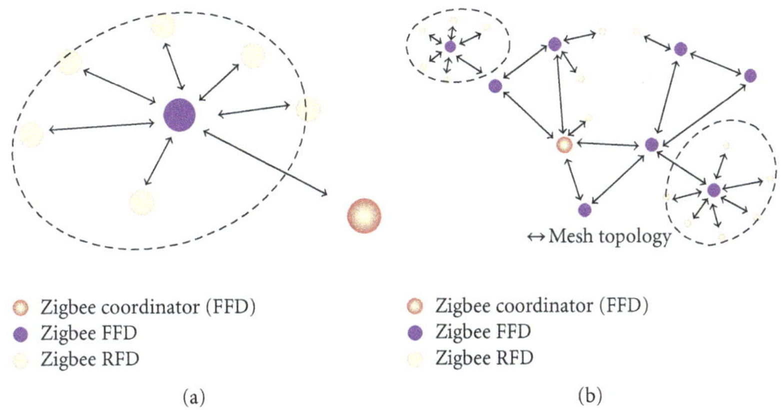

FIGURE 4: ZigBee (a) STAR and (b) MESH network topologies.

TABLE 2: Comparison of three kinds of topology for WSN.

	Networking features	Advantage	Disadvantage
Star	Supporting point-to-multipoint and multipoint-to-point communication. Need for a concentrator or hub (coordinator for ZigBee network). All data are transited by hub. Suited for networking in which the distance between equipments is small and scattered in a circular fashion. Suitable for massive, low-speed, and non-real-time applications, such as distortion monitoring.	Brief structure. Protocol is simple by using a polling mode.	The entire spatial span of the transmission network is equal to the span of the physical layer. Concentrator can carry out the role of data fusion. Its reliability has the decisive effect regarding the performance of the entire network.
Cluster tree	Network is composed of many star topologies, using a star to connect the concentrator. Suitable for situations where data needs to be converged, such as in a structure monitoring network that has an intelligent analysis ability.	A leveled and clear structure. Transmission distance is greater than the star topology. Protocol is relatively simple and clearer than with mesh networks.	All the root nodes of the cluster tree (concentrator or the convergent point) are essential nodes, upon which there is also the demand of a high ability in terms of data fusion. Energy consumption and reliability as well as the influence on the entire network are a major issue.
Mesh	All entities can communicate with each other as long as they are within the scope of the network and if there is no direct path; they also can communicate in "multihop" fashion. They can form highly complex and large capacity networks. The network can cover a large physical space so it is suitable for installation on scattered structures. The network also has self-organisation and self-healing capabilities.	The behaviour of the single node is relatively consistent. The capacity of the network is large. Generally it does not have any essential node. There are many available communication routes and the success ratio of the transmission is increased in comparison to those gained with other networks.	The behaviour of the entire network is not easily controlled. With an increasing number of nodes, the complexity of the network behaviour increases greatly and the efficiency is reduced.

issue in the application of WSN. Impacting factors upon the lifetime of nodes and the entire network include the following.

(1) The Working Principle and Mode of Sensing System. Each of the many sensors that adopt different physical principles to detect the different phenomena of strain, vibration, obliquity, cracks, the sampling frequency, working mode, and working energy consumption has a very different working principle. A resistance strain sensor operational current needs range from about $200 \sim 400$ uA, while pulsed eddy current sensors require an excitation current of up to 1A or more. Low-speed, low real-time detection usually operates in a larger working cycle, and vibration signals are usually sampled by continuous measurements. The working mode of each sensing system has a decisive influence on the energy consumption of the overall WSN.

(2) The Low-Power and Dormancy Wake-Up Mechanism of Node. Each node employs its own technology to achieve

low power consumption, whilst ensuring the monitoring application. This has to be achieved via the work/sleep mode along with a low duty cycle [12]. Currently the dormancy mode of low-power nodes has reduced current consumption to 1 uA. Generally, the ZigBee node designed for low-speed and real-time applications can run for a period of 6 months to 2 years, with the power supplied by a single AAA battery. In SHM applications such as strain testing, the duty cycle is usually low and so the life of nodes is long.

(3) The Impact of Network Protocols and Calculated Capacity. The calculated capacity requirements of hardware and for the node itself will have a direct impact upon the function of the node. Low-power requirements often conflict with the processing protocol of the WSN, its secure computational ability, and so on. Each function of a WSN, such as self-organisability, adaptability, signal sampling, and information fusion, requires energy consumption. How to ensure the above-mentioned aspects of performance under the condition of low power consumption is always a WSN most crucial issue.

In addition, reasonable network protocols will also play an enhanced role in the reliability of the entire network. For example, in a tree structure, based on the remaining node energy consumption, the key nodes (such as the tree root node and concentrators) are selected in turn. This approach greatly enhances the reliability of the system.

(4) The Self-Powering Capacity of the Node. Currently, with the tendency towards green energy conservation technology development, the use of self-power-generation technology to support nodes has become a research hotspot. Typical self-generation technologies include solar energy technology, vibration and wind-power generation based on the principle of electromagnetic induction, vibration generation based on the principle of the piezoelectric effect, electrostatic collection techniques, and thermoelectric technology based on the temperature effect [13–16]. Amongst these, ZigBee blends the self-generation node into a self-supported system.

In some SHM applications, vibration is the object of measurement as well as the potential power supply source, such as with the vibration of motors and bridges or the spin of blades on a wind farm. In other circumstances there is a large range of solar or temperature change, especially in the west of China, where wind farms are built on a large scale.

3. Conclusion

Wireless sensor networks are one of the supporting technologies in structural health monitoring. Through intelligent, self-organising means, they connect sensor nodes, with a variety of different test objects and working principles into a network along with functions of data processing and integration. Structural health monitoring is a convergence area, with a variety of sensor and information processing technologies. Sampling frequencies, operational modalities of different sensors, their respective transmission bandwidth, and real-time monitoring including interference, wireless operation frequency problems, topology, data integration, and energy issues have been laid out as the key components that will need to be considered carefully when designing the wireless sensor network. In this FP7 project HEMOW, "Health Monitoring of the Offshore Wind Farm Monitoring," our international consortium including research teams from China, India, Poland, and the UK will address the above issues in great detail over the next few years.

Acknowledgments

The research of this paper is supported by FP7 project "HEMOW," the National Science Foundation of China (50907032/E070104), Key Project of Technology of Jiangsu (SBE200900338), and the Ph.D. Programs Foundation of Ministry of Education of China (20093218120019).

References

[1] http://www.jlhlabs.com/jhill_cs/jhill_thesis.pdf.

[2] Part 15.4: Wireless Medium Access Control (MAC) and Physical Layer (PHY) Specifications for Low-Rate Wireless Personal Area Networks (WPANs), IEEE 802.15.4, 2009.

[3] http://www.ti.com/.

[4] H. Q. Si and T. G. Wang, "Calculation of the unsteady airloads on wind turbine blades under yawed flow," *Modern Physics Letters B*, vol. 23, no. 3, pp. 493–496, 2009.

[5] S. Yuan, X. Lai, X. Zhao, X. Xu, and L. Zhang, "Distributed structural health monitoring system based on smart wireless sensor and multi-agent technology," *Smart Materials and Structures*, vol. 15, no. 1, pp. 1–8, 2006.

[6] Fundamental study of Large Size Wind Turbine Aerodynamics, funded by the National Basic Research Program of China; funding: 35 million RMB; project leader: Tongguang Wang, duration: 2007–2011.

[7] J. W. Wilson, G. Y. Tian, and S. Barrans, "Residual magnetic field sensing for stress measurement," *Sensors and Actuators A*, vol. 135, no. 2, pp. 381–387, 2007.

[8] C. Shen, C. Srisathapornphat, and C. Jaikaeo, "Sensor information networking architecture and applications," *IEEE Personal Communications*, vol. 8, no. 4, pp. 52–59, 2001.

[9] J. Yick, B. Mukherjee, and D. Ghosal, "Wireless sensor network survey," *Computer Networks*, vol. 52, no. 12, pp. 2292–2330, 2008.

[10] ZibBee Alliance, ZigBee Specification, ZigBee Document 053474r06 Version 1.0., 2004.

[11] IEEE Standards 802.15.4 TM-2003.Wireless medium access control (MAC) and physical layer (PHY) specifications for low-rate wireless personal area networks (LR-WPANs), 2003.

[12] K. Vijayaraghavan and R. Rajamani, "Active control based energy harvesting for battery-less wireless traffic sensors: theory and experiments," in *Proceedings of the American Control Conference (ACC '08)*, pp. 4579–4584, June 2008.

[13] A. Hande, T. Polk, W. Walker, and D. Bhatia, "Indoor solar energy harvesting for sensor network router nodes," *Microprocessors and Microsystems*, vol. 31, no. 6, pp. 420–432, 2007.

[14] Y. K. Tan, K. Y. Hoe, and S. K. Panda, "Energy harvesting using piezoelectric igniter for self-powered radio frequency (RF) wireless sensors," in *Proceedings of the IEEE International Conference on Industrial Technology (ICIT '06)*, pp. 1711–1716, Mumbai, India, December 2006.

[15] L. Mateu, C. Codrea, N. Lucas, M. Pollak, and P. Spies, "Energy harvesting for wireless communication systems using thermogenerators," in *Proceedings of the IEEE XXI Conference on Design of Circuits and Integrated Systems (DCIS '07)*, Barcelona, Spain, November 2006.
[16] M. T. Penella and M. Gasulla, "A review of commercial energy harvesters for autonomous sensors," in *Proceedings of the Instrumentation and Measurement Technology Conference (IMTC '07)*, pp. 1–5, Warsaw, Poland, May 2007.

8

Delamination Detection of Reinforced Concrete Decks Using Modal Identification

Shutao Xing,[1,2] Marvin W. Halling,[2] and Paul J. Barr[2]

[1] *DeepSea (US) Inc., Houston, TX 77042, USA*
[2] *Department of Civil and Environmental Engineering, Utah State University, Logan, UT 84322, USA*

Correspondence should be addressed to Marvin W. Halling, marv.halling@usu.edu

Academic Editor: Andrea Cusano

This study addressed delamination detection of concrete slabs by analyzing global dynamic responses of structures. Both numerical and experimental studies are presented. In the numerical examples, delaminations with different sizes and locations were introduced into a concrete slab; the effects of presence, sizes, and locations of delaminations on the modal frequencies and mode shapes of the concrete slab under various support conditions were studied. In the experimental study, four concrete deck specimens with different delamination sizes were constructed, and experimental tests were conducted. Traditional peak-picking, frequency domain decomposition, and stochastic subspace identification methods were applied to the modal identification from dynamic response measurements. The modal parameters identified by these three methods correlated well. The changes in modal frequencies, damping ratios, and mode shapes that were extracted from the dynamic measurements were investigated and correlated to the actual delaminations and can indicate presence and severity of delamination. Finite element (FE) models of reinforced concrete decks with different delamination sizes and locations were established. The modal parameters computed from the FE models were compared to those obtained from the laboratory specimens, and the FE models were validated. The delamination detection approach was proved to be effective for concrete decks on beams.

1. Introduction

Delamination occurs when the corrosion in the steel rebar induces cracks and the cracks joined together to cause the concrete cover to separate from the substrate concrete. It results in the loss of structural strength and facilitates a rapid deterioration of the deck [1]. The delamination impairs both the appearance and the serviceability of the structure, and repairs can be very costly. It is estimated that annual maintenance and repair costs related to corrosion for concrete infrastructure approach $100 billion worldwide [2, 3]. As a corrosion-induced problem, delamination is of great concern for bridges, and routine inspection is necessary. Many methods have been developed to detect concrete delamination. These methods include the conventional chain drag method, impact-echo, ultrasonic tests, ground penetrating radar, imaging radar, and infrared thermography [1, 4]. Efforts have been made to expand, improve, and combine currently available techniques. While the conventional methods have been successfully used in the past, they require deployment of professional people with devices to field sites and traffic control which can be very costly. With the expansion of structural health monitoring systems, increasing numbers of real-time monitoring systems are being deployed on actual bridges and buildings. Taking advantage of permanently installed sensors can be useful in delamination detection. Vibration sensors can be flexibly deployed and located in-situ for long-term monitoring applications that include delamination detection.

Various analytical, numerical, and experimental studies have addressed delamination detection in composite structures by identification from the measurements of vibration sensors. Zou et al. [5] provided a review of vibration-based model-dependent delamination identification for composite structures. Valdes and Soutis [6] conducted experiments to study the effects of delamination in laminated beams on the changes in modal frequencies. Racliffe and Bagaria [7] used curvature mode shapes to locate delaminations in a composite beam. Wei et al. [8] evaluated delamination of multilayer composite plates using model-based neural networks. Yan and Yam [9] employed energy distribution of dynamic responses decomposed by wavelet analysis to detect the delaminations in composite plates and reported that this method is capable of detecting localized damage. Among these studies, most are on simple composite structures such as beams, with very few, if any, studies performed on plates. For civil engineering concrete structures, studies on delamination detection by vibration measurements from structural health monitoring systems are rare. Xing et al. [10] investigated delamination detection by using vibration measurements by sensors for civil engineering concrete plates through numerical studies. In this numerical study, finite element models of the concrete plates were modeled using ANSYS. The modal analyses were performed for examination of the effect of delamination parameters on the modal characteristics of the models. The effectiveness of using changes in modal frequencies and mode shapes as damage indicators of the delamination was studied. The proposed methodology in this paper provides the advantage of a nonsubjective means to quantitatively evaluate bridge decks.

The current study presents an expansion of the previous numerical studies [10] with a more complete parametric study. Additionally, experimental studies of four reinforced concrete slabs with different delamination areas were tested dynamically to verify the numerical results. Delaminations were simulated by embedding plexiglass inside the concrete plates during casting. The dynamic tests were conducted approximately four months after placement of the concrete. The experiments were original and significant since the models were relatively large-scale concrete plates. The models were intended to simulate concrete decks on girder bridges. Initial experimental results were discussed in [11], and a more comprehensive study is addressed in this paper. No similar experimental testing was encountered in the literature.

The primary purpose of this paper is to investigate the applicability of delamination detection of concrete plates by modal identification using dynamic responses measured by vibration sensors. The fundamental principle is that the delaminations decrease the stiffness and consequently the modal frequencies. Additionally, damping increases and mode shapes become irregular; the amplitude of the mode shapes around the delamination area is changed. Taking advantage of the changes in modal characteristics can avoid dealing with the complicated delamination mechanisms, such as the random occurrence and irregular shape and distribution of concrete delamination, which need many parameters to evaluate.

After numerical studies of several finite element models, the experiments were conducted. Since different types of input excitation may influence the dynamic properties estimated from their responses; the experimental testing consisted of dynamic tests using random, swept sine, and impact excitation. The dynamic characteristics estimated from the responses due to these excitations were presented to provide reference for output only systems (ambient vibration). Modal frequencies, modal damping ratios, and mode shapes were extracted from the measured velocity responses. The differences in modal characteristics between the various delaminated models were compared and indicated presence and severity of delaminations. Finite element models of the concrete slabs were also established using ANSYS software. The concrete, steel reinforcement, wood supports, delaminations, and boundary conditions were included in the model. The modal characteristics computed from the finite element models were compared with those from dynamic tests for validation. The finite element results could be used as reference for modal identification from dynamic response measurements.

2. Modal Identification Methods for the Output Only Systems

The classical peak-picking (PP), frequency domain decomposition (FDD), and stochastic subspace identification (SSI) methods were used in this study. In the PP method, the peaks of the power spectra of time histories measurements were used to determine modal frequencies. FDD and SSI methods were also adopted to extract modal characteristics from measurements of the dynamic tests.

For lightly damped structures, [12] derived a relationship between response spectral and modal parameters, which provide a basis for FDD method. In application of FDD algorithm, the power spectral density (PSD) of output measurements $\widehat{G}_{yy}(j\omega)$ is estimated and then decomposed at $\omega = \omega_i$ by taking the singular value decomposition (SVD) of the PSD matrix:

$$\widehat{G}_{yy}(j\omega_i) = U_i S_i U_i^{H}, \qquad (1)$$

where the unitary matrix $U_i = [u_{i1}, u_{i2}, \ldots, u_{im}]$ holds singular vectors u_{ij} and the diagonal matrix S_i holds singular values s_{ij}. If only a kth mode is dominating at selected frequency ω_i, there would be only one singular value in (1). Hence the 1st singular vector u_{i1} is an estimate of the kth mode shape $\widehat{\phi} = u_{i1}$. Damping can be obtained from the correlation function of the SDOF system [13].

The SSI method is a time-domain identification method proposed in [14]. It has been applied effectively for various types of structures. It can obtain linear models from column and row spaces of the matrices computed from input-output data [15]. This paper used data-driven SSI that does not need computation of output covariance. An enhancement of it, called reference-based SSI method, was presented in [16]. Discrete-time stochastic state-space model is a normal model without input terms [17]:

$$x_{k+1} = Ax_k + w_k; \qquad y_k = Cx_k + v_k, \qquad (2)$$

where w is process noise vector, v is observation noise vector; they are independent of each other. The SSI is used to compute A and C from output-only measurements y_k.

The identification steps are briefly summarized below, for details, refer to [15, 16].

(a) Construct Hankel matrix $Y_{0|2i-1}$ from the output measurements:

Hankel matrix $Y_{0|2i-1}$ is a matrix that has the same elements in every antidiagonal as shown in (3);

$$Y_{0|i} = \begin{bmatrix} y_0 & y_1 & \cdots & y_{j-1} \\ y_1 & y_2 & \cdots & y_j \\ \cdots & \cdots & \cdots & \cdots \\ y_{i-1} & y_i & \cdots & y_{i+j-2} \end{bmatrix};$$

$$Y_{i|2i-1} = \begin{bmatrix} y_i & y_{i+1} & \cdots & y_{i+j-1} \\ y_{i+1} & y_{i+2} & \cdots & y_{i+j} \\ \cdots & \cdots & \cdots & \cdots \\ y_{2i-1} & y_{2i} & \cdots & y_{2i+j-2} \end{bmatrix}; \quad (3)$$

$$Y_{0|2i-1} = \left[\frac{Y_{0|i}}{Y_{i|2i-1}}\right] = \left[\frac{Y_P}{Y_f}\right],$$

where subscript of y denotes time instant, i is the total number of block rows and must be greater than the system order n, which is the dimension of state matrix A, j is the number of columns, y_k is a vector with l rows, which are the degrees of freedom to be measured. $Y_{0|2i-1}$ is composed of the past output part Y_p and the future output part Y_f; each part has i block rows and j columns. When SSI is used, the system order can be over specified, and the spurious modes can be eliminated by investigating stabilization diagrams. Due to space limitations, these diagrams are not presented in this paper; the results are verified to be correct with an appropriate system order value. Let ny denote the number of time samples of output y_k; to guarantee y_k populate Hankel matrix, use $j = ny - 2i + 1$.

(b) Orthogonally, project the row space of future outputs Y_f on the row space of past outputs Y_p:

$$T_i = Y_f Y_p^T \left(Y_p Y_p^T\right)^{\dagger} Y_p, \quad (4)$$

where ^{T,†} denote transpose and pseudoinverse, respectively.

(c) Apply SVD to the orthogonal projection:

$$T_i = USV^T = \begin{pmatrix} U_1 & U_2 \end{pmatrix} \begin{pmatrix} S_1 & 0 \\ 0 & S_2 \end{pmatrix} \begin{pmatrix} V_1^T \\ V_2^T \end{pmatrix} \approx U_1 S_1 V_1^T, \quad (5)$$

where U and V are orthonormal matrices, S is a diagonal matrix containing singular values in descending order, S_2 is a block containing neglected values.

(d) Calculate the discrete-time system matrices A and C using (6), which compute the extended observability matrix from SVD of (5) and the definition of it:

$$O_i = U_1 S_1^{1/2}; \qquad O_i = \begin{pmatrix} C & CA & \cdots & CA^{i-1} \end{pmatrix}^T. \quad (6)$$

(e) Postprocess to extract modal parameters.

The matrices Λ and ψ containing eigenvalues and eigenvectors of A are computed using $A = \psi \Lambda \psi^{-1}$. The state matrix of the continuous-time system is $A_C = \psi_C \Lambda_C \psi_C^{-1}$. The relationships are shown below:

$$A = e^{A_C \Delta t}; \qquad \psi_c = \psi; \qquad \lambda_{C_q} = \ln \frac{(\lambda_q)}{\Delta t};$$
$$\lambda_{C_q}, \lambda_{C_q}^{*} = -\xi_q \omega_q \pm j\omega_q \sqrt{1 - \xi_q^2}; \qquad \Phi = C\psi, \quad (7)$$

where Δt is time step, ω_q is modal frequency, ξ_q is modal damping ratio, Φ contains mode shapes.

When actual measurements are processed, they result in complex frequencies and complex mode shapes. For lightly damped systems, the magnitudes of the complex mode shapes can be plotted as real mode shapes.

3. Numerical Studies

In this section, a finite element model of a reinforced concrete plate was created and then several delaminations of various sizes and locations were introduced into the model, separately. The modal analyses were performed on the model, and the effects of delamination on modal characteristics of the reinforced concrete finite element model were studied to provide useful reference for further numerical studies as well as for the laboratory experiments. The effects of support conditions on the applicability of the proposed method were also examined.

3.1. Description of the Numerical Examples. The reinforced concrete plate of the numerical studies is illustrated in Figure 1. The width a is 4200 mm along the X direction, the length b is 8000 mm along the Z direction, and thickness h is 200 mm along the Y direction, the origin is defined at the bottom corner node; the coordinate system is shown in Figure 1(a). The concrete's elastic modulus is E_c = 32500 MPa, Poisson's ratio ν_c = 0.24, ultimate uniaxial compressive strength is σ_c = 25.5 MPa, ultimate uniaxial tensile strength is σ_t = 2.56 MPa, and density is ρ_c = 2400 kg/m^3. The steel rebar's size is #6, nominal diameter is d = 19 mm, elastic modulus is E_s = 200 GPa, yield stress is f_y = 410 MPa; Poisson's ratio is ν_s = 0.3; and the density is ρ_s = 7850 kg/m^3. Two layers of steel rebar are placed in the concrete plate at the horizontal planes of Y = 50 mm and Y = 150 mm and rebar are along both X and Z directions as exhibited in Figure 1(b). The space between all adjacent rebar is 200 mm on center. The concrete plate is simply supported at the two opposite edges X = 0 mm and X = 4200 mm and free at the other two opposite edges Z = 0 mm and Z = 8000 mm. This will be symbolized by SS-F-SS-F.

3.2. Finite Element Modeling. The finite element software, ANSYS, was used to perform the finite element modeling. Solid65 and link8 elements were selected to represent concrete and steel rebar, respectively. Solid65 element has

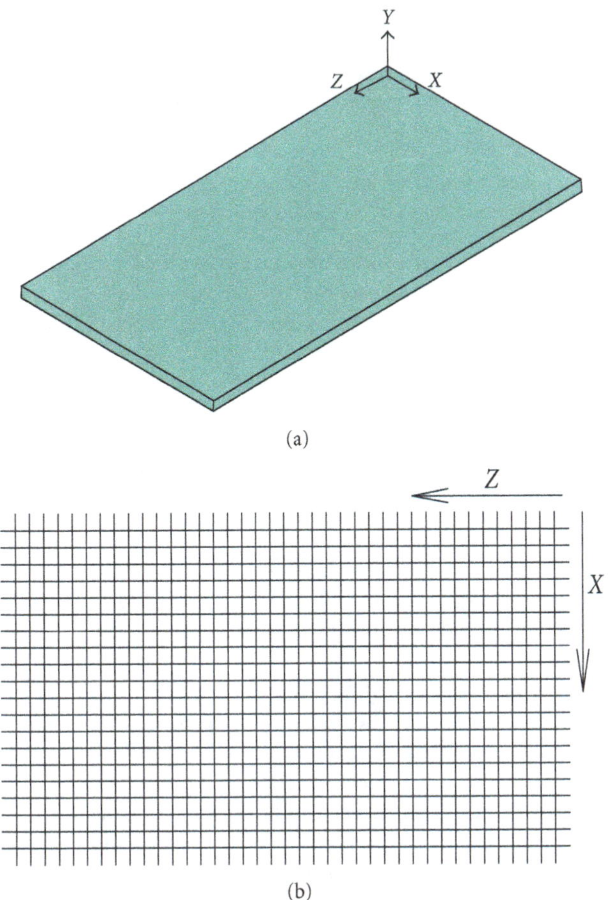

FIGURE 1: The concrete slab model, (a) isometric view of the concrete model, (b) plane view of the steel rebar.

eight nodes with three translational degrees of freedom at each node. The reason to use solid65 element is that its capability in modeling concrete may be used in the future studies. The 3D spar, Link8, element is a uniaxial tension-compression element with three translational degrees of freedom at each node [18]. In this paper, all analyses were restricted to linear elastic response and the elements were configured accordingly. When the mesh size of concrete was changed from 50 mm to 25 mm, the same modal analysis results were obtained without convergence problems.

Delamination was modeled using a similar method to [19]. First, two separate decks that were located above and below the delamination plane were attached together. The nodes in the undamaged area were declared as coupled nodes utilizing coupling/constraint equations and the nodes in delaminated area were uncoupled. The delamination locations and areas were adjusted for different damage degrees. All the delaminations were on the XZ plane at a vertical elevation of Y equal to 150 mm, which was purposely to simulate a bridge deck in which delamination often occurs in a horizontal plane at the top layer of steel reinforcement. In the following sections, the ratio of the delamination area to the total area $A_{\text{delam}}/A_{\text{tot}}$ is used to denote severity of delamination.

TABLE 1: Comparison of frequencies by analytical and finite element model for a numerical example.

Mode	1	2	3	4	5
Analytical solution	20.91	27.99	49.49	84.32	87.02
ANSYS solution	20.96	27.12	50.32	81.49	85.73
Difference (%)	0.24	−3.11	1.68	−3.36	−1.48

3.3. Modal Frequencies Validation of the Finite Element Models. The analytical solutions for the natural frequencies were computed to validate the modeling. For convenience of using the analytical formula from [20], the Poisson's ratio of the concrete was changed to $\nu_c = 0.3$, the dimensions were changed to $a = 4000$ mm and $b = 6000$ mm. The other parameters take same values as those of the model in the above section. In order to use the analytical solution, the equivalent elastic modulus and density were used, which were computed by (8) and (9):

$$E(A_c + A_s) = E_c A_c + E_s A_s, \tag{8}$$

$$\rho(V_c + V_s) = \rho_c V_c + \rho_s V_s, \tag{9}$$

where E is the equivalent elastic modulus, ρ is the equivalent density, V_c and V_s are the volumes of concrete and steel rebar, respectively, A_c, ρ_c, ρ_s, A_s have same meanings as those in the above section.

Since minimum of $(a, b)/h$ is equal to 1/20, it can be analyzed as thin plate and Kirchhoff assumptions are applicable. The analytical method follows [20]. The governing equation is

$$D\nabla^4 w + \rho \frac{\partial^2 w}{\partial t^2} = 0, \tag{10}$$

where w is transverse deflection, ∇^4 is biharmonic differential operator in rectangular coordinates, $D = Eh^3/12(1-\nu^2)$ is the flexural rigidity. The boundary conditions for the simply supported and free edges are (11) and (12) accordingly:

$$w = 0; \qquad \frac{\partial^2 w}{\partial x^2} + \nu \frac{\partial^2 w}{\partial z^2} = 0, \tag{11}$$

$$\frac{\partial^2 w}{\partial x^2} + \nu \frac{\partial^2 w}{\partial z^2} = 0; \qquad \frac{\partial^3 w}{\partial x^3} + (2-\nu) \frac{\partial^3 w}{\partial x \partial z^2} = 0. \tag{12}$$

Combining (10)–(12), characteristic equations that can numerically solved by Newton's method to yield the frequency parameter can be written as

$$\lambda = \omega a^2 \sqrt{\frac{\rho}{D}}. \tag{13}$$

Equation (13) can be used to calculate the natural frequencies. The comparison results are listed in Table 1. The percent differences are within 3.4% for the first five frequencies, and the difference for the first frequency is only 0.24%. It is shown that most frequencies by analytical solution were a little larger than those by modeling. This is reasonable, because thin plate theory usually overestimates the natural

Table 2: The locations and sizes of the delaminated areas for numerical example.

A_{delam}/A_{tot}	1/8	1/6	1/5	1/4	1/3 unsymmetrical	1/3 symmetrical	1/2
X (mm)	100~2149	100~2466	100~2692	100~2998	100~3447	427~3773	100~4199
Z (mm)	100~2149	100~2466	100~2692	100~2998	100~3447	2327~5673	100~4199
Y (mm)	150	150	150	150	150	150	150

frequencies [21] and Mindlin Plate theory would be more accurate for a thick plate. Therefore, based on the results, it was concluded that the finite element model accurately produced the correct modal frequencies.

3.4. Modal Analysis of the Finite Element Models. The modal analysis was carried out to correlate the delamination with modal characteristics. The analysis was confined to the first 6 modes. Table 2 provides the sizes and locations for the delamination cases of this study. Table 3 shows the comparisons of modal frequencies of the undamaged model and those of different delaminated models under SS-F-SS-F support. The percent differences reported in Table 3 were calculated relative to the undamaged model. Figures 2-3 compare the 3rd-mode shapes from the undamaged and delaminated models. The modal amplitudes shown in the figure have been scaled by the same factor. The modal order is the subnumber in the figures minus one for this analysis.

Based on the frequency variation listed in Table 3, it is clear that the higher the modal order is, the larger the reduction of the modal frequency is. The changes in the 4th modal frequency are much larger than their neighboring modes, which indicate that specific modes are more sensitive to delamination than other modes. Table 3 shows that the larger the delamination areas, the larger percentage reduction of the corresponding frequencies. It is also found that when $1/8 \le A_{delam}/A_{tot} \le 1/6$, the changes in the 4th frequencies can indicate the delamination quite well with frequency difference from 8.42% to 26.59%, while the changes in the first three frequencies are not greater than 3%. For $1/6 < A_{delam}/A_{tot} \le 1/4$, the changes in the 3rd frequencies can indicate delamination with frequency difference from 6.63% to 19.57%. When $1/3 \le A_{delam}/A_{tot} \le 1/2$ without consider the 1/3 symmetrical case, the changes in the 1st frequencies can indicate delamination with frequency difference from 6.39% to 29.64%. However, when $A_{delam}/A_{tot} < 1/8$, the changes in first four frequencies are within 3%, which can not indicate delamination, and the exhaustive results for the cases of these small delamination areas are not listed.

Similar trends were found to be present in the mode shapes shown in Figures 2-3. While the mode shapes are more sensitive to the delamination and are able to show the location from the irregular curves at the delaminated locations. There are also larger amplitudes at the delaminated locations than the corresponding part of the undamaged model. Comparing the delamination locations in Table 2 with the mode shapes in Figure 2, the irregular parts correspond almost exactly to where the delamination location occurs.

In Table 3, the frequencies of 1/3-unsymmetrical-delaminated and 1/3-symmetrical-delaminated models are listed. The frequency differences between the delaminate models and the undamaged model show that the more unsymmetrical the delamination is, the larger the modal frequencies decrease. For example, the frequencies difference for the 1st frequency of 1/3-symmetrical-delaminated model is 3.10%, but the difference increases to 6.39% for the 1/3-unsymmetrical-delaminated model. This is also demonstrated by the mode shapes in Figure 3, in which the 1st mode shapes of the two 1/3-delaminated models are illustrated. It is shown that the mode shapes of the unsymmetrical-delaminated model are more irregular and have larger amplitude at the delaminated location than that of the symmetrical delaminated model.

3.5. The Applicability of the Proposed Approach to Different Support Conditions. The support conditions of the above numerical examples were changed to two different conditions to examine the effectiveness of the proposed approach for concrete slabs under various boundary conditions.

For the first case, the SS-F-SS-F was changed to C-F-C-F, which symbolizes clamped-free-clamped-free support. The concrete deck was clamped at the two opposite edges $X = 0$ and $X = 4200$ mm and free at the other two opposite edges. The modal analyses were performed for the undamaged and delaminated models under C-F-C-F support condition. The modal frequencies for these models and the percent difference were listed in Table 4. By comparing the results listed in Tables 4 and 3, it is observed that for the first three or four modes, the C-F-C-F case has larger difference, and for the higher modes, the simply supported have larger difference. This indicates that the performance of the proposed approach is better when the restraints increase.

For the second case, the concrete deck was supported on steel beams at the two opposite edges $X = 0$ and $X = 4200$ mm and free at the other two opposite edges. The isometric view of the model is shown in Figure 4. The beams were supported over columns. The beams were clamped at the corner of $X = 4200$ and $Z = 0$ mm, free along Z at the corner of $X = 0$ and $Z = 0$ mm, free along X and Z at the corner of $X = 0$ and $Z = 8000$ mm, free along X at the corner of $X = 4200$ and $Z = 8000$ mm. Each beam has a length of 8000 mm and a rectangular section of 400×400 mm. The steel beams had same material properties with the steel rebar. This support condition is closer to the actual bridge decks supported on steel girders. The modal frequencies related to bending of the decks for the undamaged and delaminated models under this support condition, and the percent difference were listed in Table 5. Figure 5 shows the typical mode shapes for the undamaged and the delaminated models. It is observed from these results that the changes in modal characteristics can indicate delamination.

Table 3: Comparison of natural frequencies of the undamaged and delaminated concrete decks for numerical example under SS-F-SS-F support condition.

Mode	Undamaged		1/8 delaminated		1/6 delaminated		1/5 delaminated		1/4 delaminated		1/3 unsymmetrical delaminated		1/3 symmetrical delaminated		1/2 delaminated	
	Mode Shape	Freq (Hz)	Freq (Hz)	Difference (%)	Freq (Hz)	Difference (%)	Freq (Hz)	Difference (%)	Freq (Hz)	Difference (%)	Freq (Hz)	Difference (%)	Freq (Hz)	Difference (%)	Freq (Hz)	Difference (%)
1	1st bending	18.88	18.70	−0.94	18.63	−1.32	18.57	−1.65	18.43	−2.40	17.67	−6.39	18.29	−3.10	13.28	−29.64
2	2nd bending	22.69	22.42	−1.20	22.30	−1.70	22.19	−2.18	21.90	−3.46	20.60	−9.19	22.08	−2.68	18.67	−17.72
3	3rd bending	37.01	36.50	−1.38	35.91	−2.96	34.56	−6.63	29.77	−19.57	24.35	−34.21	26.87	−27.39	21.81	−41.06
4	4th bending	57.64	52.79	−8.42	42.31	−26.59	37.87	−34.29	35.76	−37.96	33.28	−42.26	38.76	−32.76	27.53	−52.24
5	5th bending	73.47	56.89	−22.57	55.36	−24.66	54.07	−26.40	51.22	−30.28	43.50	−40.79	47.21	−35.75	27.89	−62.04
6	6th bending	76.44	73.42	−3.95	71.58	−6.35	68.91	−9.84	59.98	−21.53	45.70	−40.21	51.12	−33.12	37.55	−50.87

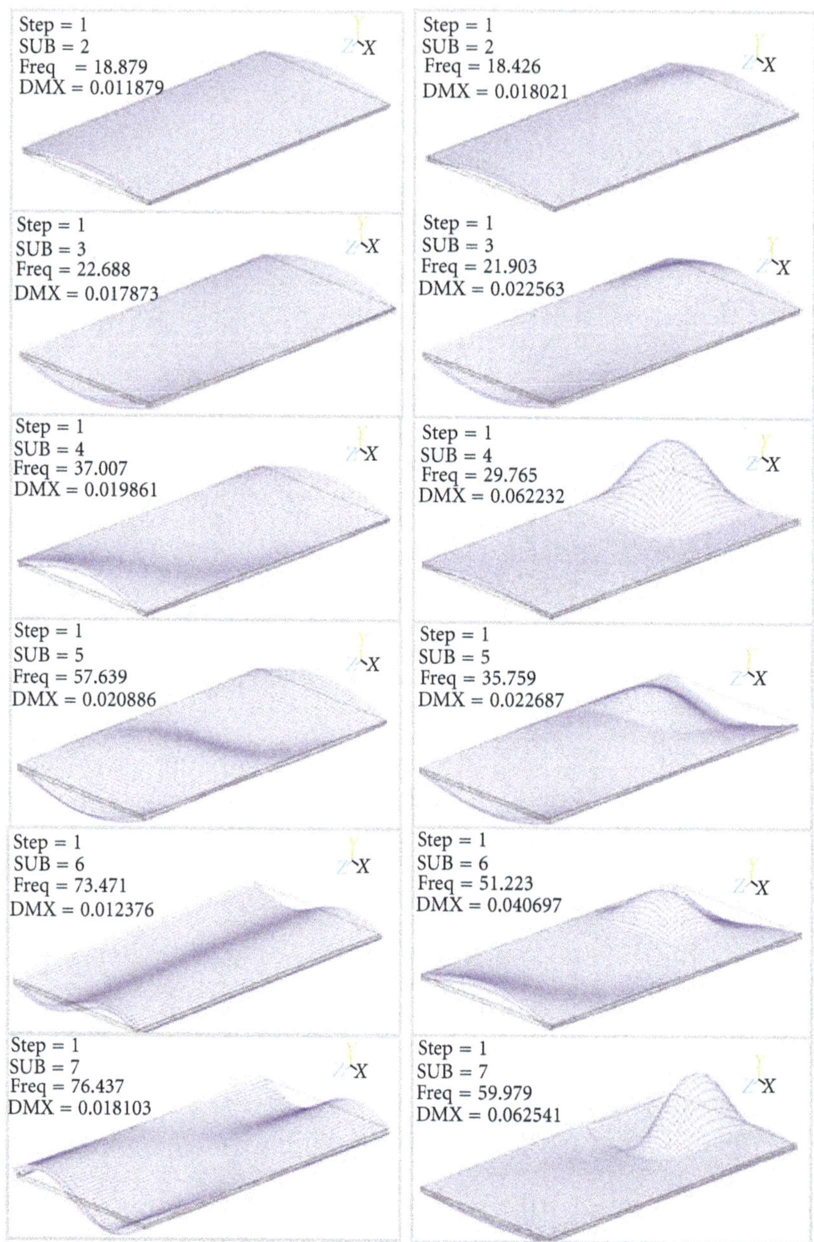

FIGURE 2: The comparison of the first six mode shapes of two models (left is for the undamaged and right is for the 1/4 delaminated models).

It is found that the general conclusions from the case of SS-F-SS-F still apply to the case of C-F-C-F and the case of supported on beams. While the most sensitive mode and the percentage differences vary. When the proposed method for delamination detection is used, it is necessary and beneficial to perform modal analysis for the specific structures.

4. Experimental Studies

The objective of the experimental studies is to investigate the feasibility of delamination detection of concrete structures from real vibration measurements. The general conclusions from this section can be used to verify the previous numerical studies. In the experimental studies, four reinforced concrete plates with simulated delaminations were constructed in the Systems, Materials, and Structural Health Laboratory (SMASH) at Utah State University. Modal characteristics were extracted from the dynamic test data using three modal identification methods. The changes in modal characteristics were used for delamination detection. The finite element models, in this section, were developed for the experimental concrete slabs and the modal analyses based on the structural properties were carried out. Some useful conclusions were drawn from the studies.

4.1. Experimental Setup and Dynamic Tests. Four reinforced concrete plates were constructed in the SMASH lab. Each concrete plate has the same size 1.83 m × 2.74 m × 0.14 m

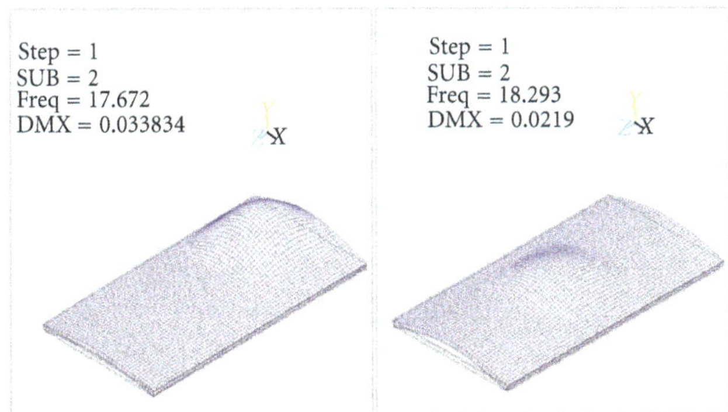

FIGURE 3: Comparison of the 1st mode shapes of the two 1/3 delaminated models (left and right are for a unsymmetrical and a symmetrical model, respectively).

and same layout of #6 steel rebar as illustrated in Figure 6. All the concrete was placed from the same ready mix concrete batch in order to achieve similar concrete strengths between test specimens. The only differences between the models were the different delamination scenarios that were introduced into the concrete plates.

In the delamination detection studies of composites [6, 7], to generate the effects of delamination, Teflon films were inserted into the composite beams/plates to prevent layers from bonding together. In this study, plexiglass sheets (1.57 mm thick) were used to simulate the effect of a delamination. Plexiglass is unaffected by moisture and offers a high strength-to-weight ratio, which is good for preventing the bonding of concrete above and below it. One layer of plexiglass was placed into the horizontal plane of each delaminated concrete plate that was 139.7 mm from the bottom of the plate during the concrete pouring. The horizontal areas of the delaminations were adjusted in each specimen. The first specimen had no delamination and is referred to as undamaged. The other three specimens had delamination sizes of 914 mm × 1219 mm (22.2% delamination), 1219 mm × 2134 mm (51.9% delamination) and 1524 mm × 2438 mm (74.1% delamination), respectively. Figure 7 shows the plan and elevation views of the undamaged and all delaminated specimens. Four months after concrete placing, the dynamic tests were performed. During the tests, the two long edges of each concrete plate were placed on timber supports, while the two short edges were unsupported. All of the concrete slabs were tested with the same boundary conditions.

Based on the 28 day concrete compression and tensile tests, the concrete's elastic modulus is $E_c = 24000$ MPa, ultimate uniaxial compressive strength is $\sigma_c = 27.5$ MPa, ultimate uniaxial tensile strength is $\sigma_t = 14.69$ MPa. The concrete density is $\rho_c = 2300\,\text{kg/m}^3$. Poisson's ratio of concrete is $\nu_c = 0.15$. The steel rebar's size is #6, its nominal diameter is $d = 19$ mm, elastic modulus is $E_s = 200$ GPa, yield stress is $f_y = 410$ MPa, Poisson's ratio is $\nu_s = 0.3$, and the density is $\rho_s = 7850\,\text{kg/m}^3$. It needs to be noted that Poisson's ratios of wood are difficult to be measured accurately for they vary within and between species and are affected by moisture content and specific gravity [22]. According to the literature [22–24] and finite element model updating, the timber's anisotropic properties are as follows. The timber's density is $430\,\text{kg/m}^3$. The elastic moduli along the longitudinal, radial, and tangential axes of timber are denoted by E_Z, E_X, and E_Y; they are $E_Z = 8$ GPa, $E_X = 0.068 \times E_Z$, and $E_Y = 0.05 \times E_Z$. The values of Poisson's ratios $\nu_{ZX} = 0.496$, $\nu_{XY} = 0.56$ and $\nu_{YZ} = 0.274$ were used in this study.

The same dynamic tests were performed on the four specimens for convenience of comparisons. Figure 8 illustrates the layout of excitation sources and sensors. An electromagnetic shaker was employed to generate excitations. Because hammer is convenient and more often used in field tests, the hammer was also used to simulate more practical situations. Based on the identified results from the impulse excitations by shaker and hammer, the two impulses don't make difference. Six velocity transducers ($V1 \sim V6$) were used to measure the vertical responses and one velocity transducer ($V7$) was used to measure the horizontal responses. An accelerometer was attached to the shaker to measure the real excitation inputs, and the input acceleration of the instrumented hammer was also measured. These measurements were recorded by a data physics vibration controller/signal analyzer, in which the antialiasing filter was integrated. For the dynamic tests, the APS shaker generated swept sine, random and impulse excitations on the specimens in sequence. Then the instrumented hammer applied impact excitations a few times on two locations. The goal was to detect delamination from the ambient vibration data. Therefore only response measurements were used for modal identification, and frequency response functions (FRFs) were not calculated. The typical swept sine, random, and impact acceleration inputs and the power spectral density (PSD) of them are shown in Figure 9 for the 74.1% delaminated specimen. The PSD plots in this figure show that all the excitations have frequency bands from 0 to 400 Hz and similar power distributions over the bands, which ensure that each excitation method has consistent frequency band and power over it. The inputs for the other specimens were similar. The velocity responses and the acceleration inputs were recorded for subsequent analyses. The signal-to-noise ratios (SNRs) for the response measurements by sensors $V1 \sim V7$ were estimated to be 54∼58 db under hammer impact excitation, and the numbers were decreased to 18∼30 db or lower when the specimens were under shaker excitation. The modal identification showed that same results were obtained from these different responses, which means that the different SNR values had little influence on modal identification results, and all the response measurements were good to provide modal information. The recorded duration for each input by the shaker and hammer was 32 seconds. The sampling frequency was 1024 Hz, theoretically modal frequencies as high as 512 Hz could be identified; in reality the lower frequencies are critical.

4.2. Modal Identification of the Dynamic Measurements. Classical peak-picking, frequency domain decomposition, and stochastic subspace identification methods were used to

TABLE 4: Comparison of natural frequencies of the undamaged and delaminated concrete decks for numerical example under C-F-C-F support condition.

Mode	Undamaged		1/6 delaminated		1/5 delaminated		1/4 delaminated	
	Mode shape	Freq (Hz)	Freq (Hz)	Difference (%)	Freq (Hz)	Difference (%)	Freq (Hz)	Difference (%)
1	1st bending	29.19	28.06	−3.87	27.50	−5.79	25.54	−12.51
2	2nd bending	31.96	31.02	−2.96	30.74	−3.81	30.05	−5.98
3	3rd bending	41.76	39.82	−4.63	36.82	−11.83	33.23	−20.43
4	4th bending	60.66	43.60	−28.13	41.32	−31.89	39.89	−34.23
5	5th bending	73.71	57.95	−21.38	56.44	−23.43	53.08	−28.00
6	6th bending	76.94	72.13	−6.25	69.77	−9.32	60.70	−21.11

FIGURE 4: The isometric view of the concrete deck supported on beams.

obtain the modal parameters from the velocity responses. Typical time signals and PSD of them for the PP method are shown in Figure 10 for the vertical responses due to swept sine, random, and impact excitations, respectively. Typical singular value plots by FDD for the undamaged model are shown in Figure 11.

The time histories of responses in Figures 10(b) and 10(c) are associated with same external excitations, and therefore they can be compared. It is obvious that the amplitudes of responses of the delaminated model are larger than those of the undamaged model. For example, in Figure 10(b), the response of the delaminated model is about 2 times of the response of the undamaged model.

Applying the SSI method in this study, the system order was $n = 60$, and the number of block rows was $i = 400$; however, when taking $n = 30$ and $i = 150$ there was an insignificant change in the results. The issue on how to select the system order is not studied deeply in this paper. The modal frequencies extracted by using PP, FDD, and SSI methods are listed in Table 6 and the damping ratios identified by the SSI method are listed in Table 7. Figure 12 includes comparisons of the mode shapes of a specified mode for all the concrete slabs obtained by the FDD and SSI methods. For the purpose of convenient comparison, the so-called mode shapes in Figure 12 are plotted with respect to the velocity transducer numbers $V1 \sim V6$ instead of the real locations of the $V1 \sim V6$ in three dimensions.

The modal frequencies extracted from transducers $V1$–$V6$ vary by a maximum of only 2% for the presented results; only the results from $V4$ and $V7$ are shown in the figures for illustration. The frequencies listed in the tables are also identified from these selected sensors. In Table 6, the frequencies identified by the three methods agree very well for most modes, the differences between the 1st bending frequencies by these methods for all the specimens are less than 2%. From the comparisons in Figure 12, it is evident that the mode shapes extracted by the FDD and SSI methods yield consistent results. While this paper mainly demonstrates analyses on the responses due to random and impact excitations, the modal characteristics obtained from responses due to all three different excitations match very well, for example, the differences between the 1st frequencies obtained from responses due to random and impact excitations are within 2.5%. In other words, excellent agreements are obtained for modal frequencies and mode shapes regardless of the different identification methods and different excitation inputs, and this validates that the extracted modal characteristics are correct for the dynamic test data.

The effectiveness of using changes in modal frequencies and mode shapes as damage indicators of delaminations is examined in Figures 8, 9, 10, 11 and Tables 4-5. Figure 10 shows that the fundamental frequencies of the delaminated specimens are decreased compared with that of the undamaged specimen. It can also be observed that vibration energy of the response of the undamaged specimen is concentrated in the first few modes; however, the delaminated specimens exhibit a relatively higher level of response in the higher modes. The same conclusions can be drawn from the results for the 51.9% and 74.1% delaminated specimens. From Table 6, it can be seen that the corresponding frequencies decrease with the increase of delamination size, among which the 1st bending frequency decreases from 31.77 Hz to 24.01 Hz. The 22.2% delaminated specimen was supposed to have higher frequencies than the corresponding ones of 51.9% delaminated specimen, while they are slightly lower than expected values. This is because the 22.2% delaminated specimen was the first model to have plexiglass placed, and the building process was slower than the other delaminated models, which slightly reduced the integral strength.

TABLE 5: Comparison of natural frequencies of the undamaged and delaminated concrete decks for numerical example supported on beams.

Mode	Undamaged		1/4 delaminated		1/3 delaminated	
	Mode shape	Freq (Hz)	Freq (Hz)	Difference (%)	Freq (Hz)	Difference (%)
1	1st bending	11.51	11.47	−0.37	11.43	−0.68
2	2nd bending	17.91	17.80	−0.62	17.65	−1.45
3	3rd bending	22.19	21.26	−4.19	19.38	−12.70
4	4th bending	24.33	23.86	−1.92	23.34	−4.07
5	5th bending	37.39	30.05	−19.63	25.00	−33.13
6	6th bending	47.81	35.97	−24.77	33.57	−29.78

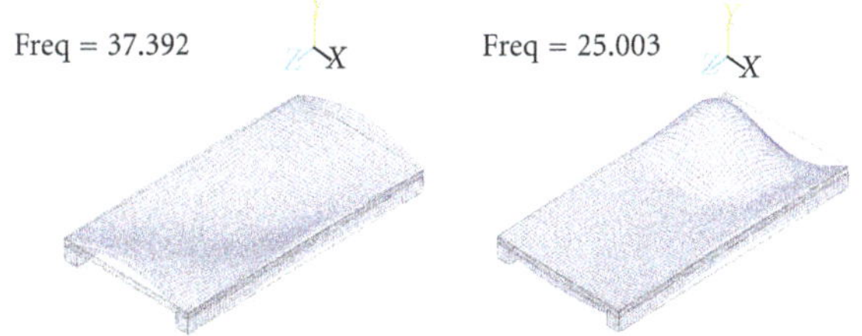

FIGURE 5: Comparison of the 5th mode shapes of the undamaged (left) and 1/3 delaminated (right) models supported on beams.

From Table 7, it is observed that generally the listed damping ratios of the corresponding mode show a trend of increase with the increase of delamination size from 1.05% (1.25%) to 1.71% (1.43%). The damping ratios can assist in identifying delamination, but it is not recommended to rely on damping alone due to the complexity of damping mechanism and difficulty in obtaining accurate damping values in experimental studies. Figure 12 displays the changes in corresponding mode shapes for different delaminated models based on responses due to random excitation. The mode shapes demonstrate that when delamination develops to an undamaged area, the relative amplitude of this area with respect to its neighboring areas becomes larger. Because of the insufficient number of sensors in this study, detailed results are not very accurate regarding the use of mode shapes for the detection of delaminations. It is recommended that additional work is to be performed with denser array of sensors in order to obtain more conclusive results regarding mode shapes.

For real bridges, different types of damage or combination of them may exist. It is challenging to distinguish the changes in modal characteristics caused by the different types of damage. If delamination area is small, it may be hard to determine whether the changes in modal characteristics attribute specifically to delamination instead of concrete spalling or part loss of section stiffness. Further investigations of damage characteristics of delamination may be needed to address this issue.

4.3. Finite Element Modeling of the Experimental Models. The finite element models of the reinforced concrete specimens were created using ANSYS software. The origin of the coordinate system is at the left bottom corner as shown in Figure 7. The longitudinal steel rebar were placed in the model at the plane of $Y = 50.8$ mm. The transversal rebar were placed immediately above the longitudinal rebar. The two long edges were supported on timber, while the two short edges were unsupported. Solid65, link8, and solid45 elements were selected to represent concrete, steel rebar, and timber, respectively. The laboratory floor was modeled using solid65 elements with infinite strength. Dynamic characteristics are sensitive to the boundary conditions, so the timber supports were modeled realistically instead of as ideal simple supports. Contact and target elements were used to simulate the contact between the concrete and timber and between the timber and the concrete lab floor. The contact and target elements used in this study were conta173 and targe170 elements, respectively. Coefficient of friction was an important parameter and was the only parameter needed to be adjusted for the different delaminated models. Its values range between 0.28 and 0.32. Normal penalty stiffness factor FKN can be a small value less than 0.1, and the penetration tolerance factor FTOLN can be 0.1. The properties of the contact elements were updated to match the results of modal analysis from real measurements. The delaminations were modeled by reducing the elastic modulus of the delaminated parts to very small numbers. In the modeling process, the mass of the shaker was also included in some models and found to be negligible, particularly for the lower modes.

Modal analyses of all of the four reinforced concrete slab specimens were performed. The computed modal frequencies and mode shapes were compared with those from experimental tests. Comparisons of frequencies are listed in Table 8 and one mode shape of the undamaged and delaminated models are shown in Figure 13. Figure 13 shows that the irregular part of the mode shape locates where delamination occurs. Table 8 demonstrates that the lowest and even relatively high-order modal frequencies calculated by ANSYS model match those from dynamic testing well; the difference between the ANSYS model and FDD method from test data for the first three frequencies is maximally 7%. Figure 13 shows that the frequencies of the delaminated models decrease and the mode shapes have abrupt changes at the delaminated areas. The comparison of mode shapes by the Ansys modeling with those by the FDD and SSI methods from the response measurements due to random excitation is shown in Figure 12. It is observed that the mode shapes of

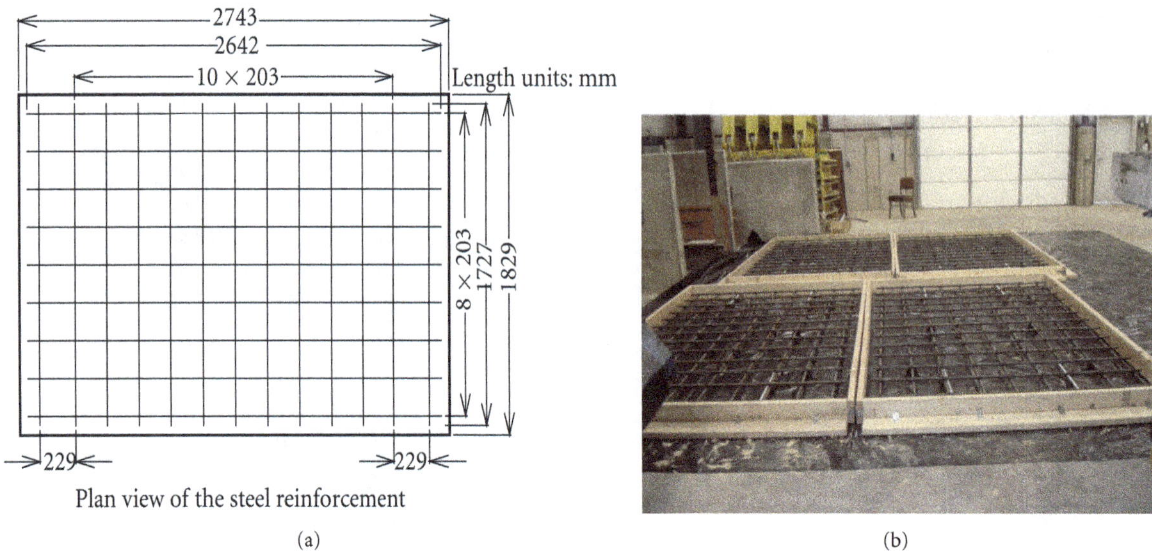

FIGURE 6: Plan view of steel reinforcement. (a) Drawing of the plan view of steel reinforcement; (b) formwork.

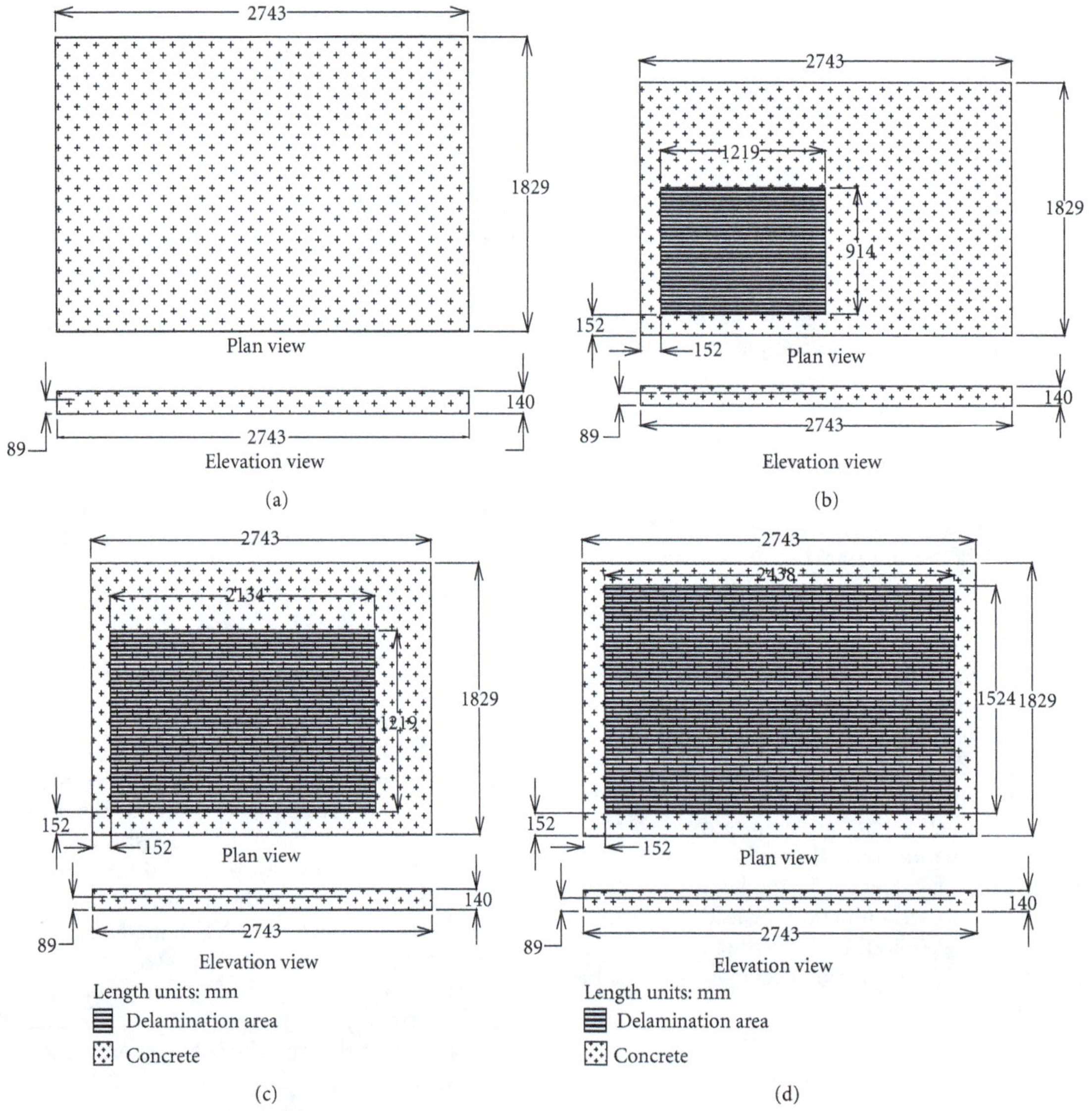

FIGURE 7: The locations of delamination areas (plexiglass) for the experimental concrete plates: (a) undamaged model; (b) 22.2% delaminated model; (c) 51.9% delaminated model; (d) 74.1% delaminated model.

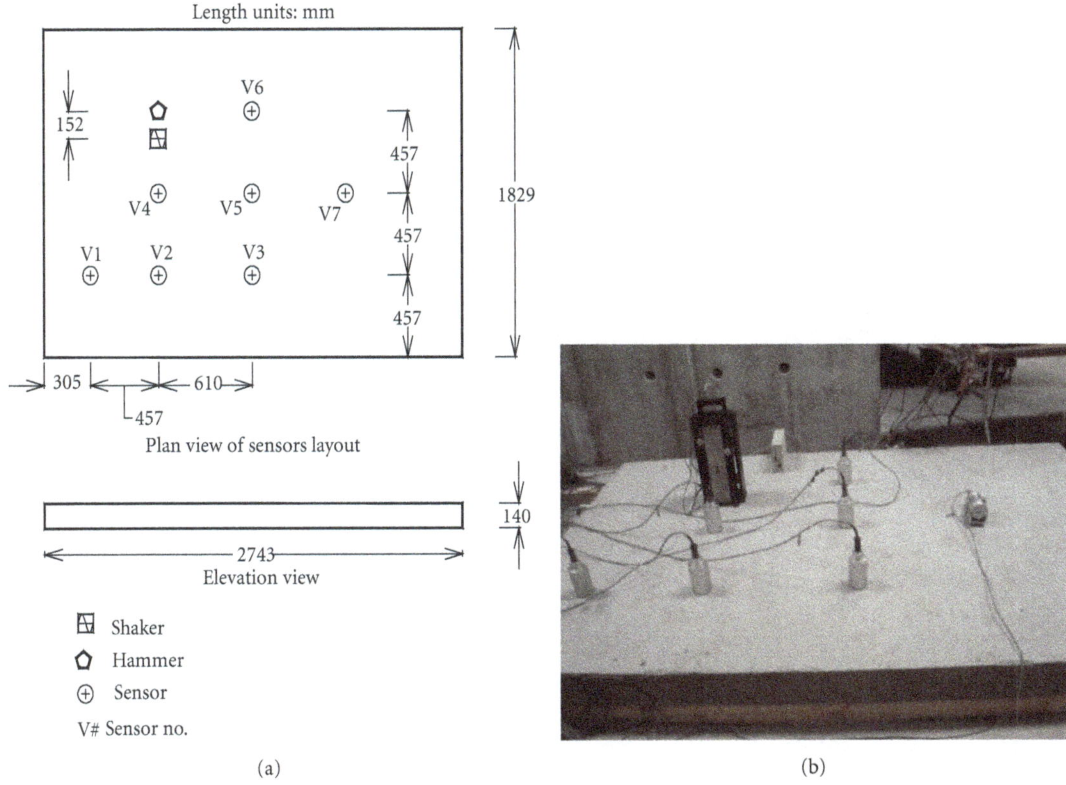

FIGURE 8: Layout of excitation sources and sensors for dynamic tests: (a) layout of excitation sources and sensors, (b) one of the models for dynamic test.

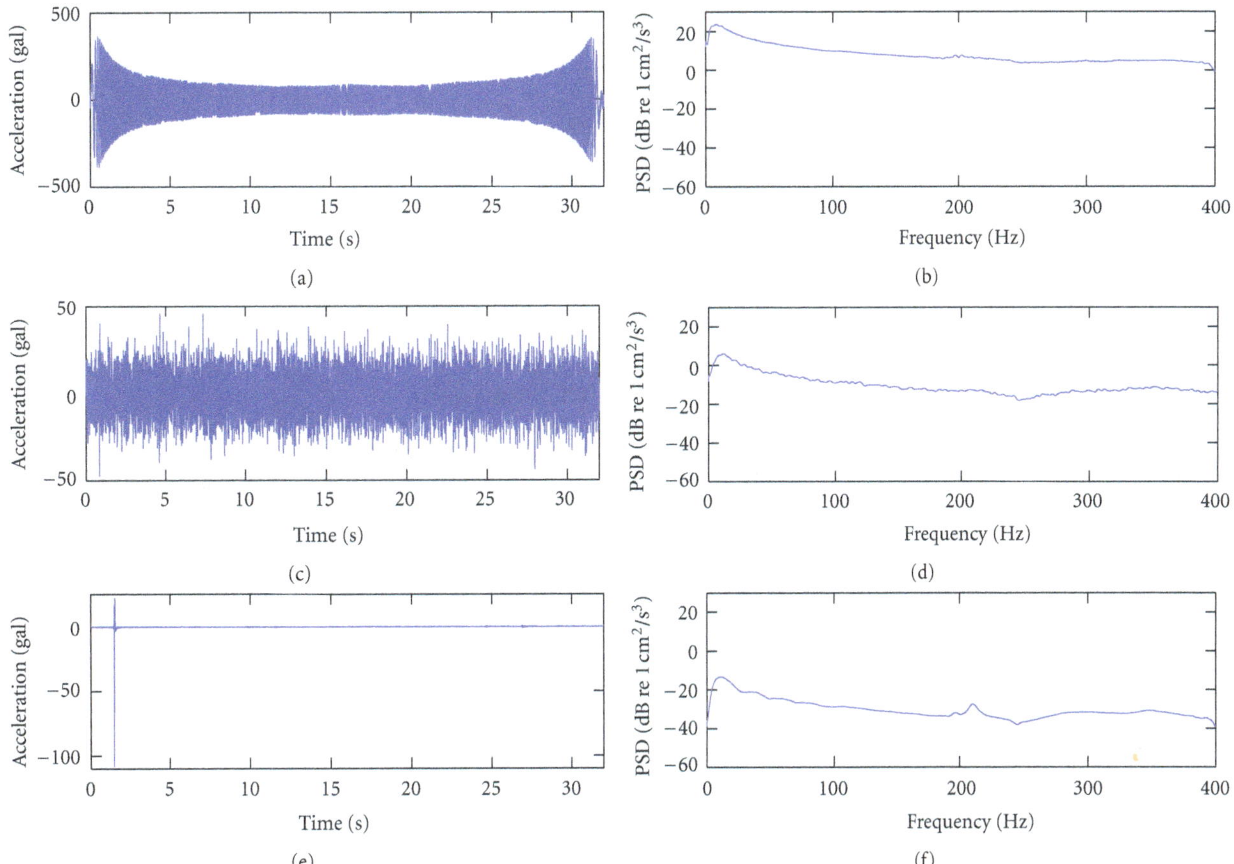

FIGURE 9: The swept sine (upper), random (middle) and impact (lower) excitations applied on the 74.1% delaminated specimen and the PSD estimates of the accelerations.

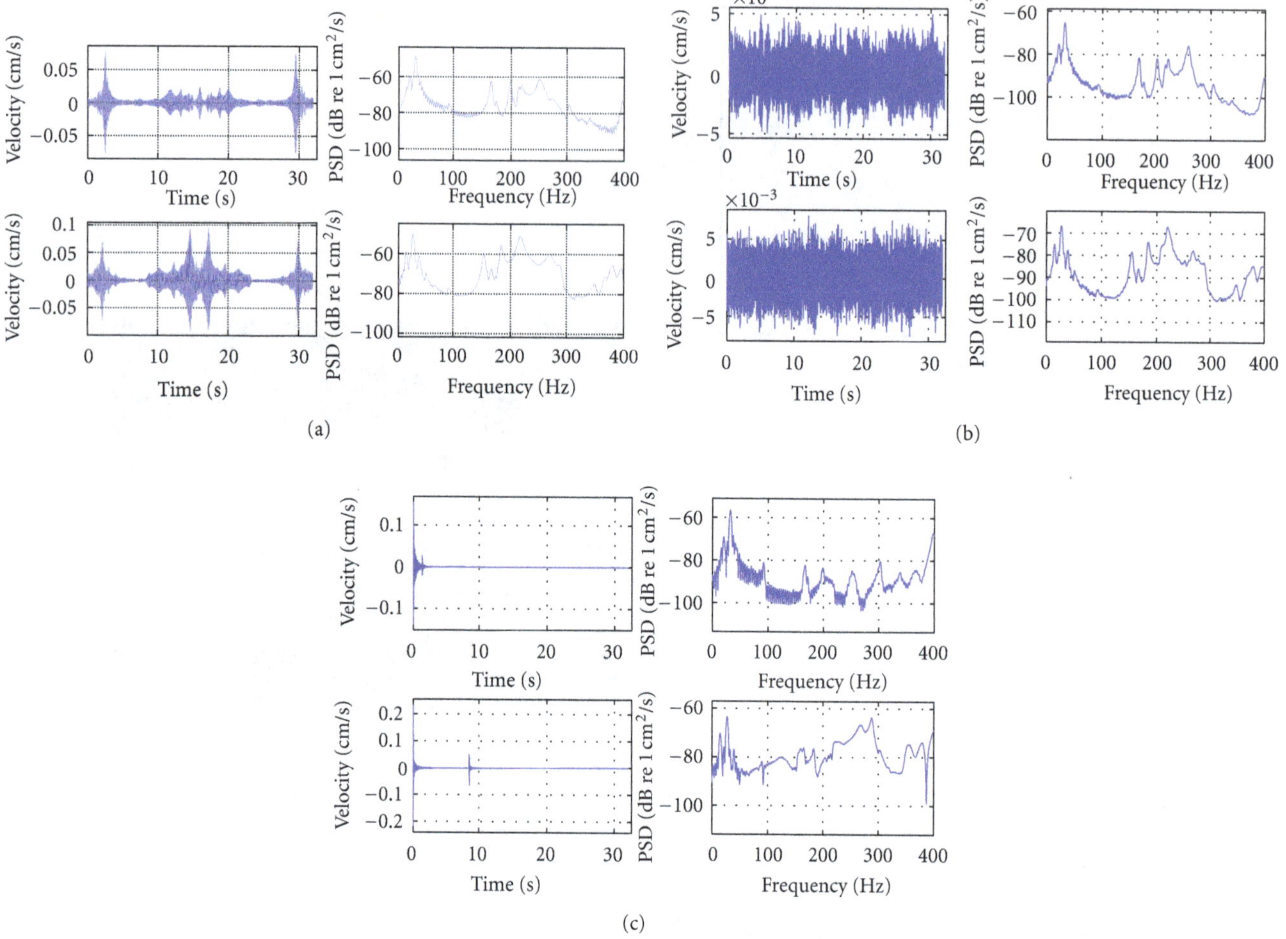

FIGURE 10: Typical time signals of vertical sensor $V4$ and PSD estimates of them: upper row is for undamaged model, lower row is for 22.2% delaminated model: (a) due to swept sine excitations; (b) due to random excitations; (c) due to impact excitations.

the ANSYS models are consistent with those obtained from real measurements; this could validate the ANSYS model to some degree.

It is concluded that the finite element modeling can approximate the undamaged and delaminated concrete slabs for these experimental studies for modal analysis. The results computed by the finite element analysis can be used as supplements for modal identification of the dynamic test data to determine the modal characteristics. The finite element modeling can provide initial investigation for further experimental tests and can also be used effectively for parameter studies.

4.4. Field Test of the Proposed Approach to the Actual Delamination. In the experimental modal testing, the vibration sensors were installed at the top surface. Since vertical modes were concerned for detecting delamination, it does not affect the identification results by placing the sensors either above or under the deck. Therefore, the proposed method will be applicable to actual delamination of concrete decks, in which the sensors were usually installed under the decks. For the next phase of this study, it will benefit to investigate the effectiveness of the method in delamination detection of real structures.

5. Conclusions

This paper presents study on delamination detection of concrete plates by modal identification of output only vibration measurements. The feasibility is examined through numerical as well as experimental studies.

Parameter studies on the effects of delamination on modal characteristics were performed through finite element modeling. The delaminations with different sizes and locations were introduced into the concrete plates separately, which were simply supported on two opposite edges and free on the others.

(a) For the concrete slabs, modal characteristics are dependent on the size of the delamination. The changes in mode shapes were sensitive to delaminations and can indicate and locate the development of delamination. Based on the results, the effects

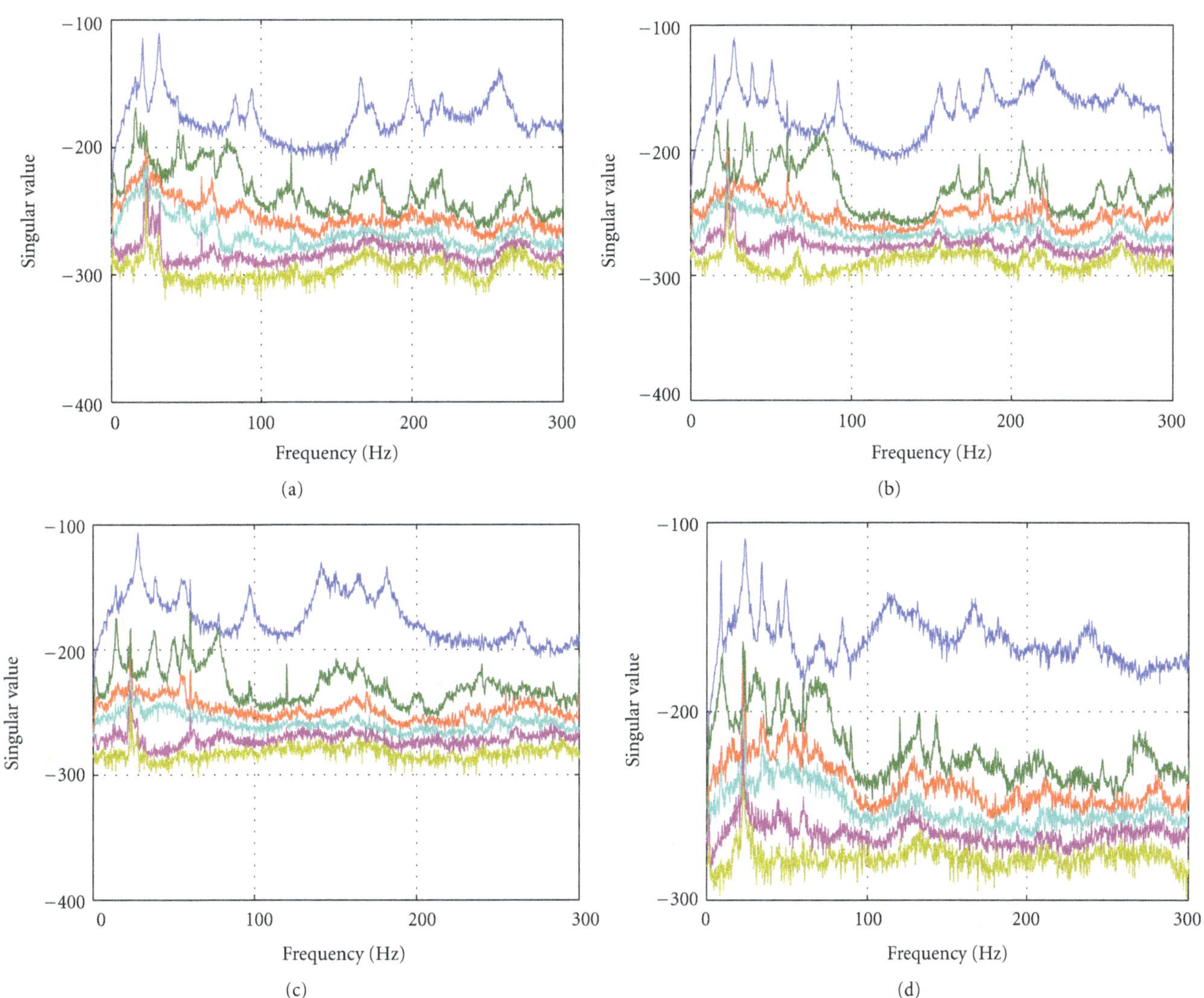

Figure 11: Singular values of PSD of the responses ($V1 \sim V6$) due to random excitation for the tested models: (a) undamaged model; (b) 22.2% delaminated model; (c) 51.9% delaminated model; (d) 74.1% delaminated model.

Table 6: Modal frequencies identified by PP, FDD, and SSI ($i = 400$, $n = 60$) methods from experimental specimens under random excitation.

—	—	1st bending	2nd bending	1st bending/torsion	3rd bending
Undamaged	PP (Hz)	31.60	68.88	84.66	165.95
	FDD (Hz)	31.77	68.03	83.04	166.08
	SSI (Hz)	31.61	67.62	82.60	166.37
22.2% delaminated	PP (Hz)	26.69	51.00	—	92.07
	FDD (Hz)	27.01	50.77	—	91.79
	SSI (Hz)	26.83	50.67	—	91.74
51.9% delaminated	PP (Hz)	27.60	55.94	—	96.79
	FDD (Hz)	27.51	54.53	—	96.80
	SSI (Hz)	27.56	55.52	—	96.94
74.1% delaminated	PP (Hz)	24.41	44.69	—	84.38
	FDD (Hz)	24.01	44.77	—	84.54
	SSI (Hz)	24.31	44.37	—	84.63

Table 7: The damping ratios identified by using SSI method from experimental specimens.

—		Undamaged	22.2% delaminated	51.9% delaminated	74.1% delaminated
From random responses	Frequency (Hz)	31.61	26.83	27.56	24.31
	ξ (%)	1.05	1.34	1.68	1.71
From impact responses	Frequency (Hz)	31.62	26.87	27.44	9.15
	ξ (%)	1.25	1.46	1.32	1.43

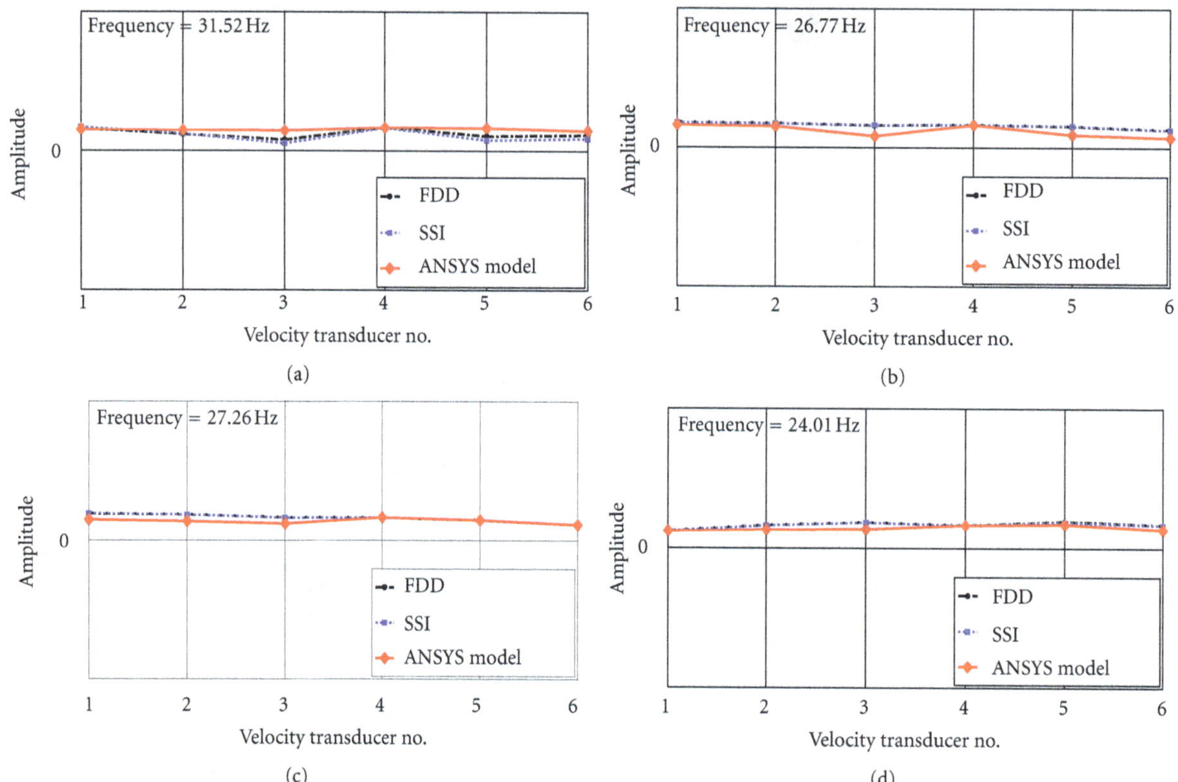

Figure 12: Comparison of mode shapes by FDD, SSI from responses due to random excitation and ANSYS model: (a) undamaged model; (b) 22.2% delaminated model; (c) 51.9% delaminated model; (d) 74.1% delaminated model.

of the delamination on the modal frequencies can be summarized as follows. The higher the modal order, the larger frequency difference between the delaminated models and the undamaged model. The frequency difference also increases with the increase of delamination area. When $1/8 \leq A_{\text{delam}}/A_{\text{tot}} \leq 1/6$, the changes in the 4th mode shape can indicate the presence of delamination. When $1/6 < A_{\text{delam}}/A_{\text{tot}} \leq 1/4$, the reductions in the 3rd frequency range can indicate the presence of delamination, and the corresponding mode shapes can be used to locate the delaminations. When $A_{\text{delam}}/A_{\text{tot}} \geq 1/3$ the reduction in the 1st frequency was greater than 6%, when $A_{\text{delam}}/A_{\text{tot}} = 1/2$ the reduction in the 1st frequency was greater than 29.64%, so the changes in the 1st frequency and mode shape may indicate delamination.

From the results of the 1/3 delaminated models at an unsymmetrical location and a symmetrical location, it is observed that the location of the delamination can have significant effect on modal parameters. The more unsymmetrical the delamination area was with respect to the total area, the bigger the changes in the frequencies and mode shapes. For example, for the cases in this paper, the changes in the 1st frequencies are 6.39% and 3.10% for the unsymmetrical and symmetrical delaminated models, respectively.

Some specific modes were more sensitive to delamination than their adjacent modes and are shown to be excellent indicators of delamination. For example, the changes in the 4th modal frequencies of the $A_{\text{delam}}/A_{\text{tot}} = 1/6, 1/5, 1/4, 1/3$ unsymmetrical 1 delaminated models are bigger than their neighboring modes.

(b) For the delaminated models, it was shown that amplitudes of delaminated areas in the mode shapes were significantly irregular compared to the undamaged models. This feature can be used to detect and locate

TABLE 8: Comparison of modal frequencies by ANSYS modeling and FDD of the responses of experimental specimens under random excitation.

—	—	1st bending	2nd bending	1st bending/torsion	3rd bending
0% delaminated	ANSYS (Hz)	31.46	69.13	81.68	165.80
	FDD (Hz)	31.77	68.03	83.04	166.08
22.2% delaminated	ANSYS (Hz)	27.38	51.99	65.63	91.71
	FDD (Hz)	27.01	50.77	—	91.79
51.9% delaminated	ANSYS (Hz)	26.98	50.81	65.27	96.99
	FDD (Hz)	27.51	54.53	—	96.80
74.1% delaminated	ANSYS (Hz)	24.91	42.78	60.47	81.39
	FDD (Hz)	24.01	44.77	—	84.54

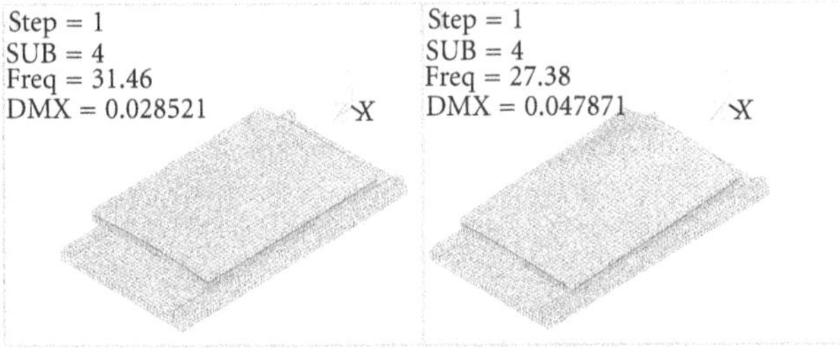

FIGURE 13: The comparison of the 3rd mode shapes for the undamaged (left) and 22.2% delaminated (right) specimens.

delamination. The limitation is that it requires a dense layout of sensors to obtain mode shapes and quantify the irregular shapes. With the development of new sensor technology and algorithms, this problem may be solved satisfactorily.

(c) The effects of the support conditions on the applicability of the proposed approach were studied. It was concluded that the proposed approach applies to various support conditions that include concrete decks supported on beams.

Experimental studies were conducted on reinforced concrete plates with different delamination sizes. The modal characteristics were extracted from the dynamic responses and analyzed. Finite element models of concrete slabs were developed using ANSYS. Conclusions are drawn as follows.

(a) The response measurements show that, if given same external excitation, the amplitudes of the delaminated models were obviously larger than those of the undamaged model.

(b) Changes in frequencies and mode shapes can indicate the occurrence and degree of delamination. The frequencies decrease with the increase of delamination sizes. The 1st modal frequency decreased 43% for the 74.1% delaminated specimen when compared to the undamaged specimen. The changes in higher modal frequencies are larger. The experimental results show that the damping ratios increase with the increase of delamination sizes. Damping can assist in the delamination identification, but it alone is not suitable as a delamination indicator. It demonstrated that the mode shapes have abrupt changes in the delamination areas of the specimens. When other damage exists, for example, concrete spalling and rock pocket, more effort is needed to determine whether the changes in modal characteristics can be attributed specifically to delamination. Further investigations of damage characteristics of delamination may be needed.

(c) It is beneficial to use multiple identification methods to ensure correct identification results. The frequencies identified by the PP, FDD, and SSI methods agree well; the maximum difference between the 1st frequencies by these methods is less than 4.8%. The mode shapes that are extracted by FDD and SSI are shown to be identical.

(d) In the finite element modeling, the frequencies and mode shapes calculated by the FE models agreed well with those identified from test data, among which the differences between the first three modal frequencies by ANSYS model and FDD method from test data are within 7%. Therefore finite element models can approximately model delaminated concrete slabs for modal analysis.

This work provides useful information for practical delamination detection of bridge decks by vibration-based health monitoring systems. The experimental as well as numerical models were intended to simulate a portion of real concrete deck on beams. Therefore, the results were applicable to the concrete panels on girders, while the delamination detection approach in this study is also an alternative way for other structural forms.

For the future work, experimental modal testing with more stable and practical boundary conditions is needed. Stabilization diagrams are suggested to be used for obtaining more accurate modal characteristics. Further work is necessary for developing more sensitive delamination indicators and more accurate finite element models. Additionally, field tests for actual delamination detection using the actual vibration sensors under bridge deck are highly recommended. It is anticipated that the proposed new methodology can be applied to an actual bridge deck in five years.

Acknowledgment

The research work was financially supported by the Utah Transportation Center (UTC) and the Utah Department of Transportation (UDOT).

References

[1] J. P. Warhus, J. E. Mast, and S. D. Nelson, "Imaging radar for bridge deck inspection," in *Proceedings of the Nondestructive Evaluation of Aging Bridges and Highways*, pp. 296–305, Oakland, Calif, USA, June 1995.

[2] J. Broomfield, *Corrosion of Steel in Concrete, Understanding, Investigating, and Repair*, E & FN Spon, London, UK, 2nd edition, 1997.

[3] C. Q. Li, J. J. Zheng, W. Lawanwisut, and R. E. Melchers, "Concrete delamination caused by steel reinforcement corrosion," *Journal of Materials in Civil Engineering*, vol. 19, no. 7, pp. 591–600, 2007.

[4] N. Gucunski, Z. Wang, T. Fang, and A. Maher, "Rapid bridge deck condition assessment using three-dimensional visualization of impact echo data," in *Proceedings of the Non-Destructive Testing in Civil Engineering (NDTCE '09)*, Nantes, France, 2009.

[5] Y. Zou, L. Tong, and G. P. Steven, "Vibration-based model-dependent damage (delamination) identification and health monitoring for composite structures—a review," *Journal of Sound and Vibration*, vol. 230, no. 2, pp. 357–378, 2000.

[6] S. H. Diaz Valdes and C. Soutis, "Delamination detection in composite laminates from variations of their modal characteristics," *Journal of Sound and Vibration*, vol. 228, no. 1, pp. 1–9, 1999.

[7] C. P. Ratcliffe and W. J. Bagaria, "Vibration technique for locating delamination in a composite beam," *AIAA Journal*, vol. 36, no. 6, pp. 1074–1077, 1998.

[8] Z. Wei, L. H. Yam, and L. Cheng, "Delamination assessment of multilayer composite plates using model-based neural networks," *JVC/Journal of Vibration and Control*, vol. 11, no. 5, pp. 607–625, 2005.

[9] Y. J. Yan and L. H. Yam, "Detection of delamination damage in composite plates using energy spectrum of structural dynamic responses decomposed by wavelet analysis," *Computers and Structures*, vol. 82, no. 4-5, pp. 347–358, 2004.

[10] S. Xing, M. W. Halling, and P. J. Barr, "Delamination detection and location in concrete deck by modal identification," in *Proceedings of the Structures Congress*, pp. 741–751, Orlando, Fla, USA, 2010.

[11] S. Xing, M. W. Hailing, and P. J. Barr, "Delamination detection in concrete plates using output-only vibration measurements," in *Proceedings of the 29th International Modal Analysis Conference (IMAC '11)*, pp. 255–262, Jacksonville, Fla, USA, February 2011.

[12] R. Brincker, L. Zhang, and P. Andersen, "Modal identification from ambient responses using frequency domain decomposition," in *Proceedings of the International Modal Analysis Conference (IMAC '00)*, pp. 625–630, San Antonio, Tex, USA, 2000.

[13] R. Brincker, C. E. Ventura, and P. Anderson, "Damping estimation by frequency domain decomposition," in *Proceedings of the International Modal Analysis Conference (IMAC '01)*, pp. 698–703, Kissimmee, Fla, USa, 2001.

[14] P. Van Overschee and B. De Moor, "Subspace algorithms for the stochastic identification problem," *Automatica*, vol. 29, no. 3, pp. 649–660, 1993.

[15] P. Van Overschee and B. De Moor, *Subspace Identification for Linear Systems: Theory, Implementation*, Kluwer Academic, Dordrecht, The Netherlands, 1996.

[16] B. Peeters and G. De Roeck, "Reference-based stochastic subspace identification for output-only modal analysis," *Mechanical Systems and Signal Processing*, vol. 13, no. 6, pp. 855–878, 1999.

[17] T. Katayama, *Subspace Methods for System Identification*, Springer, Englewood Cliffs, New Jersey, USA, 1st edition, 2005.

[18] AnsysH Release 10.0 ANSYS theory reference, 2005.

[19] S. Rajendran and D. Q. Song, "Finite element modeling of delamination buckling of composite panel using Ansys," in *Proceedings of the 2nd Asian ANSYS User Conference*, Singapore, 1998.

[20] A. Leissa, "The free vibration of rectangular plates," *Journal of Sound and Vibration*, vol. 31, no. 3, pp. 257–293, 1973.

[21] K. M. Liew, Y. Xiang, S. Kitipornchai, and C. M. Wang, *Vibration of Mindlin Plates: Programming the P-Version Ritz Method*, Elsevier Science, Oxford, UK, 1998.

[22] Forest Products Laboratory, *Wood Handbook: Wood as an Engineering Material*, U.S. Department of Agriculture, Forest Service, Forest Products Laboratory, Madison, Wis, USA, 2010.

[23] R. S. McBurney and J. T. DROW, "The elastic properties of wood Young's moduli and poisson's ratios of douglas-fir and their relations to moisture content," Report no. 1528-D, Forest Products Laboratory, Forest Product Service, U.S. Department of Agriculture, 1962.

[24] A. Sliker, "Measuring Poisson's ratios in wood—a method for measuring Poisson's ratios along with Young's moduli in wood is described by the author," *Experimental Mechanics*, vol. 12, no. 5, pp. 239–242, 1972.

9

Low-Cost MEMS-Based Pedestrian Navigation Technique for GPS-Denied Areas

Abdelrahman Ali and Naser El-Sheimy

Department of Geomatics Engineering, The University of Calgary, 2500 University Drive N.W., Calgary, AB, Canada T2N 1N4

Correspondence should be addressed to Abdelrahman Ali; asali@ucalgary.ca

Academic Editor: Kai-Wei Chiang

The progress in the micro electro mechanical system (MEMS) sensors technology in size, cost, weight, and power consumption allows for new research opportunities in the navigation field. Today, most of smartphones, tablets, and other handheld devices are fully packed with the required sensors for any navigation system such as GPS, gyroscope, accelerometer, magnetometer, and pressure sensors. For seamless navigation, the sensors' signal quality and the sensors availability are major challenges. Heading estimation is a fundamental challenge in the GPS-denied environments; therefore, targeting accurate attitude estimation is considered significant contribution to the overall navigation error. For that end, this research targets an improved pedestrian navigation by developing sensors fusion technique to exploit the gyroscope, magnetometer, and accelerometer data for device attitude estimation in the different environments based on quaternion mechanization. Results indicate that the improvement in the traveled distance and the heading estimations is capable of reducing the overall position error to be less than 15 m in the harsh environments.

1. Introduction

Personal navigation requires technologies that are immune to signal obstructions and fading. One of the major challenges is obtaining a good heading solution in different environments and for different user positions without external absolute reference signals. Part of this challenge arises from the complexity and freedom of movement of a typical handheld user where the heading observability considerably degrades in low-speed walking, making this problem even more challenging. However, for short periods, the relative attitude and heading information is quite reliable. Self-contained systems requiring minimal infrastructure, for example, inertial measurement units (IMUs), stand as a viable option, since pedestrian navigation is not only focused on outdoor navigation but also on indoor navigation.

Nowadays, most of the smartphones are programmable and equipped with self-contained, low cost, small size, and power-efficient sensors, such as magnetometers, gyroscopes, and accelerometers. Hence, integrating IMUs navigation solution with a magnetometer-based heading can play an important role in pedestrian navigation in all environments. In the current state of the art in MEMS technology, the accuracy of gyroscopes is not good enough for deriving an absolute heading or relative heading over longer durations of time. However, for short periods, the relative attitude information is quite reliable. Magnetometers, on the other hand, provide absolute heading information once calibrated. However, they can easily be disturbed by ferrous objects nearby, making them unreliable for brief intervals. This calls for the investigation of possible sources of heading error in complementary sensors such as a gyroscope and a magnetometer and improving the accuracy of the result based on an improved Kalman filter design.

Much research towards the heading estimation for personal positioning applications has been conducted in the recent years. Some approaches use magnetometers exclusively for heading estimation [1] while others integrate it tightly with an IMU [2, 3]. One commercially available personal locator system based on this principle is the Dead Reckoning Module DRM-4000 made by Honeywell [4]. A quaternion-based method to integrate IMU with magnetometer is presented by [5]. Three body angular rates and four quaternion elements were used to express attitude and were selected as the states of the Kalman filter. The method needs to model the angular motion of the body. In [6], a linear system error model based on the Euler angles errors expressing the local frame errors is developed, and the corresponding system observation model is derived. The proposed method does not need to model the system angular motion and also avoids the nonlinear problem which is inherent in the customarily used methods. A similar technique is proposed by [7] where the angular rates were modeled to be a constant. A nonlinear derivative equation for the Euler angle integration kinematics is investigated in [8]. Work in [9, 10] presented an Euler angle error based method to integrate IMU with magnetometer data where three Euler angle errors and three gyroscope biases were used as states for the Kalman filter. The estimated states were used to correct the Euler angles and to compensate gyroscope drifts, respectively. The work at [11] presented a mathematical model for compass deviation by creating an a priori look-up table for heading corrections. A Kalman filtering approach was investigated by [12] to estimate the angular rotation from the input of a magnetometer compass and three gyroscopes. References [13, 14] presented a least squares technique with improvement which is used for the estimation of the compass deviation model. In addition, much research has been conducted to use the 3D magnetometer-based heading for personal navigation applications in the recent years [15].

The magnetometer cannot be used as standalone source for heading information in the harsh environments, especially indoor [16]. In addition, it is required to have knowledge about the preexisted magnetic anomalies resulted from some of the man-made infrastructure [17]. Using magnetic field measurements in heading estimation for indoor navigation also has some limitations as the magnetic field signal needs to be strong enough. Also, the mobile navigation device should be away from any source of disturbances to avoid any perturbation effect [18]. Besides that, the magnetic field during the indoor environment is not completely constant due to the presence of the electronic and electrical devices everywhere. To avoid the problem of magnetometer anomaly, arising out of ferrous materials in the vicinity of the magnetometers, a perturbation detection technique is required. In such scenario, the filter works only in the propagation mode without any update for the attitude. Also the gyroscope bias drifts with time and temperature can be compensated by magnetometers. In this paper, a method is presented to obtain seamless attitude information by integrating the heading outputs based on magnetometer, accelerometer, and gyroscope measurements using the Kalman filter (KF).

2. Pedestrian Dead Reckoning (PDR)

Pedestrian dead reckoning is a relative means of positioning where the initial position and heading of the user are supposed to be known. The basic concept and components of the proposed PDR algorithm are shown in Figure 1. Generally, steps of the user are detected based on the accelerometer data. To get the travelled distance, the total number of steps is multiplied by the step length. With known heading and reference point, the user can be tracked by successive steps count.

Step detection is a basic step for any PDR algorithm. Usually, the accelerometer signal is used to detect the steps of the person. Once the step is detected, the total number of steps for a pedestrian can be counted. As a result, the total travelled distance can be estimated by multiplying the step length with the total number of steps. Using travelled distance and heading, user can be located during a typical trip. Step detection algorithm can be performed based on different kinds of sensors, that is, not only accelerometers but even gyroscopes and magnetometers. However, our step event detection scheme is based on using the accelerometer signal. The norm of the three accelerometers is used as in (1), where it is possible to clearly identify the steps by observing, for example, the signal over time:

$$\text{accel}_{\text{norm}} = \sqrt{(Fx^2 + Fy^2 + Fz^2)}. \quad (1)$$

Steps are detected as peaks in the resulting norm, where the step is the highest local maximum in the norm acceleration between the current peak and the previous step peak.

3. Sensors Performance

The used device in the test, Samsung Galaxy Nexus smartphone, is shown in Figure 2 with axes definition.

Besides other sensors, the device is equipped with triad magnetometer (M), triad gyroscope (G), and triad accelerometer (A). The manufactures and the ranges of the main used sensors are listed in Table 1.

Table 1: Sensors manufacture and ranges.

Sensor	Manufacture	Range
Gyroscope	mpu-3050	±250 to ±2000°/sec
Accelerometer	BMA220	±2, ±4, ±8, ±16 g
Magnetometer	YAS530	±800 μT

3.1. Gyroscope Drift. Gyroscopes are mainly used to determine device attitude. The output of such sensor is rotational rate, and performing a single integration on the gyroscopes outputs is necessary to obtain a relative change in angle. Due to the integration process which is very sensible to the systematic errors of the gyroscopes, the bias introduces a quadratic error in the velocity and a cubic error in the position [19]. Gyroscopes measurements can generally be described using

$$I_\omega = \omega + b_\omega + S_\omega + N_\omega + \varepsilon(\omega), \quad (2)$$

where I_ω is the measured angular rate, ω is the true angular rate, b_ω is the gyroscope bias, S is the linear scale factor matrix, N is the nonorthogonality matrix, and $\varepsilon(\omega)$ is the sensor

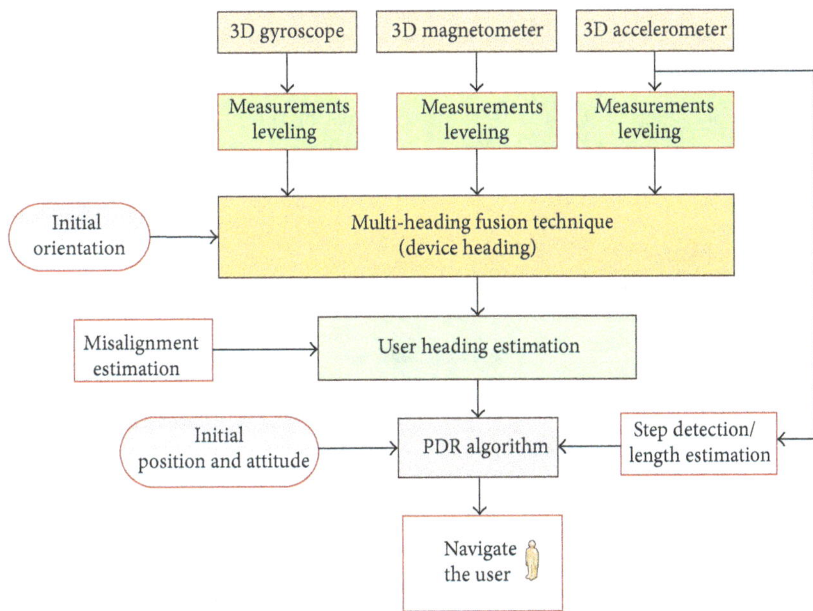

FIGURE 1: The main concept of the PDR algorithm.

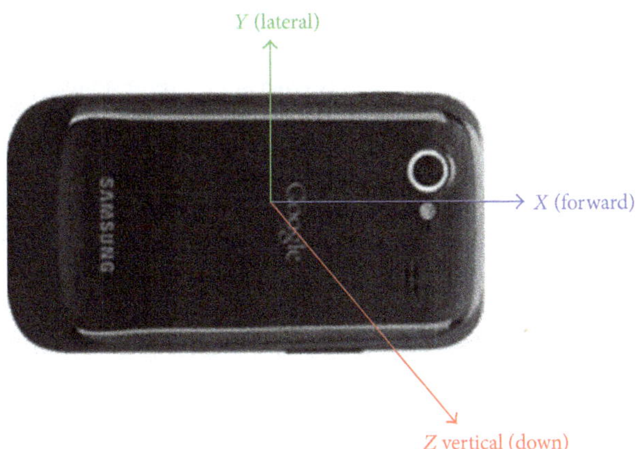

FIGURE 2: Device axis definition.

noise. With integration, the gyroscope bias will introduce an angle error in pitch or roll proportional to time; that is, $\delta\theta = \int b_\omega dt = b_\omega t$; this small angle will cause misalignment of the IMU. Therefore, when projecting the acceleration from for example the gravity vector g, from the body frame to the local-level frame, the acceleration vector will be incorrectly projected due to this misalignment error. This will introduce an error in one of the horizontal acceleration; that is, $\delta a = g\sin(\delta\theta) \approx g\delta\theta \approx gb_\omega t$. Consequently, this leads to an error in velocity $\delta v = \int b_\omega gt\, dt = (1/2)\, b_\omega gt^2$ and in position $\delta p = \int \delta v\, dt = \int (1/2) b_\omega gt^2 dt = (1/6) b_\omega gt^3$. To overcome the problem of error drift, a bias compensation of gyroscopes is required. When the device is in stationary mode, the deterministic bias of the gyroscope can be estimated by calculating the average gyroscope output during that time.

3.2. Magnetometer Perturbation. During operation, especially inside buildings, a magnetometer is subject to many external disturbances such as large metal objects [20]. Other objects like steel structures and electromagnetic power lines can affect the solution of the magnetometer. These kinds of disturbance lead to unpredictable performance of the magnetometer which is a major drawback of using magnetic sensors. In the meanwhile, the magnetic field parameters such as strength, horizontal and vertical magnetic field, and change in the inclination angle can be checked to detect the perturbation in magnetometer measurement by comparing to the reference values which can be found at [21]. Figure 3 shows an example for harsh environment as the magnetometer is totally disturbed due to the steel constructions. The figure shows the total magnetic field in a perturbed area during a walking test compared to the reference value which is 570 mGauss for Calgary [21].

4. Multisensors Heading Fusion Filter

The attitude of the device is commonly estimated using the inertial sensor. There are three main approaches for attitude representation: DCM, Euler angle, and quaternion. Among the three techniques quaternion algebra is the preferred. However, the estimated attitude from the gyroscope is very noisy leading to unbounded growth in the heading errors. An integration scheme for the gyroscope, accelerometer, and magnetometer data is proposed to estimate the device attitude and the gyroscope bias. The proposed scheme is a quaternion-based KF as shown in Figure 4.

In order to use the KF-based estimator for quaternion parameters and gyroscope biases estimation for a device which is carried by a pedestrian, the required model for the states and measurements and their respective system and measurement error models are presented in this section.

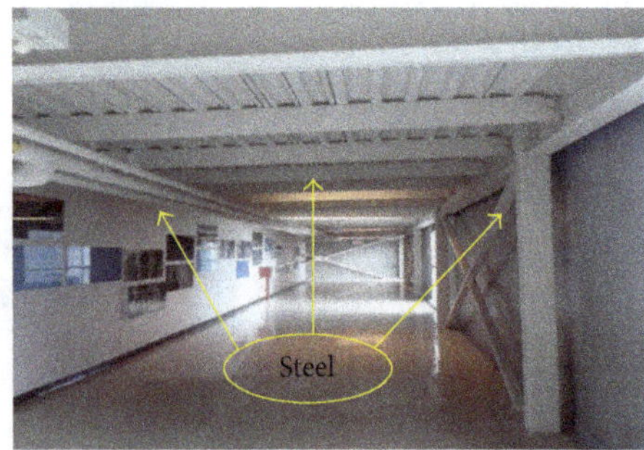

(a) The structure at the CCIT 2nd floor

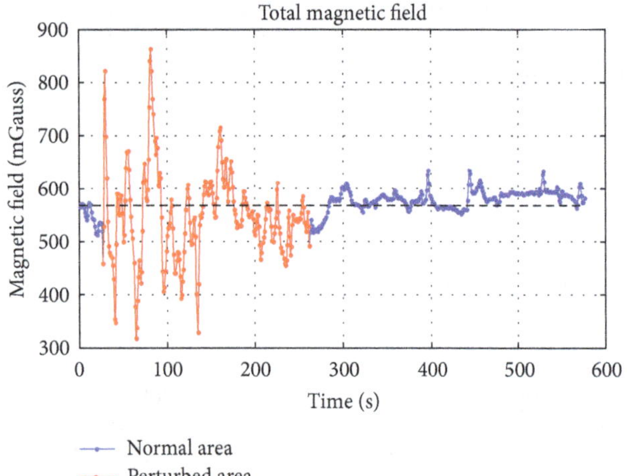

(b) Total magnetic field

FIGURE 3: Magnetic field perturbation.

4.1. Quaternion Mechanization. The relationship between the direct cosine matrix (DCM) and Euler angles is given in (3) with the sequence of azimuth, pitch, and roll ($\psi\theta\phi$) or ($R_\phi^x R_\theta^y R_\psi^z$) [22]:

$$C_l^b = \begin{bmatrix} c\theta c\psi & c\theta s\psi & -s\theta \\ s\phi s\theta c\psi - c\phi s\psi & s\phi s\theta s\psi + c\phi c\psi & s\phi c\theta \\ c\phi s\theta c\psi + s\phi s\psi & c\phi s\theta s\psi - s\phi c\psi & c\phi c\theta \end{bmatrix}. \quad (3)$$

A quaternion is a four-dimensional vector which is defined based on a vector q and a rotation angle. The vector q is given as

$$q = (q_1, q_2, q_3, q_4). \quad (4)$$

The DCM matrix in terms of quaternion vector components can be obtained from using

$$C^n(q) = \begin{bmatrix} 2q_1^2 - 1 + q_2^2 & 2q_2q_3 + 2q_1q_4 & 2q_2q_4 - 2q_1q_3 \\ 2q_2q_3 - 2q_1q_4 & 2q_1^2 - 1 + q_3^2 & 2q_3q_4 + 2q_1q_2 \\ 2q_2q_4 + 2q_1q_3 & 2q_3q_4 - 2q_1q_2 & 2q_1^2 - 1 + q_4^2 \end{bmatrix}, \quad (5)$$

where C^n represents the DCM matrix in terms of the quaternion vector.

The matrix transforms from the body frame to the local level (navigation) frame. The roll, pitch, and azimuth values can be obtained by using the $a\tan 2$ math function on the values of the C^n propagated inside the sensors navigation equations

$$\phi = \tan^{-1}\left(\frac{C^n(2,3)}{C^n(3,3)}\right),$$
$$\theta = -\sin^{-1}\left(C^n(1,3)\right), \quad (6)$$
$$\psi = \tan^{-1}\left(\frac{C^n(1,2)}{C^n(1,1)}\right).$$

4.2. Filter States. Basically, the target of the proposed filter is to estimate the device attitude based on the quaternion technique. Consequently, any improvement in the quaternion estimate leads to improving the estimated attitude values. The implementation of the KF is optimal for linear systems driven by additive white Gaussian noise (AWGN). The state model can be written in the following form:

$$\dot{x} = Fx + Gw, \quad (7)$$

where x is the state vector, F is the state transition matrix, and Gw represents the covariance matrix of the applied state model. The measurement system can be represented by a linear equation of the following form:

$$Z = Hx + v, \quad (8)$$

where Z is the vector of measurement updates, H is the design (observation) matrix that relates the measurements to the state vector, and v is the measurement noise.

The nonlinear form of the system model in the absence of the known input can be written as

$$\dot{x}(t) = F(x(t), t) + G(t)W(t), \quad (9)$$

where $F(x(t), t)$ is now a nonlinear function describing the time evolution of the states. Consider a nominal trajectory, $x^{\text{nom}}(t)$, related to the actual trajectory, $x(t)$, as

$$\delta x(t) = x(t) - x^{\text{nom}}(t), \quad (10)$$

where $\delta x(t)$ is a perturbation from nominal trajectory. performing a Taylor series expansion equation (10) about the nominal trajectory yields

$$\begin{aligned} \dot{x}(t) &\approx F(x^{\text{nom}}(t), t) \\ &\quad + \left.\frac{\partial F(x(t),t)}{\partial x(t)}\right|_{x(t)=x^{\text{nom}}(t)} \delta x(t) \\ &\quad + G(t)W(t) \\ &= \dot{x}^{\text{nom}}(t) + F\delta x(t) + G(t)W(t), \\ \dot{x}(t) - \dot{x}^{\text{nom}}(t) &= F\delta x(t) + G(t)W(t), \\ \delta\dot{x}(t) &= F\delta x(t) + G(t)W(t), \end{aligned} \quad (11)$$

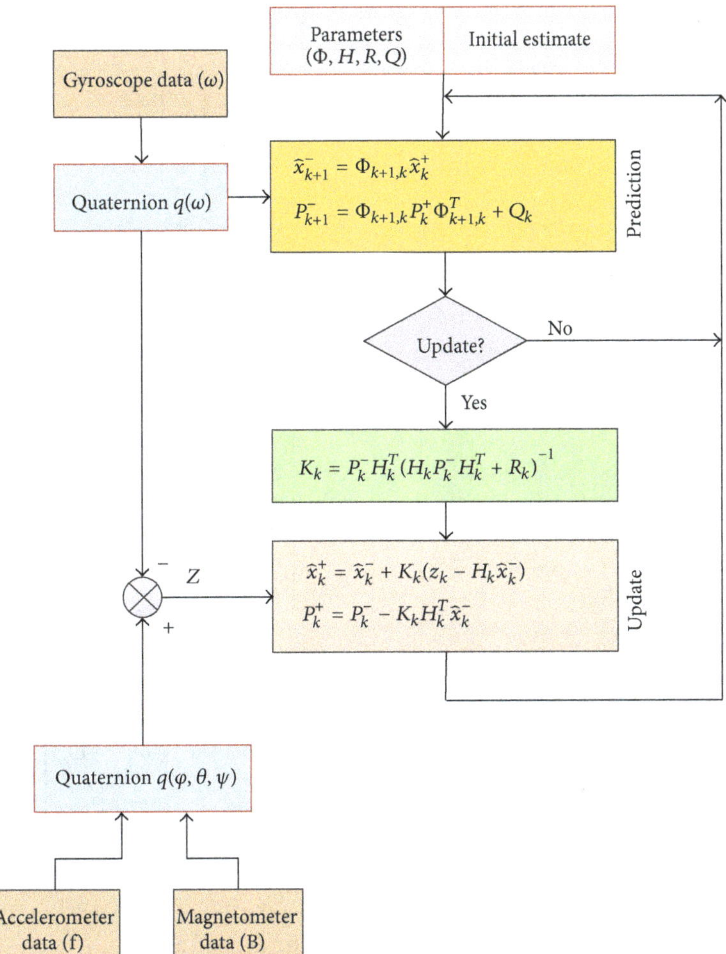

FIGURE 4: Flow of the Kalman filter process.

where F is now the dynamic matrix for a system with state vector which consists of the perturbed states, δx. Thus, the main states to be estimated are the errors in the quaternion parameters given by:

$$\delta q = \left[\delta q_1 \; \delta q_2 \; \delta q_3 \; \delta q_4\right]^T, \tag{12}$$

where δq_i is the error in the ith quaternion parameter.

The quaternion parameters are primarily determined using the angular rates obtained from gyroscopes' measurements. The deterministic errors associated with the gyroscope can be compensated using data from static interval at the beginning of the test while the stochastic errors in biases are given by

$$b_\omega = \left[b_{\omega x} \; b_{\omega y} \; b_{\omega z}\right]^T, \tag{13}$$

where $b_{\omega x}$, $b_{\omega y}$, and $b_{\omega z}$ are the gyroscope biases.

The complete state vector is defined as a 7-dimensional vector with the first four components being errors in the elements of the quaternion and the last three being the elements of the gyroscope biases.

$$x = \left[\delta q_1 \; \delta q_2 \; \delta q_3 \; \delta q_4 \; b_{\omega x} \; b_{\omega y} \; b_{\omega z}\right]^T. \tag{14}$$

4.3. The State Transition Model. The angular rate is linked to the quaternion parameters as in the following:

$$\dot{q} = \frac{1}{2} \cdot q \otimes \omega = \frac{1}{2} \begin{bmatrix} -q_2 & -q_3 & -q_4 \\ q_1 & -q_4 & q_3 \\ q_4 & q_1 & -q_2 \\ -q_3 & q_2 & q_1 \end{bmatrix} \begin{bmatrix} \omega_x \\ \omega_y \\ \omega_z \end{bmatrix}, \tag{15}$$

where q is the attitude quaternion ω_x, ω_y, and ω_z represent angular rate measurements in the sensor frame obtained using the rate gyroscopes. Quaternion is used to represent attitude in the filter design because it does not have the singularity problem associated with Euler angles. The previous equation can be rewritten as

$$\begin{bmatrix} \dot{q}_1 \\ \dot{q}_2 \\ \dot{q}_3 \\ \dot{q}_4 \end{bmatrix} = \frac{1}{2} \begin{bmatrix} -q_2\omega_x - q_3\omega_y - q_4\omega_z \\ q_1\omega_x - q_4\omega_y + q_3\omega_z \\ q_4\omega_x + q_1\omega_y - q_2\omega_z \\ -q_3\omega_x + q_2\omega_y + q_1\omega_z \end{bmatrix}. \tag{16}$$

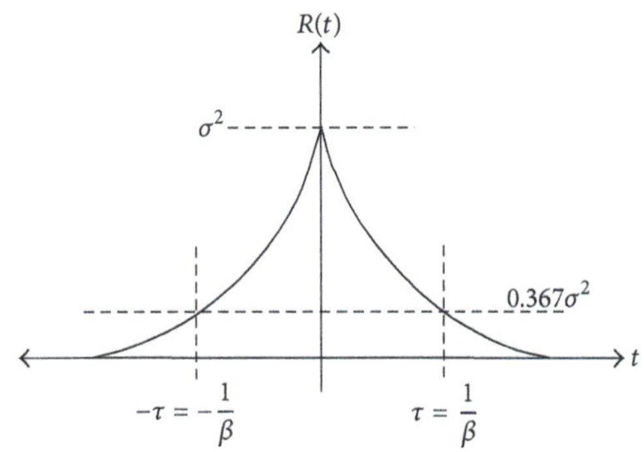

(a) Gauss Markov parameters

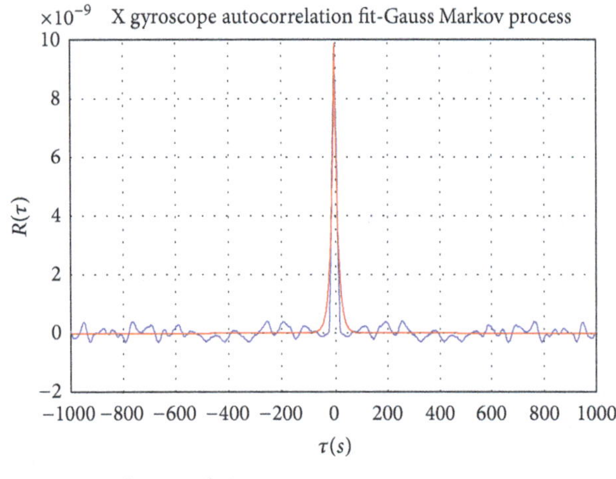

(b) Autocorrelation fit for the x gyroscope

FIGURE 5: Gauss-Markov model representation.

The Taylor series expansion to first order is shown in the following:

$$\delta\dot{r} = \frac{\partial\dot{r}}{\partial r}\delta r, \tag{17}$$

where r is the state vector δr is the error in the state. Thus, the quaternion parameters error can be obtained as

$$\begin{bmatrix}\delta\dot{q}_1\\ \delta\dot{q}_2\\ \delta\dot{q}_3\\ \delta\dot{q}_4\end{bmatrix} = \frac{1}{2}\begin{bmatrix}0 & -\omega_x & -\omega_y & -\omega_z\\ \omega_x & 0 & \omega_z & -\omega_y\\ \omega_y & -\omega_z & 0 & \omega_x\\ \omega_z & \omega_y & -\omega_x & 0\end{bmatrix}\begin{bmatrix}\delta q_1\\ \delta q_2\\ \delta q_3\\ \delta q_4\end{bmatrix}. \tag{18}$$

A general equation for the 1st order Gauss-Markov model is given as

$$\dot{b} = -\beta b + \sqrt{2\beta\sigma^2}\, w(t), \tag{19}$$

where β is the reciprocal of the correlation time and σ^2 is the variance of the gyroscope signal. The different parameters of the Gauss-Markov model can be determined as shown in Figure 5.

According to (19), the gyroscopes bias can be modeled as

$$\dot{b}_\omega = \begin{bmatrix}\dot{b}_{\omega x}\\ \dot{b}_{\omega y}\\ \dot{b}_{\omega z}\end{bmatrix} = \begin{bmatrix}-\beta_x & 0 & 0\\ 0 & -\beta_y & 0\\ 0 & 0 & -\beta_z\end{bmatrix}\begin{bmatrix}b_{\omega x}\\ b_{\omega y}\\ b_{\omega z}\end{bmatrix} + \begin{bmatrix}\sqrt{2\beta_x\sigma_x^2}\\ \sqrt{2\beta_y\sigma_y^2}\\ \sqrt{2\beta_z\sigma_z^2}\end{bmatrix} w. \tag{20}$$

The complete state model can be written as the following:

$$\begin{bmatrix}\delta\dot{q}_1\\ \delta\dot{q}_2\\ \delta\dot{q}_3\\ \delta\dot{q}_4\\ \dot{b}_{\omega x}\\ \dot{b}_{\omega y}\\ \dot{b}_{\omega z}\end{bmatrix} = \begin{bmatrix}F_\omega & 0_{4\times 3}\\ 0_{3\times 4} & F_b\end{bmatrix}\cdot\begin{bmatrix}\delta q_1\\ \delta q_2\\ \delta q_3\\ \delta q_4\\ b_{\omega x}\\ b_{\omega y}\\ b_{\omega z}\end{bmatrix} + \begin{bmatrix}0\\ 0\\ 0\\ 0\\ \sqrt{2\beta_x\sigma_x^2}\\ \sqrt{2\beta_y\sigma_y^2}\\ \sqrt{2\beta_z\sigma_z^2}\end{bmatrix}\cdot w, \tag{21}$$

where

$$F_\omega = \begin{bmatrix}0 & -\omega_x & -\omega_y & -\omega_z\\ \omega_x & 0 & \omega_z & -\omega_y\\ \omega_y & -\omega_z & 0 & \omega_x\\ \omega_z & \omega_y & -\omega_x & 0\end{bmatrix}, \qquad F_b = \begin{bmatrix}-\beta_x & 0 & 0\\ 0 & -\beta_y & 0\\ 0 & 0 & -\beta_z\end{bmatrix}. \tag{22}$$

Using the dynamic matrix in (21), the state transition matrix can be defined as

$$\Phi_{k+1,k} = e^{(F\cdot\Delta t)} = I + F\cdot\Delta t + \frac{(F\cdot\Delta t)^2}{2!}. \tag{23}$$

4.4. Measurement Models. The magnetometer measurements along with the accelerometer data are used as main source of update. The measured magnetic field is tested for perturbation. Once the magnetic field is free of disturbances, the geomagnetic heading is estimated from the calibrated data. Also, the accelerometer measurements are experienced to be noisy as low cost MEMS sensors are used and usually measured at higher rate. Therefore, the average of the measured data over the step time is used to estimate the roll and pitch values. The roll, pitch, and heading estimates are used to calculate the quaternion parameters.

The only source of update is the quaternion parameters. Thus, the design matrix can be set as

$$\delta z = \begin{bmatrix}\delta q_1\\ \delta q_2\\ \delta q_3\\ \delta q_4\end{bmatrix} = \begin{bmatrix}1 & 0 & 0 & 0 & 0 & 0 & 0\\ 0 & 1 & 0 & 0 & 0 & 0 & 0\\ 0 & 0 & 1 & 0 & 0 & 0 & 0\\ 0 & 0 & 0 & 1 & 0 & 0 & 0\end{bmatrix}\cdot\begin{bmatrix}\delta q_1\\ \delta q_2\\ \delta q_3\\ \delta q_4\\ b_{\omega x}\\ b_{\omega y}\\ b_{\omega z}\end{bmatrix}. \tag{24}$$

4.5. Modeling of Process and Measurement Noises. In order to complete the design of the KF, it is necessary to define the noise covariance matrices, the process noise covariance matrix Q and the measurement noise covariance matrix R. These matrices reflect the confidence in the system model and the measurements, respectively. The covariance of w_k is often called the process noise matrix, Q_k, and can be computed as

$$Q_k = E\left[w_k \cdot w_k^T\right]. \quad (25)$$

The Q_k matrix is a 7-dimension square matrix which can be computed using as follows [23, 24]:

$$Q_k = \int_{t_k}^{t_{k+1}} \Phi_{t_{k+1},\tau} \cdot G(\tau) \cdot Q(\tau) \cdot G^T(\tau) \Phi_{t_{k+1},\tau}^T d\tau. \quad (26)$$

The measurement noise covariance matrix R is also known as the covariance matrix for v. The R_k matrix represents the level of confidence placed in the accuracy of the measurements and is given by

$$R_k = E\left[v_k \cdot v_k^T\right]. \quad (27)$$

The R_k matrix is a 4-dimension diagonal square matrix. The diagonal elements are the variances of the individual measurements, which can be determined experimentally using measurement data from the used sensors.

4.6. Filter State Initialization. The state vector should be initialized at the beginning of the process. For the gyroscope bias states, all biases are initialized as zeros. The quaternion states can be initialized from the DCM matrix using the Euler angles. The mean of the accelerometer calibrated data during a stationary period can be used to estimate the initial roll and pitch using the following relationships [25]:

$$\begin{aligned} \phi_o &= \tan^{-1}\left(\frac{-\overline{f}_y}{\sqrt{\overline{f}_x^2 + \overline{f}_z^2}}\right), \\ \theta_o &= \tan^{-1}\left(\frac{\sqrt{\overline{f}_x^2 + \overline{f}_y^2}}{-\overline{f}_z}\right), \end{aligned} \quad (28)$$

where $\overline{f}$ is the mean of the accelerometer data. During the same interval, the roll and pitch estimates are used for leveling the magnetometer data to be in the navigation frame. The calibrated magnetometer data is used to estimate the initial azimuth as

$$\psi_o = \tan^{-1}\left(\frac{H_y}{H_x}\right). \quad (29)$$

The DCM is calculated using the initial Euler angles values as in (3). Then, the following relation between the quaternion and the DCM [22] is used to calculate the initial quaternion vector:

$$q = \begin{bmatrix} \cos\frac{\phi}{2}\cos\frac{\theta}{2}\cos\frac{\psi}{2} + \sin\frac{\phi}{2}\sin\frac{\theta}{2}\sin\frac{\psi}{2} \\ \sin\frac{\phi}{2}\cos\frac{\theta}{2}\cos\frac{\psi}{2} - \cos\frac{\phi}{2}\sin\frac{\theta}{2}\sin\frac{\psi}{2} \\ \cos\frac{\phi}{2}\sin\frac{\theta}{2}\cos\frac{\psi}{2} + \sin\frac{\phi}{2}\cos\frac{\theta}{2}\sin\frac{\psi}{2} \\ \cos\frac{\phi}{2}\cos\frac{\theta}{2}\sin\frac{\psi}{2} - \sin\frac{\phi}{2}\sin\frac{\theta}{2}\cos\frac{\psi}{2} \end{bmatrix}. \quad (30)$$

The initial quaternion vector, q_o, is calculated using the initial Euler angles values (ϕ_o, θ_o, ψ_o). Equation (30) is also used to get the updated quaternion parameters.

5. Results and Discussion

In this section, the performance of the proposed attitude algorithm is assessed. The device is held in the texting mode. All tests start with around 30 seconds of stationary period to calibrate for the gyroscope while the magnetometer is calibrated by moving the device in the 3D space. The first scenario is conducted to assess the performance of the proposed technique in the indoor environments while the second scenario is performed in the downtown area.

5.1. Environment Change Test. In this scenario, the test is started outside the building close to the Olympic Oval at the University of Calgary. The PDR solution is initialized using initial position from GPS and initial heading from the magnetometer. The device is held in stationary at the beginning of the test for about 40 s to calibrate for the gyroscope deterministic bias and to estimate the initial orientation, roll and pitch, of the device. The device is held in the compass, texting mode, and the user keeps going through the first floor with a long corridor inside the building. The trajectory is ended outside the McEwan Student Center in front of Taylor Family Digital Library (TDFL). The time for this trajectory was about 5.5 minutes taken in 594 steps with a total travelled distance of 461 m. Various reasons made this place a candidate for the test.

(i) It is a popular and attractive place for students' activities.

(ii) It has a long corridor which is a challenging area for magnetometer.

(iii) At the starting point, in front of the building, there is a huge metal structure which can affect the magnetometer performance.

(iv) This place is full of students and simulates the normal walking scenario of a smartphone user.

The derived heading form the magnetometer is totally affected at the beginning due to the presence of a huge metal object in the surrounding area as shown in Figure 6. The total magnetic field is distorted yielding incorrect heading estimation from magnetometer during the first 40 seconds.

(a) Area at the starting point

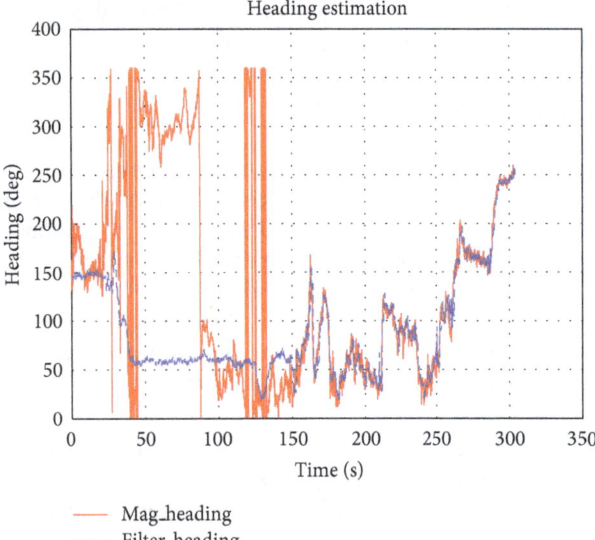

(b) Heading estimate

FIGURE 6: Heading estimation and the effect of the surrounding environment.

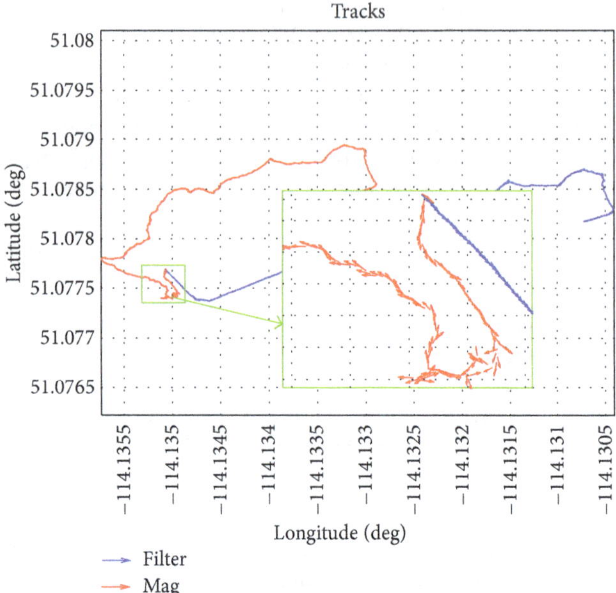

FIGURE 7: Magnetometer heading direction during perturbation area.

FIGURE 8: PDR trajectory compared to a reference trajectory.

In addition, the calibration process is performed based on data taken outside the building, and so, once the user is moved to be inside the building, the distribution of the magnetic field is different from that outside as shown in Figure 6. Thus, magnetometer-based heading is still not accurate. However, after around 2 minutes the heading from the magnetometer starts to be the main source of update for the attitude filter. As a result, during the perturbation interval the attitude filter does not perform the update stage and keeps propagating the attitude of the device based on the gyroscopes' measurements.

The magnified part in Figure 7 shows the magnetometer heading during a perturbation area. The figure shows that the heading is diverted and scattered when the magnetic field is distorted in contrast to the attitude filter heading during the same interval.

To evaluate the overall accuracy of the PDR algorithm, a reference trajectory plotted on the map of the test site is used. A reference trajectory for the actual direction of the test is plotted on the map of the first floor of the University of Calgary. As shown in Figure 8, the maximum error is less than 5 m during the test. Also, the trajectory is finished at the correct point with position drift of less than 4 m. The error in the distance is about 1.1% of the total traveled distance.

5.2. Downtown Test. Downtown is an attractive place for tourists in addition to most people in the city. It has the major attractive sites such as shopping centers, administration offices, museums, restaurants, cafes, and theaters. So, it is important to have a good navigation system which is able to help the pedestrians to their destinations. Some conventional techniques such as GPS suffer from the multipath and signal destruction due to the high buildings. The test is conducted in downtown Calgary to assess the performance of the proposed algorithm in the harsh environments. The performance of the magnetometer is totally affected due to the distortion of the magnetic field in the downtown area. The high buildings, cars, and traffic signals add more complications for the heading estimation based on the magnetometer. The test is started at the intersection of the 7th street SW and the 8th avenue SW downtown Calgary as shown in Figure 9. The selected trajectory is a square starting in the east direction followed by north, west, and south directions. The length of the trajectory is 490 m taking about 6 minutes of walking.

FIGURE 9: Test starting point.

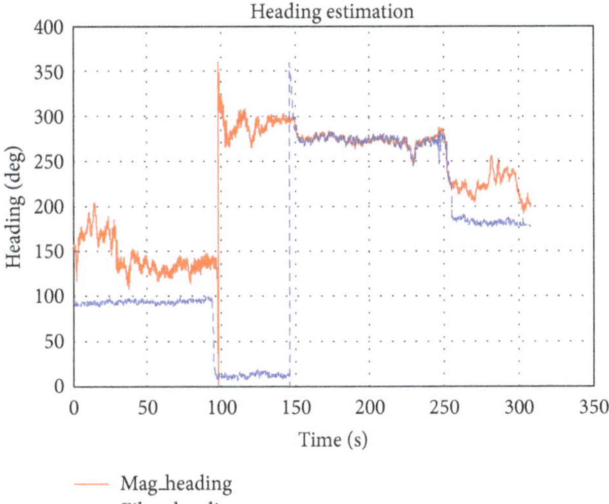

FIGURE 10: Heading estimation.

Although the magnetometer gets good maneuvering at the start of the test the presence of a strong perturbation source affected the heading estimate. As shown in Figure 10, the estimated heading from magnetometer is totally unused for major part of the test due to the distortion in the magnetic field. It shows that the heading update is available only for two minutes during the interval of 140 s to 260 s.

Figure 11 shows the PDR solution using the attitude filter compared to the same solution based on the heading from the magnetometer stand-alone, gyroscope heading stand-alone, and GPS solution. As shown in the figure, there is no accurate stand-alone solution for the test trajectory.

(i) The gyroscope stand-alone solution gives the accurate direction with position drift due to the uncompensated bias.

(ii) The magnetometer stand-alone solution is diverted at many parts of the test; however, in certain parts it performs well and provides the correct heading.

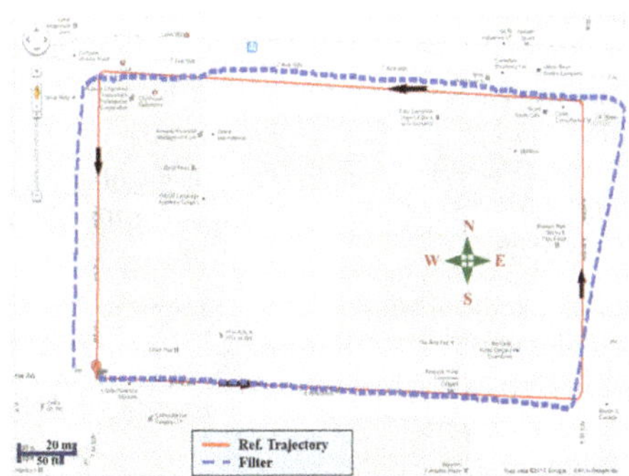

FIGURE 11: PDR trajectory compared to a reference trajectory.

(iii) The GPS is considered the poorest solution as it does not provide any correct information at all during the test due to the satellites unavailability.

However, the PDR algorithm based on the attitude filter, in corporation with the magnetometer anomaly detection technique, provides an acceptable solution.

The overall accuracy of the PDR algorithm is evaluated by comparing the PDR solution with a reference trajectory on Google map. As shown in Figure 11, the maximum error in position happens at the north direction side of the trajectory where no update is available yet, and it was around 10 m. Also, the trajectory is finished at the correct place with position drift of less than 4 m. The overall error in distance is about 1% of the total traveled distance.

6. Conclusion

An enhanced attitude estimation technique based on two different attitude sensors is presented. The algorithm is based on the quaternion mechanization to estimate the device attitude. The filter is working as a complementary technique; the Earth's magnetic field and the angle rate are integrated to estimate the device heading. The filter is propagated in the prediction mode using the gyroscope measurements while the magnetometer and accelerometer measurements are used for the update stage. The improvement in the attitude estimation leads to an improved PDR result. The PDR-based estimated trajectory results are compared to reference trajectories to show the accuracy of the proposed technique. The results show that the presented algorithm is able to provide the necessary navigation information accurately even in the harsh environments.

Acknowledgments

This work was supported in part by research funds from TECTERRA Commercialization and Research Centre, the Canada Research Chairs Program, and the Natural Science and Engineering Research Council of Canada (NSERC).

References

[1] S. Y. Cho, K. W. Lee, C. G. Park, and J. G. Lee, "A personal navigation system using low-cost MEMS/GPS/Fluxgate," in *Proceedings of the 59th Annual Meeting of the Institute of Navigation and CIGTF 22nd Guidance Test Symposium*, Albuquerque, NM, USA, 2003.

[2] C. Aparicio, *Implementation of a Quaternion-Based Kalman Filter for Human Body Motion Tracking Using MARG Sensors*, California Naval Postgraduate School, Monterey, Calif, USA, 2004.

[3] X. Yun and E. R. Bachmann, "Design, implementation, and experimental results of a quaternion-based Kalman filter for human body motion tracking," *IEEE Transactions on Robotics*, vol. 22, no. 6, pp. 1216–1227, 2006.

[4] Honeywell, "DRM 4000 Dead Reckoning Module," 2009, http://www51.honeywell.com/aero/common/documents/myaerospacecatalog-documents/Missiles-Munitions/DRM4000.pdf.

[5] J. L. Marins, X. Yun, E. R. Bachmann, R. B. McGhee, and M. J. Zyda, "An extended Kalman filter for quaternion-based orientation estimation using MARG sensors," in *Proceedings of IEEE/RSJ International Conference on Intelligent Robots and Systems*, vol. 4, pp. 2003–2011, November 2001.

[6] S. Han and J. Wang, "A novel method to integrate IMU and magnetometers in attitude and heading reference systems," *Journal of Navigation*, vol. 64, no. 4, pp. 727–738, 2011.

[7] S. Emura and S. Tachi, "Compensation of time lag between actual and virtual spaces by multi-sensor integration," in *Proceedings of IEEE International Conference on Multisensor Fusion and Integration for Intelligent Systems (MFI '94)*, pp. 463–469, October 1994.

[8] J. M. Cooke, M. J. Zyda, D. R. Pratt, and R. B. McGhee, "NPSNET: flight simulation dynamic modeling using quaternions," *Presence*, vol. 1, pp. 404–420, 1992.

[9] E. Foxlin, "Inertial head-tracker sensor fusion by a complementary separate-bias Kalman filter," in *Proceedings of the IEEE Virtual Reality Annual International Symposium*, pp. 185–194, April 1996.

[10] P. Setoodeh, A. Khayatian, and E. Farjah, "Attitude estimation by separate-bias Kalman filter-based data fusion," *Journal of Navigation*, vol. 57, no. 2, pp. 261–273, 2004.

[11] S.-W. Liu, Z.-N. Zhang, and J. C. Hung, "High accuracy magnetic heading system composed of fluxgate magnetometers and a microcomputer," in *Proceedings of the IEEE National Aerospace and Electronics Conference (NAECON '89)*, pp. 148–152, May 1989.

[12] B. Hoff and R. Azuma, "Autocalibration of an electronic compass in an outdoor augmented reality system," in *Proceedings of IEEE and ACM International Symposium on Augmented Reality*, pp. 159–164, Munich, Germany, October 2000.

[13] D. Gebre-Egziabher, G. H. Elkaim, J. D. Powell, and B. W. Parkinson, "A non-linear, two step estimation algorithm for calibrating solid-state Strapdown magnetometers," in *Proceedings of the 8th International St. Petersburg Conference on Navigation Systems*, IEEE/AIAA, St. Petersburg, Russia, 2001.

[14] C. L. Tsai, *Motion-based differential carrier phase positioning: heading determination and its applications [Ph.D. dissertation]*, National Taiwan University of Science and Technology, 2006.

[15] S. Kwanmuang, L. Ojeda, and J. Borenstein, "Magnetometer-enhanced personal locator for tunnels and GPS-denied outdoor environments," in *Sensors, and Command, Control, Communications, and Intelligence (C3I) Technologies for Homeland Security and Homeland Defense X*, Proceedings of SPIE, Orlando, Fla, USA, April 2011.

[16] L. Xue, W. Yuan, H. Chang, and C. Jiang, "MEMS-based multi-sensor integrated attitude estimation technology for MAV applications," in *Proceedings of the 4th IEEE International Conference on Nano/Micro Engineered and Molecular Systems (NEMS '09)*, pp. 1031–1035, January 2009.

[17] W. F. Storms and J. F. Raquet, "Magnetic field aided vehicle tracking," in *Proceedings of the 22nd International Technical Meeting of the Satellite Division of the Institute of Navigation (ION GNSS '09)*, pp. 1767–1774, September 2009.

[18] E. R. Bachmann, X. Yun, and C. W. Peterson, "An investigation of the effects of magnetic variations on inertial/magnetic orientation sensors," in *Proceedings of IEEE International Conference on Robotics and Automation (ICRA '04)*, pp. 1115–1122, May 2004.

[19] N. El-Sheimy, *Inertial Techniques and INS/DGPS Integration*, Lecture Notes ENGO 623, Department of Geomatics Engineering, the University of Calgary, Calgary, Canada, 2012.

[20] A. S. Ali, S. Siddharth, Z. Syed, C. L. Goodall, and N. El-Sheimy, "An efficient and robust maneuvering mode to calibrate low cost magnetometer for improved heading estimation for pedestrian navigation," *Journal of Applied Geodesy*, vol. 7, no. 1, pp. 65–73, 2013.

[21] C. C. Finlay, S. Maus, C. D. Beggan et al., "International geomagnetic reference field: the eleventh generation," *Geophysical Journal International*, vol. 183, no. 3, pp. 1216–1230, 2010.

[22] J. B. Kuipers, *Quaternions and Rotation Sequences*, Princeton University Press, Princeton, NJ, USA, 1999.

[23] R. G. Brown and P. Y. Hwang, *Introduction to Random Signals and Applied Kalman Filtering*, Wiley, New York, NY, USA, 1992.

[24] A. Gelb, *Applied Optimal Estimation*, MIT Press, 1974.

[25] H. J. Luinge and P. H. Veltink, "Measuring orientation of human body segments using miniature gyroscopes and accelerometers," *Medical and Biological Engineering and Computing*, vol. 43, no. 2, pp. 273–282, 2005.

10

Brillouin Distributed Fiber Sensors: An Overview and Applications

C. A. Galindez-Jamioy and J. M. López-Higuera

Photonic Engineering Group of the University of Cantabria, R&D&i Telecommunication Building, Avenida Los Castros, 39005 Santander, Spain

Correspondence should be addressed to J. M. López-Higuera, higuera@teisa.unican.es

Academic Editor: Romeo Bernini

A review focused on real world applications of Brillouin distributed fiber sensors is presented in this paper. After a brief overview of the theoretical principles, some works to face the two main technical challenges (large dynamic range and higher spatial resolution) are commented. Then an overview of some real and on-field applications is done.

1. Introduction

Optical fibers have commonly used as communication channels where light waves propagate along haul distances. In this situation the fibers are isolated from the external perturbations by means of cabling techniques.

However, by enhancing the environment influences on the properties of the light that travels into the waveguide, the fibers can be used to detect, to monitor, and even to measure external perturbations (measurands) in an integral or distributed format [1].

When the optical power overpass a given power threshold, nonlinear phenomena such as Brillouin scattering can be forced inside the core fiber. Furthermore, Brillouin scattering of light in optical fibers can be used as a basis to develop optical devices such as fiber lasers [2], optical filters [3], and due to its strong dependence on environmental variables (strain and temperature), it is successfully employed in distributed fiber sensor systems [4].

In these cases, the optical fiber constitutes the medium where the interaction takes place, acting at the same time as distributed transducer and optical channel. These sensors are capable of measuring the change of a specific parameter (measurand) along the entire fiber transducer. Hence, the dynamic range (correlated with the maximum fiber-length of the transducer) and the spatial resolution (minimum fiber-length required to measure to consecutive perturbations or events) are key factors, whose values still remain as challenges (they must be improved).

2. Background: Brillouin Frequency Shift in Monomode Optical Fibers

The interaction between an electromagnetic wave and matter can generate variations in the molecular structure of the material. Classically, the incident light wave generates acoustic waves through the electrostriction effect (electrostriction is the tendency of materials to become compressed in the presence of an electric field) and induces a periodic modulation of the refractive index of material that provokes a light-backscattering like a Bragg grating. This scattered light is down-shifted in frequency due to the Doppler shift associated with the grating moving at the acoustic velocity. From the point of view of quantum physics, when the intensity of light can modify locally the density of the solid, a scattering process can appear. In this process the material absorbs part of the energy from the electromagnetic wave. This energy is used for generating a periodic structure, while the remaining energy is reemitted as a wave of lower frequency, provided that conditions of resonance between the light wave and the phonon are met. The scattering associated to this process was named as "Brillouin scattering" after Leon Brillouin did for the first time the theoretical description in 1910.

Spontaneous scattering of light is mostly caused by thermal excitation of the medium, and it is proportional to the incident light intensity. On the other hand, the scattering process becomes stimulated if fluctuations in the medium are stimulated by the presence of another electromagnetic wave that reinforces the spontaneous scattering. The scattering process is in the stimulated regime provided that the intensity of the input light has a value above a level known as the threshold, which is lower than the threshold of the spontaneous regime. The stimulated scattering process is readily observed when the light intensity reaches a range between 10^6 and $10^9\,\mathrm{Wcm}^{-2}$ [5] and is capable of modifying the optical properties of the material medium.

Stimulated Brillouin Scattering (SBS) can be achieved by using two optical light waves. In addition to the optical pulse, usually called the pump, a continuous wave (CW), the so-called probe signal, is used to probe the Brillouin frequency profile of the fiber [6]. A stimulation of the Brillouin scattering process occurs when the frequency difference of the pulse and the CW signal corresponds to the Brillouin shift and provided that both optical signals are counter-propagating in the fiber. The interaction leads to a larger scattering efficiency, resulting in an energy transfer from the pulse to the probe signal and an amplification of the probe signal.

The stimulated Brillouin interaction in single mode fibers can be modeled by the three-wave transient equations for the pump (subscript p) and Stokes (subscript s) waves with the field amplitudes $E_{p,s}(z,t)$ interacting with the acoustic wave $E_a(z,t)$ (in time t and position z along the fiber) [7]:

$$\frac{\partial E_p}{\partial z} - \frac{n_f}{c}\frac{\partial E_p}{\partial t} = E_a E_s$$

$$\frac{\partial E_s}{\partial z} + \frac{n_f}{c}\frac{\partial E_s}{\partial t} = E_a^* E_p \qquad (1)$$

$$\frac{\partial E_a}{\partial z} + \Gamma E_a = \frac{1}{2}\Gamma_1 g_B E_p E_a^*$$

$$\Gamma = \Gamma_1 + i\Gamma_2,$$

where n_f is the fiber refractive index, c is the velocity of light in vacuum, and g_B is the Brillouin gain factor. Γ is the damping constant of the acoustic wave, Γ_1 is the damping time of the phonon, Γ_2 is the detuning angular frequency given by $\Gamma_2 = 2\pi(\nu - \nu_B)$ and ν is the beat frequency between the probe wave and the pump wave. The sensor system focuss on measuring the Brillouin backscattered light, which gives information about changes in temperature or strain experienced by the fiber. Considering pump pulses larger (time-width) than the phonon lifetime and no pump depletion, the Brillouin backscattered light power $P_B(z,\nu)$ detected at the receiver can be given by [5, 7]

$$P_s(z=0,\nu,t) = |E_s|^2,$$

$$P_S = g_B(\nu,\nu_B)\left(\frac{c}{2n}\right)P(t)\exp(-2\alpha z), \qquad (2a)$$

$$g_B = g_0\frac{(\Delta\nu_B/2)^2}{(\nu-\nu_B)^2 + (\Delta\nu_B/2)^2},$$

$$g_0 = \frac{2\pi n^7 {p_{12}}^2}{\lambda_p^2 \rho_0 v c \Delta\nu_B}, \qquad (2b)$$

where $P(t)$ is the total power of the launched pulsed light, α is the attenuation coefficient of the fiber, p_{12} is the electooptic coefficient, v is the acoustic velocity, λ_p is the pump wavelength, ρ_0 is the material density, and the Brillouin gain spectrum g_B in this case is given by a Lorentzian function and is assumed not to depend on z. The parameter ν_B is the frequency at which g_B has a peak value g_0, and $\Delta\nu$ is the full width at half-maximum (FWHM). Also, it is important to remark that there is an exponential relationship between the Brillouin power gain and the Brillouin gain spectrum, which is maximized at ν_B value.

According to the mathematical description presented in (1), Brillouin scattering is strongly dependent on thermodynamical variables [8–12]. The dielectric constant varies according to the pressure wave that is generated and which travels along the medium. Then, the Brillouin shift frequency is a function of the acoustic phonon, as well as the medium structure and its constituents. The material structure is clearly perturbed by changes on environmental temperature or by strong alterations on its density distribution; such is the case when a longitudinal force or a stress is applied. Those transversal or longitudinal forces relay on a shrinking or an enlargement of the original size of the material.

The Brillouin frequency shift has a linear dependence (for values of strain and temperature within its tolerance ranges) on the applied strain ε and the temperature variation ΔT (at a reference temperature T_0) that can be written as [8, 9]

$$\nu_B(T,\varepsilon) = C_\varepsilon \Delta\varepsilon + C_T \Delta T + \nu_B(T_0,\varepsilon_0), \qquad (3)$$

where C_ε is the strain coefficient (MHz/$\mu\varepsilon$), C_T is the temperature coefficient (MHz/°C) and reference strain ε_0. These values are mostly determined by the fiber composition, pump wavelength, fiber coatings, and jackets.

Brillouin scattering was proposed for the first time to measure temperature in 1989 [13], and, currently, it is widely used for distributed temperature and strain sensing because the Stokes side-lobe is temperature and strain dependent. One reason of this success is that Brillouin effect can be used in long transducers (hundreds of kilometers) made of standard monomode telecommunication optical fibers, since the Brillouin frequency shift is about 10-11 GHz at 1550 nm [4]. An example of this dependence for a standard single mode fiber (SMF) is shown in Figure 1.

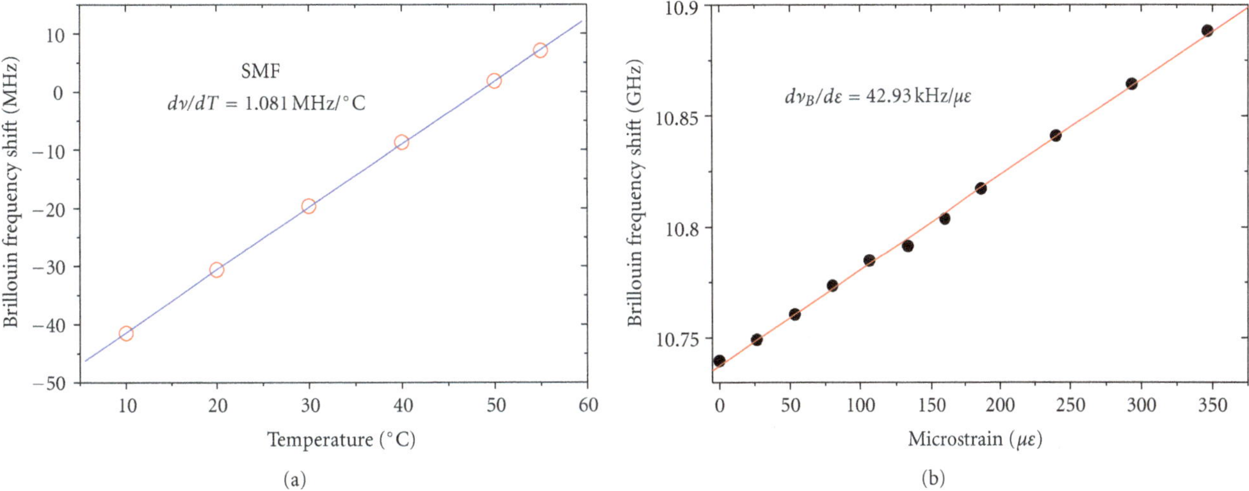

FIGURE 1: Experimental data of Brillouin shift dependence on the temperature and strain.

3. Brillouin Scattering Distributed Fiber Sensors: Quick Overview

A device or a system capable of detecting, measuring, and reproducing faithfully a particular physical or chemical variable (measurand) in the electrical domain may be recognized as a *sensor* or a *sensor system* [1]. If light is used in such sensors, and the measurand changes some of the light properties, the device is known as photonic or optical sensor. These changes on the light properties usually happen in the transducer part of the sensor. Within the photonic sensors, the fiber sensors are those who are made up with optical fibers and the technology around them [1]. Distributed fiber optic sensors are capable of detecting and measuring variables along a fiber that acts both as a distributed transducer and as an optical channel. Some examples of sensor fibers are shown in Figure 2, in this case three fibers, one for strain sensing, one for temperature sensing, and another for simultaneous strain and temperature sensing, are pictured (picture courtesy of the Photonic Engineering Group of the University of Cantabria).

As it was previously mentioned, Brillouin scattering is a phenomenon that strongly depends on the temperature in the medium, its entropy, and the material density. In most of the cases, these variables affect the frequency shift or the Brillouin spectrum linewidth. Thus, this dependence can be used for indirectly determining the influence of external or intrinsic variations in the medium where the scattering takes place. Additionally, the Brillouin shift process can be accurately (~cm) localized along the fiber, by time domain or frequency correlation techniques, among others. Hence, Brillouin scattering of light in optical fibers can be used as a basis to develop accurate distributed optical sensors and optical devices. In these cases, the optical fiber constitutes the medium where the interaction takes place, acting at the same time as distributed transducer and optical channel. These sensors are able to measure the change of a specific parameter along the entire fiber transducer within a given certitude.

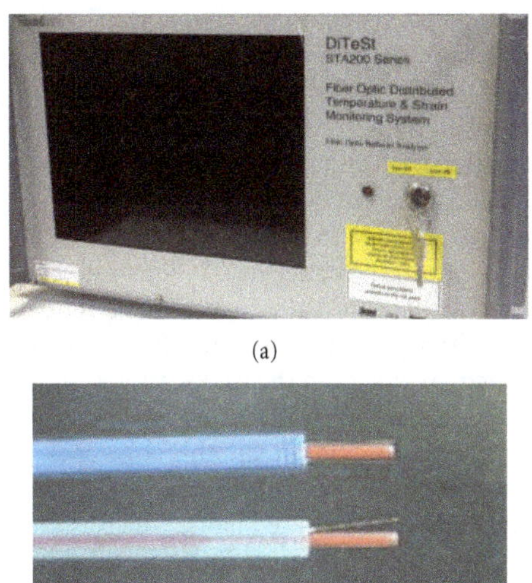

FIGURE 2: Illustration of one optoelectronic (a) and three samples of fiber transducers for both temperature and strain (b) of a stimulated Brillouin distributed sensing system, courtesy of the Photonic Engineering Group of the University of Cantabria.

Prior to talk about the sensor schemes and ultimate ranges, it is important to clarify the meaning of some key parameters commonly used to characterize the distributed fiber sensors, such as spatial resolution, dynamic range, and distance range.

The minimum change in the measurement, meaningfully detectable by the measurement system, is the *resolution*. In an optical fiber the minimum distance between two step transitions of the measurand is the *spatial resolution*. The maximum cumulated one-way (or two-way) loss in the optical link between the interrogator and the measurement point that makes possible a measurement within a specified performance is the *dynamic range*, which is measured as the ratio of the difference between the extremes of the measurement range to the resolution (in dB). The fiber length over which the measurement can be performed within the stated uncertainty and spatial resolution is denoted as *distance range*.

From a general point of view Brillouin sensors can be classified in two main kinds: *spontaneous Brillouin sensors* and *stimulated Brillouin sensors*. The methods where only the incident light is launched into the optical fiber and there is not any additional stimulus on the phonon generation are denoted as *spontaneous Brillouin sensing configurations* [14–18]. Coherent detection of spontaneous Brillouin scattering gives greater dynamic range and allows simultaneous measurements of temperature and strain. However, nonlinear effects that limit the input probe pulse power, the weakness of the Brillouin signal, and the two-way fiber loss (0.4 dB/km) lead to a rapid reduction in performance as the sensing ranges are extended beyond 50 km. Brillouin optical time domain reflectometry (BOTDR) [19, 20], the Landau-Placzek ratio [21] method, and Brillouin optical correlation domain reflectometry (BOCDR) [22] are part of this classification. On the other hand, *Stimulated Brillouin sensors* are based in the additional stimulus on the phonon generation given by an additional incident light, that is, the Brillouin scattering is enhanced or amplified.

BOTDR is a coherent detection method that uses a pulsed light. This light is launched into the optical fiber to generate spontaneous Brillouin scattering. As illustrated in Figure 3, the backward light is measured with a coherent receiver by mixing the scattering signal with that from a local oscillator [19, 20]. Since the power in the backscattered signal is small, fiber attenuation can induce a negative effect on the quality of the measurement. To compensate this drawback a coherent detection is currently used. In Brillouin optical time domain reflectometry the backscattered light, which is generated by the pump light launched in a fiber, is combined with a local oscillator. Nonetheless its dynamic range decreases with the length of the fiber. More disadvantages on this method are the fact that the spatial resolution cannot be less than one meter, the frequency shift is simultaneously dependent on both the temperature and the longitudinal strain, and that is necessary to introduce electrical filtering to eliminate Rayleigh signal.

Since the Brillouin frequency shift is a function of temperature and strain simultaneously, there is no way, uniquely measure each variable unless their effects can be separated. Nonetheless, the ratio between the intensities of the Rayleigh and Brillouin backscattered light (Landau-Placzek ratio (LPR)) is only dependent on temperature, and it can be used for obtaining a distributed temperature profile from a fiber regardless of the strain distribution [21]. The standard OTDR technique is used for determining the fiber loss profile as a function of position, later this is subtracted from the intensity measurements made with the BOTDR system and the intensity measurements are cleaned out of fiber attenuation and any form of fiber losses.

The Brillouin shift on BOCDR is measured by controlling the interference of the continuous probe and pump

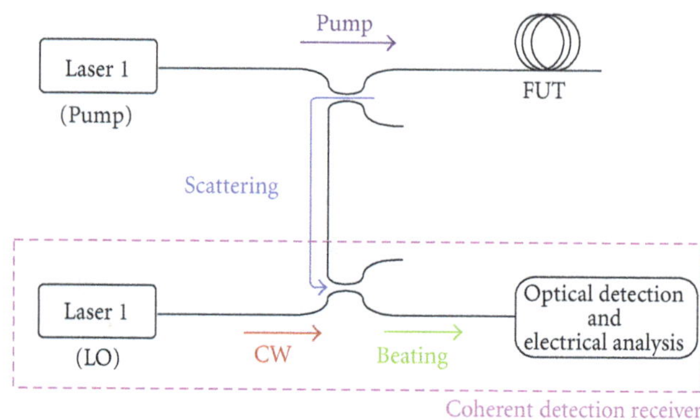

Figure 3: Typical configuration for a BOTDR system. FUT is the fiber under test or distributed transducer, LO is the local oscillator, and CW is the continuous wave.

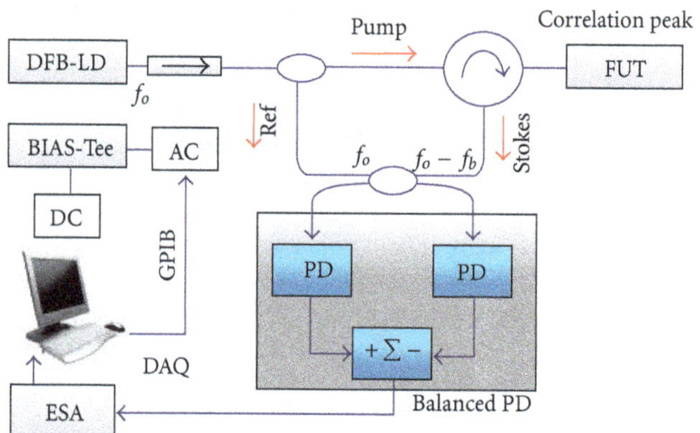

Figure 4: Schematic description of a Brillouin optical correlation domain reflectometry.

waves within the optical fiber, and the position is determined by scanning along the fiber. This technique allows the spatial range to be below the one meter range along one kilometer of optical fiber [23]. Since BOCDR is not based on an optical pulse but on a continuous wave, long integration time for the reflected signal is not needed, hence it is faster than BOTDR. However, the number of sample points on the fiber is limited by the spatial resolution and the spatial range. An experimental setup is depicted in Figure 4.

BOTDA uses Brillouin gain (or loss) spectroscopy in which a pulsed-optical wave (pump) and a counter-propagating light (probe/Stokes), normally a continuous wave, are injected into an optical fiber; the two most frequently implementaed ways are shown in Figure 5. When the frequency difference between the pulsed and the continuous light ($\Delta\nu = \nu - \nu_B$) is tuned to the Brillouin frequency ν_B of the fiber, the continuous light is amplified through the stimulated Brillouin scattering process; that is, the increasing continuous light is measured as function of time, like with OTDR [24] (Figure 6). The gathered signal using the BOTDA technique is higher than the Rayleigh backscattering power, for example, when they launched continuous light power into standard fibers which is around

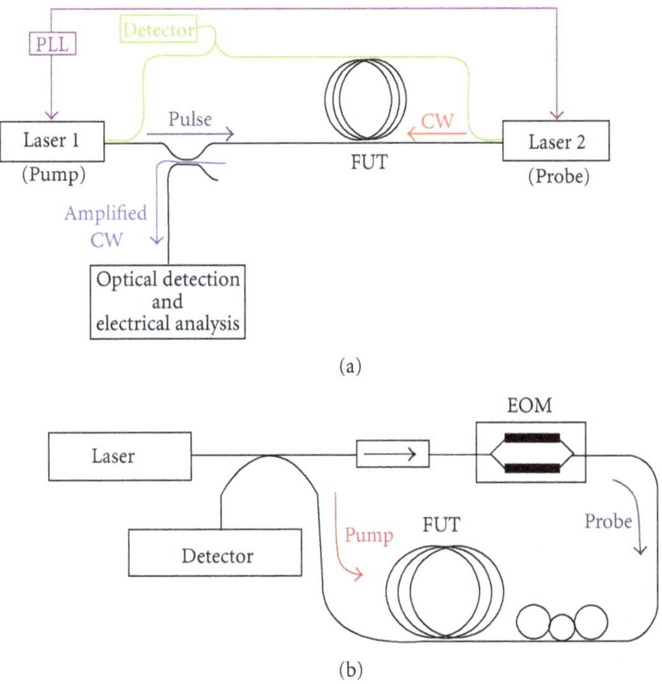

FIGURE 5: Experimental setup for BOTDA, using two sources (a) and one source (b).

0.1 mW, the backscattering can be a hundred times higher than the Rayleigh [8].

The pump and probe signals are launched into opposite ends of the fiber in order to enhance the scattering process. Thus the Brillouin gain spectrum is depicted by each value of the frequency difference, $\Delta\nu$, and is given by (2a) and (2b). Effects on the pump-probe technique have been widely studied, such as the Fresnel reflection [25], electrooptical modulation effects [26, 27], or pulse shape effects [28, 29].

In BOTDA based on pulse pumping, spatial resolution is limited to ~1 m due to the finite phonon lifetime [5, 7]. To overcome this constraint, several techniques that include variations on the probe and handling the pumping (or probe) intensity have been proposed. In the time-domain pumping schemes, the more significant methods are (1) prepumping [30, 31], (2) differential pulse-width pairs (DPP-BOTDA) with or without phase shift [32–35], (3) dark pulses [36] (by these methods, cm or even mm scale spatial resolution has been demonstrated [37, 38]), and (4) dynamic Brillouin gratings (DBGs) [39].

The mentioned methods achieve the goal of simultaneously high spatial and spectral resolution thanks to an acoustic field preactivated by a low intensity prepump pulse [31–41] or a continuous pump background [42] (see Figure 7). Even, when the pump is restored to a nonnull intensity after the pulse, it interacts with the decaying acoustic wave, producing "echoes" in the acquired signal [35]. Although the acoustic wave partially decays during the pulse duration, and a second attenuated response appears beyond the acoustic lifetime, the measurement of submeter perturbations on the fiber may be hidden. One way to suppress or attenuate the impact of this background Brillouin response is based on the subtraction of two Brillouin gain spectra obtained from pulses shifting in time or pulses with different widths [34]; in this technique, the rising and the falling times of the pulses define the spatial resolution (see Figure 8). Another proposal to overcome the appearance of ghost peaks in the Brillouin gain spectra, which makes ambiguous the determination of the Brillouin frequency shift, is addressed to the numerical correction algorithms applied to BOTDA measurements, such as the correction method based on the deconvolution of the time traces, in which each trace is inverse filtered by use of the fiber response evaluated in the resonance case [35] or the iterative numerical approaching, using the analytical model of the stimulated Brillouin scattering in the frequency domain [43].

BOTDA systems require a uniform signal-to-noise ratio (SNR) to avoid uneven Brillouin gain along the optical fiber. Thus, a low Brillouin gain across a long sensing system is needed. One compromise of low gain system over a long sensing fiber causes an increment of the spatial resolution. In order to maintain the SNR high and a low Brillouin gain, the pump power should be kept minimized and the probe power maximized, but below the influence if the modulation instability [44]. The modulation instability is a process induced mostly by the amplitude and phase modulation part, which is included in BOTDA systems to generate the probe from the same coherent pump source. To avoid the effect of the modulation instability, a dispersion shifted fiber with normal dispersion can be used [45], by coding the pulse [46]

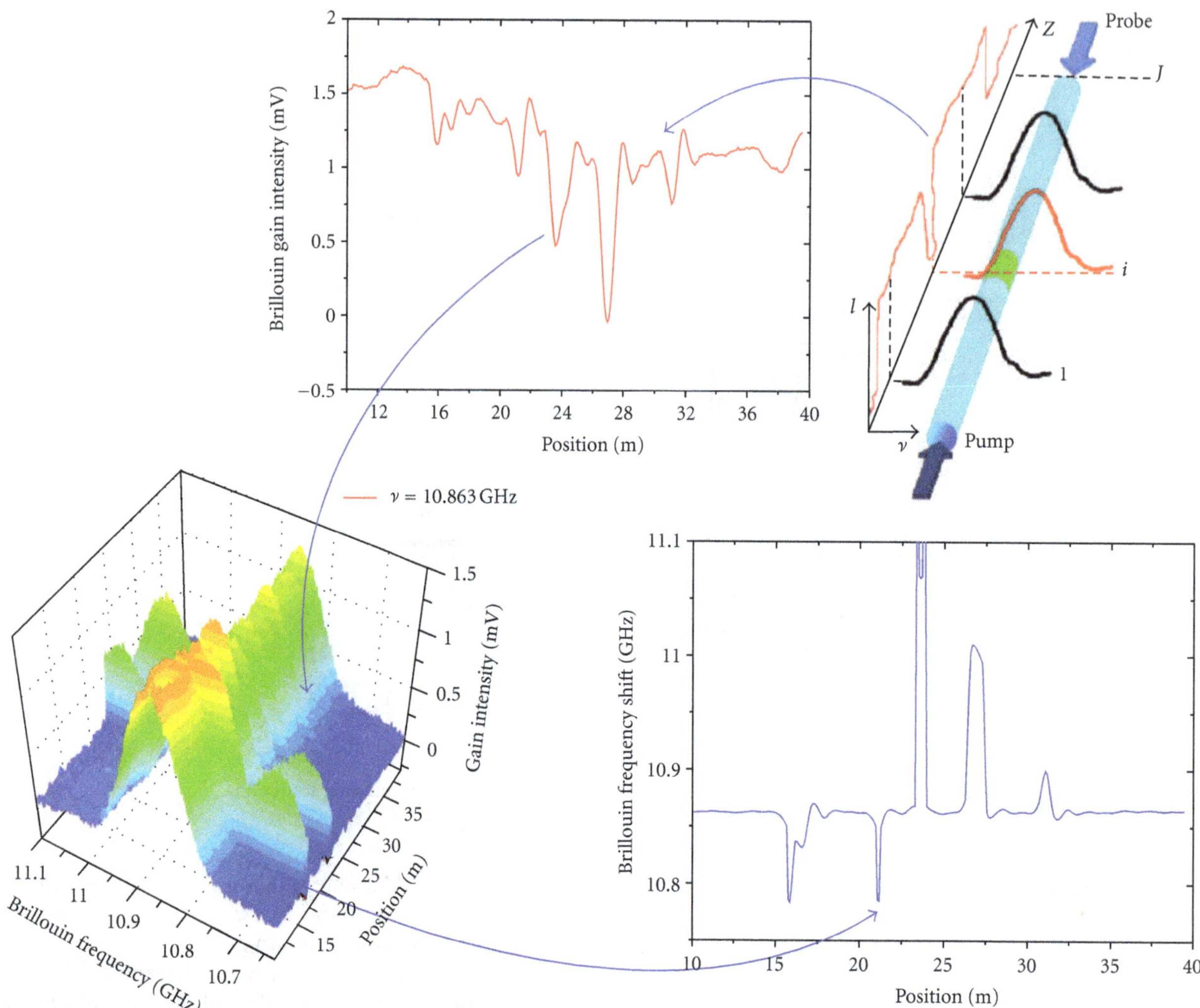

Figure 6: Traces and fiber distribution of the Brillouin frequency shift.

or an unbalanced double sideband probe [47]. The coding technique could reduce the spatial resolution, 25 cm or 2 m over a long distance range 60 km and 120 km, respectively [48, 49]. Additionally to these techniques, an interesting solution to enhance the spatial range without compromising the SNR ratio is addressed to the combination of techniques, or hybrid sensors, such as the Brillouin sensing with Raman amplification [50] to minimize the modulation instability.

The measurement time required by a traditional BOTDA system is on the order of minutes [51], time that also depends on the total length range and constitutes a serious drawback for dynamic detection in health monitoring structures. Dynamic variations of temperature or strain can be measured using Brillouin sensors by modifying the sensor technique or the system. The simplest method consists of the direct measurement of the intensity of the Brillouin peak gain/loss signal versus time, allowing the use of a BOTDA system [52, 53], which drastically reduces the measurement time, but it has a threshold detection and requires high averaging sampling as the length range is enlarged. Similar to this idea is the technique that uses two counter-propagating optical pulses with a fixed optical frequency difference; this difference is set to a spectral distance from the local Brillouin frequency shift approximately equal to half the Brillouin gain spectrum linewidth, then any vibration-induced modulation of the local Brillouin frequency shift will be measured as an intensity variation of the Stokes pulse peak intensity [54]; important changes in the setup are required in this technique and nonperiodic dynamic variations are not detected. Another method uses the polarization dependence of the Brillouin gain to avoid the need for scanning the pump-probe frequency shift [55]. To reduce the measurement time, a technique based on a multitude of pump signals in the form of a frequency-domain comb in a complex BOTDA based system is also proposed [56]. Further to these techniques, a technique for measuring dynamic variations of temperature or strain based on the well-known anomaly detection method referred to as the RX algorithm to process the data gathered from a BOTDA system [57] was presented. This technique exploits the Brillouin sensing advantages without punishing complexity or performance limitations. Most recently a method to enhance the time for measuring

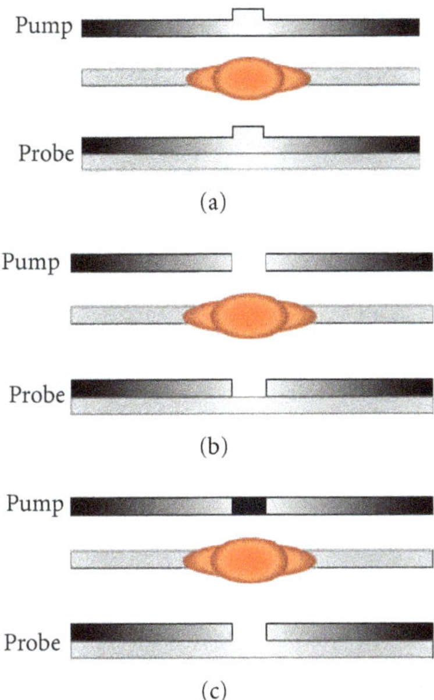

FIGURE 7: (a) Configuration of bright pulse, (b) configuration of dark pulse, and (c) configuration of phase pulse.

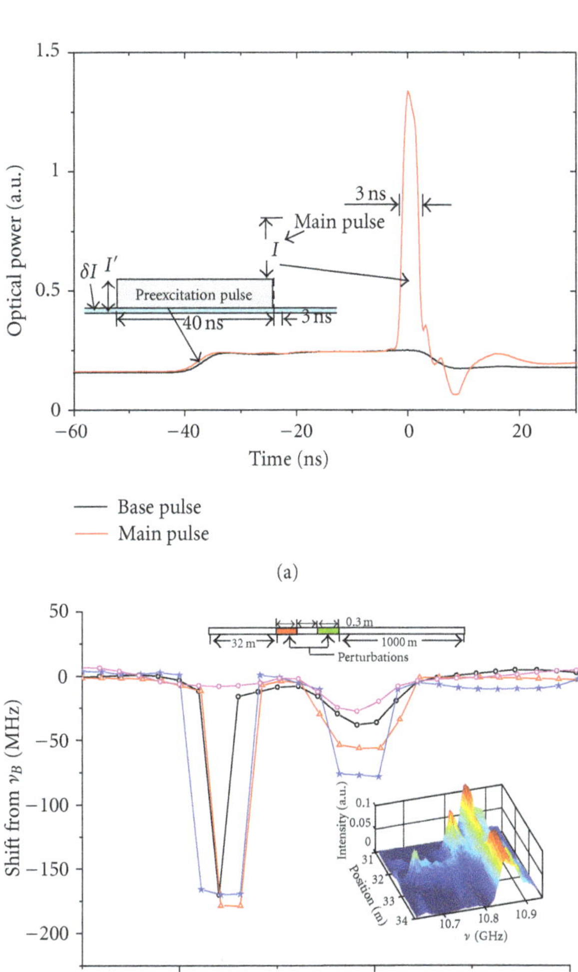

FIGURE 8: The pump light is basically formed by a pump-prepump pair pulse, and by subtracting the couple Brillouin backscattering it is possible to use larger base pulses to obtain stronger backscattered signals or to reach larger distances in the fiber. (a) Experimental data of preexcitation optical pulse. (b) Experimental data for 2 ns of main pulse and four values of P-pulse. Brillouin spectrum gain (inset figure).

the Brillouin frequency detection by generating a comb probe was proposed [58]; this method uses an arbitrary waveform generator (AWG) in the intensity modulation stage.

Proposed by Hotate [39], the BOCDA is a technique that controls Brillouin scattering using the synthesis of optical coherence function (SOCF). This technique involves phase modulation of the continuous pump wave and the probe wave. Brillouin interaction only takes place in positions where the phase of these two signals is highly correlated (Figure 9). As the phase difference between the pump and probe is changed, the position along the fiber changes where the Brillouin interaction occurs; thus providing a way to scan the length of the fiber. In conventional BOTDA techniques, if the spectral width of the pulse exceeds the Brillouin linewidth, the gain spectrum broadens and the measurement deteriorates. Nonetheless, in BOCDA if the spectral width of the pump and probe increases, the resolution of correlation technique also increases. Spatial resolution up to 1.6 mm, as well as improvements on the sampling rate of 1 kHz, and measurement range of 1 km [59, 60] were achieved.

A common drawback of the BOCDA system is its limited measurement range due to the periodic nature of sensing position and that the transducer length is shorter than other Brillouin-based sensors; additionally the transducer is much more complex and the postprocessing is more intense, which may affect negatively the measurement time.

Brillouin optical frequency domain analysis (BOFDA) is based on the measurement of a complex transfer function that relates the amplitudes of counter-propagating pump and probe waves along a fiber [61]. The continuous probe wave is modulated in intensity with a sinusoidal signal over a range of frequencies, whilst in the pump wave an intensity modulation is induced. This induced signal has an alternating component (AC component/part) due to the interaction with the counter-propagating probe wave. By measuring the changes in the AC pump wave component the Brillouin frequency shift profile is determined via the complex base-band transfer function. Once the base band transfer function is determined, the impulse response is calculated by applying the inverse Fourier transform (IFFT) to the function (Figure 10). Thus the temperature or strain can be determined from the Brillouin profile along the fiber [62, 63].

Recently, the concept of Brillouin dynamic grating (BDG) has been newly implemented in polarization maintaining fibers (PMFs) [64] and single mode fibers [65], since

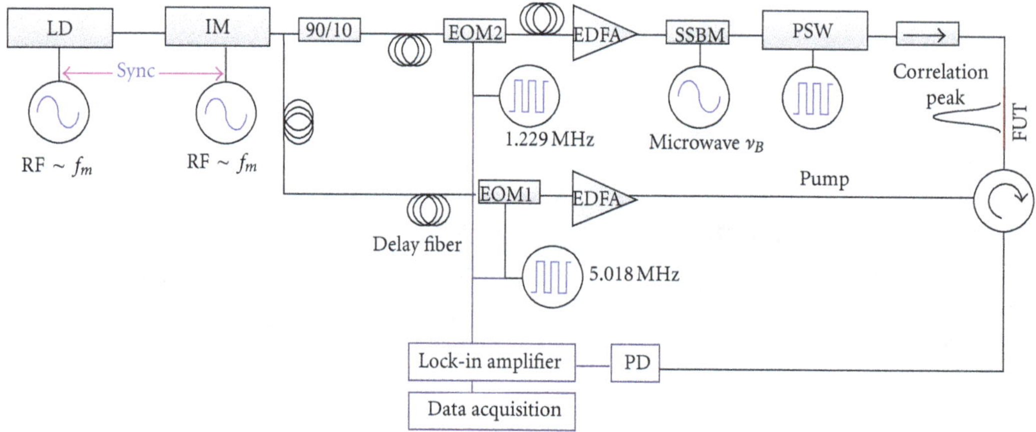

Figure 9: Experimental setup for BOCDA technique.

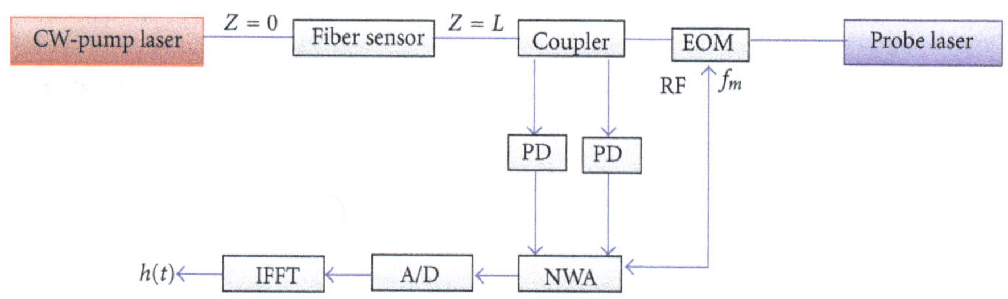

Figure 10: Basic schematic configuration of a BOFDA sensor system.

the operation principle strongly depends on the local birefringence of the medium. Acoustic waves generated during the process of stimulated Brillouin scattering in one polarization are used to reflect an orthogonally polarized wave (probe wave) at a different optical frequency from the pump. Figure 11 shows the conceptual scheme of the BDG; the frequency separation Δf between the pump and the probe waves in the BDG operation is determined by the local birefringence of the fiber and lies in the several tens of GHz in usual cases. BDG has been used as strain and temperature sensors [66, 67], birefringence sensor [68] or as tunable delay lines [69]. BDG can also be applied to enhance the spatial resolution of an ordinary Brillouin optical time-domain analysis (BOTDA) system by replacing the Brillouin probe with the reflection from the BDG [70].

4. Some Field Significant Applications

Distributed fiber sensing is a really attractive technique for structural health monitoring (SHM) [71], since it can provide information of strain and temperature about a section or the complete structure with noise electromagnetic immunity, in hazardous environments, with durability, robustness, measurement reliability, stability, good strain and temperature sensitivity ($\pm 10\,\mu\varepsilon$, $\pm 1^{\circ}$C), long range measurement (~ 100 km) and suitable spatial resolution (couple centimeters minimum) to diverse structures and shapes. Despite it is a promising diagnostic tool, it is not easy to implement outside laboratories, because each application is a unique project by itself. Each application provides new challenges on the Brillouin sensing technique adaptation and standards, as well as the transducer implementation, ranges of measurement, calibration and environmental conditions.

Typically when review articles about the distributed Brillouin sensing technique are published, authors always refer only to the laboratory implementation, but none talks about the real world challenges of the technique, that is, why in this section a review of some interesting field applications is presented, to appreciate how this technique can be adapted no matter its operating principle.

4.1. In Civil Infrastructures. A few examples of the use of Brillouin distributed sensors in infrastructures such as bridges, railways, and land monitoring are briefly commented in this section.

4.1.1. Bridges and Monitoring. Near-to-surface fiber (NSF) embedding and smart-FRP (fiber reinforced polymer) sensor bonding techniques have been experimented on small reinforced concrete (RC) bridges subject to a diagnostic load test [72]. Two bridges with similar dimensions were instrumented and tested under similar environmental conditions. Acryl cyanide and epoxy putty were used, respectively as bonding and encapsulating media for NSF. Being both fragile materials, the crack could easily propagate across the NSF section inducing a strain distribution on the sensing fiber that is mostly concentrated in the small distance between the crack edges. This very short step-like condition in the strain

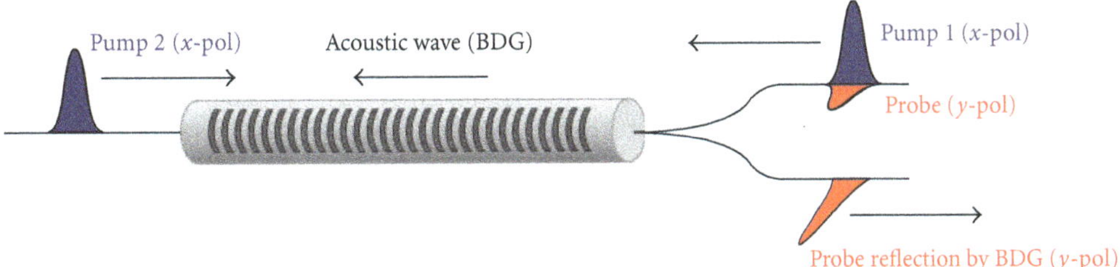

FIGURE 11: Schematic operation of BDG in an optical fiber.

distribution is extremely demanding for the performance of the BOTDR equipment, since the declared accuracy could be obtained only with a strain step length of 1 m. However, the situation is different for smart-FRP sensor, where the bridging effect due to structural fibers "spreads" the strain peak over a certain shear stress transfer length. Both the bigger fiber length that is interested by the phenomenon and the smoother transition in the strain level contribute to enhance the strain sensitivity of smart-FRP in comparison with other fiber installation technique. From the obtained results it must be remarked that NSF took about 28 man-hours to be installed on four girders, while smart-FRP took only 9 man-hours to be installed on three slabs. In addition, authors claim that NSF installation resulted to be much more problematic for the amount of dust and noise produced during groove cutting. All the results drive to the conclusion that for this kind of application in terms of performance enhancements, installation cost, and time reduction the smart-FRP system has to be considered as an optical sensor installation in bridges technique [72].

On Götaälv Bridge (built in 1939 in Sweden) severe cracking in zones above columns and a minor collapse in a structural element were observed. For safety and security reasons the bridge was monitored continuously for unusual strain changes, as well as for crack detection and localization during the refurbishing works. To do this tasks a large-scale distributed fiber optic measuring system developed by Enckell et al. [73] was implemented on the structure (Figure 12).

The project was carried out with the aims of detecting and localizing cracks that may occur due to fatigue and mediocre quality of steel; to report automatically about high strain values, high strain variation as well as temperature values in short-term and long-term perspective; and to send warnings to the traffic authorities as owner of the bridge. In Figure 13 the Brillouin gain response from a fiber transducer zone with a crack is illustrated. Results of the tests are enlightened, presented, and discussed in the reference by Enkell et al. [73].

A good example of truly-distributed sensing, based on brillouin optical time domain analysis (BOTDA)—in addition to a discrete long-gauge sensing, based on Fiber Bragg-Gratings—was installed on the Streicker Bridge at Princeton University campus during its construction. The sensors were embedded in concrete during the construction. The technique allows measurements of hydration swelling and contraction in the first stage, and posttensioning of

(a)

(b)

FIGURE 12: Götaälv Bridge with distributed Brillouin fiber sensors (a). To monitoring the structure SMARTape sensors were glued over 5 main girders (~1000 m each) (b). Courtesy of Daniele Inaudi.

concrete was registered by both systems and placed side by side in order to compare their performances. Aside from the usual behavior, an unusual increase in strain was detected by several sensors in various cross-sections. The nature of this event is still under investigation, but preliminary study indicates early-age cracking as the cause. The comparison between the two monitoring systems shows good agreement in the areas where no unusual behavior was detected, but some discrepancies are noticed at locations where unusual behavior occurred and during the early age of concrete. According to the authors, these discrepancies are attributed

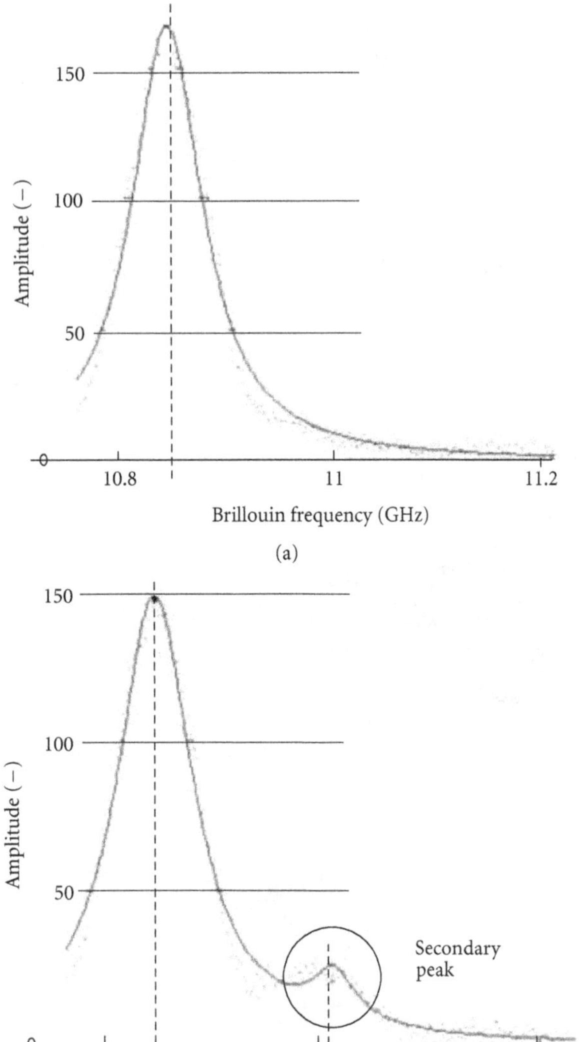

FIGURE 13: Illustration, the Brillouin gain response of fiber transducer embedded in a concrete structure with and without a crack, courtesy of Daniele Inaudy.

to the spatial resolution of the distributed monitoring system and the temperature influences at early age [74].

To measure the longitudinal strain distribution of a bridge a distributed optical fiber sensor is used [75]. A single mode optical fiber was attached on the surface using epoxy glue along the length and two ESG (electric strain gauge) were attached at a 1/4 position and 1/2 position of the length of the bridge, near the optical fiber. The optical fiber was connected to a developed fiber optic sensor system based on the concept of the Brillouin optical correlation domain analysis (BOCDA). The longitudinal strain of the bridge was directly measured at 195 points, and a spatial resolution of 14.5 cm. The measurements were made for several values of loads which remains constant during a period of time (from 0 to 50 kN), and a proportional increasing of the load from 0 to 70 kN at a rate of 0.33 kN/s. A remarkable conclusion of this work is that the strain measured by the optical fiber and the strain distribution shape of the girder coincide with the data of ESG (electric strain gauge) within the measurement error range [75].

Recently, a study of distributed sensing on 4 span model bridge was carried out. The study consists on data analysis of gathered information from BOTDA and BOTDR systems; the project was addressed to acquire useful data to identify bridge assessment conditions [76]. It can also be mentioned that employing Brillouin fiber distributed sensing techniques, the strain distribution (with a spatial resolution of 3 meters and a resolution of 15 $\mu\varepsilon$) along a supporting beam of a road-bridge with a span length of 44.40 m has been successfully carried out [77].

4.1.2. Rails Monitoring. Transverse strain of a rail can be used to determine the wheel load and lateral force, which affect the derailment coefficient and a rate of change of the wheel load. However, the longitudinal strain distribution of a rail affects on the buckling and fractures. To measure the longitudinal strain distribution of a rail section in real time, a fiber optic distributed sensor system based on the Brillouin correlation domain analysis (BOCD) was used by [78]. The experimental setup is formed by two KS60 rails with a length of 3.3 m fastened on five sleepers using the e-CLIP fastening device, a frame to add the train load evenly to the left and right side of the rail was laid in a middle position, and two vertical load actuators were placed on the rails and coupled with a frame by bolts. A single mode optical fiber with a diameter or 250 μm was attached on the surface using epoxy glue with a length of 2.8 m, 250 mm apart from both ends of the rail, and an electric strain gauge (ESG) was located in the span center of the rail, attached on the foot surface near the sensing optical fiber. The authors report that in the test measurement, the vertical load to each rail was increased to 143 kN and maintained constant for 10 min. Authors also claim that the vertical load bends the rail and that the longitudinal strain in the rail is proportional to the curvature and varies linearly with the distance from the neutral surface of the cross section. The spatial resolution for this measure was reported on 3.6 cm along the rail.

The Brillouin frequency distribution along the 2.8 m fiber (strain coefficient of 0.05 MHz/$\mu\varepsilon$) appears in a symmetrical shape with the maximum at the center position where the Brillouin frequency was reported on 10.897 GHz with an increase about 15 MHz. Additionally, in order to validate the sensor reading, the experimental data were checked with the finite element results.

4.2. Geotechnical Structures Monitoring. To understand the bearing behavior of a loaded ground anchor, the measuring and monitoring of the stress distribution in the anchor tendon is essential. To provide this information to the geotechnical engineers, novel monitoring ground anchors using embedded optical fibers along the anchor tendon were developed using a BOTDA technique [79, 80]. In a first step, optical sensors have been integrated into short tendons

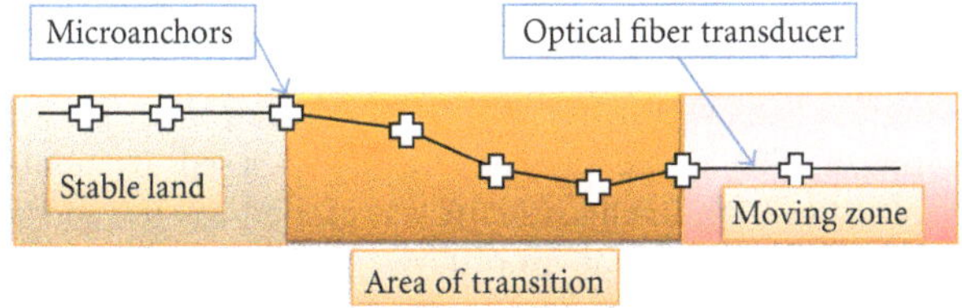

FIGURE 14: The trench cut in the hiking path: (left) attachment of the "microanchors" to the cable and, (right) compacting of the soil above the cable.

using different methods and laboratory strain testing was performed on these instrumented tendons (Figure 14). The evaluation of the laboratory testing enabled the design and development of an 8 m long monitoring ground anchor for field application. In 2009, this anchor has been placed into a wall supporting an excavation pit and, subsequently, anchor pullout test was carried out. The anchor was loaded stepwise up to 470 kN, almost reaching its ultimate bearing capacity. Optical measurements were taken successfully at each load step. Authors report comparison of the optical data with data acquired using conventional methods, and they claim good consistency of the results.

Geotextiles are generally used to perform functions of separation, reinforcement, filtration, and/or drainage, and they have been used to solve civil engineering problems over more than three decades. Since fabrication processes of geotextiles and optical fiber have substantially evolved, nowadays it is possible to combine both materials to produce a nonconventional optical sensor. Thus, several geotextiles with embedded optical fibers have been proposed and developed for optical sensing (Figure 15(a)). By using Brillouin interrogation techniques smart structures capable of monitoring and/or measuring the strain and/or temperature distribution in 2-dimentions can be developed. Employing geotextiles, distributed measurements of critical mechanical deformations (or soil displacements) on dikes with several kilometers length can be carried out. Additionally, they are commonly used in dikes for reinforcement of the dike body and erosion prevention, among other possibilities.

A BOFDA monitoring system has been optimized to fit the demands on dike monitoring: detection of mechanical deformation (strain) with a spatial resolution of 5 m over a distance range of up to 10 km. The functionality of the monitoring system and the fiber-sensors-equipped geotextiles has been proven in several installations and field tests in dikes and dams such as the gravity dam in Solina, Poland in August 2006. A thin soil layer of several 10 cm put onto the geomats after installation has been proven to be a sufficient protection of the textile-integrated glass fiber cables against heavy machinery and construction work.

It must also be noticed that on the laboratory dike (15 m long) at Hannover University, Germany, applications like the before tested were carried out. Sensor-based geotextiles were installed on top of the dike and were covered with a thin soil layer (Figure 15(b)). To simulate a mechanical deformation/soil displacement, a lifting bag was embedded into the soil and was inflated by air pressure. This induced

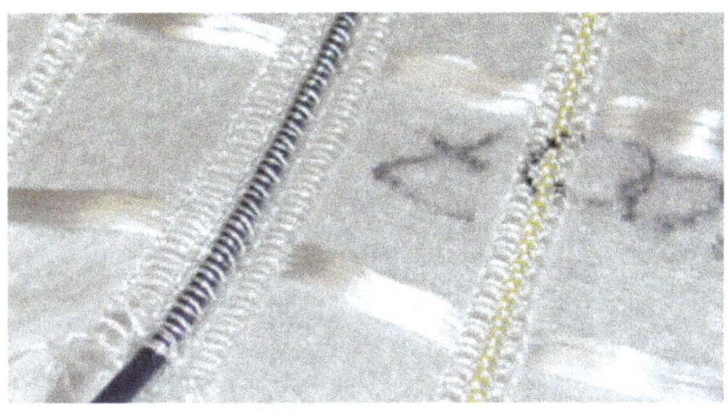

(a)

(b) (c)

FIGURE 15: Geotextile equipped with the Fiber-ware special sensing cable and the Nexans standard indoor cable (yellow). Installation in two independent scenarios: the surface of a dike foot (b) and in a soil displacement (courtesy of K. Krebber).

a break of the inner slope of the dike and a soil displacement (Figure 15(c)). The soil displacements were clearly detected and localized by the BOFDA system [81–85].

Cross-border smuggling tunnels enable unmonitored movement of people or forbidden stuff (drugs, weapons, etc.), and they may suppose a threat to homeland security. To contribute to overcome this risk, a method for detecting the excavation of small (diameter <1 m) tunnels in clayey soils using Brillouin optical time domain reflectometry (BOTDR) and a neural network was proposed [86]. The reported architecture includes two fiber optic layouts. One is a horizontal fiber buried at a shallow depth below ground surface, and the other one has a fiber embedded in vertical minipiles to detect very deep tunnels. In both configurations, strains would develop in the fiber due to the soil displacements induced by the tunnel excavation. It was demonstrated by the authors that the proposed system was capable of detecting even small tunnel, 0.5 m diameter, as deep as 20 m under the horizontal fiber or as far as 10 m aside from the minipile, if the volume loss is greater than 0.5%.

4.3. Pipelines Monitoring. By using Brillouin scattering-based fiber distributed techniques, real-time monitoring and early warning systems for liquid and gas pipes have been successfully checked in real installations in a wide set of scenarios. As illustrated in Figure 16, in buried pipes and in general terms, leaks could be detected through temperature

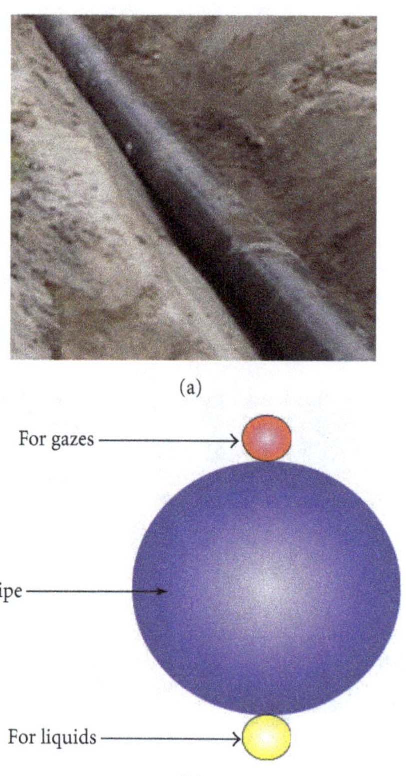

Figure 16: Illustration of a pipeline in a buried process (a). Scheme of the distributed fiber transducer placement with respect to the pipe section (b).

changes on the upper zone (gases) or the bellow zone (liquids) of the pipe; that is, the fiber optics transducer-cable of the distributed sensor system must be placed over the pipe for gases and below it for liquid leaks detection, respectively. The strain/stress state and safety of casing pipes in oil wells arenas are monitored and evaluated by using BOTDR systems such as in Daqing oilfield [87].

The brine (saturated water with salt) is a waste product derived from the mining technique for building underground caverns for gas storage in large rock-salt formation. This brine must be transported by pipelines for its processing or injected back safely into the ground. Because the brine can be harmful to the environment, pipeline small leak detection is a mandatory requirement. As in this case any change in the local temperature indicates a brine leakage in the pipeline section. As an example of this problem, a distributed temperature sensing system was integrated in the brine pipeline placed in the area of Berlin in Germany. The brine is pumped out of the underground caverns and is injected into the pipeline at a temperature of 35°C to 40°C. At normal flow rate the temperature gradient along the whole pipeline length is about 8°C. Since the pipeline is buried at a depth of approximately 2 to 3 meters, the seasonal temperature variations are quite small and the average soil temperature was measured to be around 5°C. As a result a substantial temperature increase is associated to any pipeline leak even in the case of very small leakages [88].

Offshore arctic conditions are a challenge to the safe operation of subsea pipelines exposed to seabed ice gouging, permafrost thaw settlement, strudel scour, and channel migration. The application of fiber optic-based distributed temperature monitoring systems was carried out to monitor the pipeline operational conditions in arctic pipelines [89]. As visual inspection is impossible, real-time temperature monitoring via optical fibers along the pipeline route can provide an early warning of the development of erosional events, pipeline insulation damages, and seabed soil modifications. In order to monitor two offshore pipelines in the Alaska's Beaufort Sea oil fields a distributed temperature sensor monitoring system was implemented. The pipeline installation is part of Oooguruk oil field developments in the Beaufort Sea. It is composed of 8 km of buried subsea flow-lines transporting the produced fluids from an offshore gravel island/drill site to an onshore above ground pipeline which runs to an existing transmission pipeline. A total of 14 km of pipeline distance is continuously monitored with the fiber optic communication cables installed within the pipeline bundle. The monitoring system was demonstrated to meet the monitoring performance to detect temperature events occurring over just one meter, such as leaks and erosional events. The system has been able to map seabed temperature profiles along the pipeline route and to accurately track temperature excursions as they were occurring with field verified data prior to and during pipeline operation startup. The monitoring system operates permanently and continuously with an active leak detection system based on the detection of local temperature variations [89].

4.4. Some Materials and Structures Monitoring Cases. Applications of the Brillouin distributed sensing include a large variety of structures, such as competition yachts or experimental vehicles. Monocoque structures such as those made of carbon fiber reinforced plastic (CFRP) were instrumented with this fiber distributed technology to measure and monitor strain and temperature during manufacturing and in-service structural performance like its stiffness [90]. The sensing technique was embedded in a couple of International America's Cup Class (IACC) yachts (Asura and Idaten). The yachts were equipped with a fiber-optic distributed sensor using Brillouin scattering to monitor the longitudinal and transverse strains of the yachts to assess the stiffness of the structures in America's Cup 2000 (held at Auckland in New Zealand). Another monocoque structure is a full-scale model built as a prototype to demonstrate feasibility of a Japanese experimental reentry vehicle, namely, HOPE-X (H-II Orbiting Plane-Experimental). The same type of sensor that was applied to the IACC yachts and a sensor using Raman scattering were used for strain and temperature measurements during a manufacturing process and for strain measurements in structural tests.

Another Brillouin-based system application with a spatial resolution of 10 cm was utilized for distributed strain measurement in a representative carbon fiber reinforced plastic (CFRP) stiffened panel manufactured by vacuum assisted resin transfer molding (VARTM). Strain changes induced during the manufacturing process and the impact tests were comprehensively presented and discussed. Carbon fiber reinforced plastic (CFRP) is being used on almost all modern commercial aircraft as a primary structural material, but it is still difficult to precisely manufacture cocured large-scale CFRP structures and ensure their structural integrity during operation. Hence there is an urgent need to develop innovative techniques to monitor the internal states of composite structures and utilize the obtained data to improve structural design, processing technologies, and maintenance methods. By combining all the information obtained by the fiber-optic network, the structural health can be accurately evaluated [91].

Honeycomb sandwich structures are integral constructions consisting of two thin facesheets and a lightweight honeycomb core. They are widely utilized in aircraft structures on account of their excellent mechanical properties. However, the lightweight structure absorbs a large amount of water while in service. Over time, cracks form in the facesheets and the adhesive layer as a result of mechanical and thermal loading, creating leak paths from the surface to the honeycomb core. Then, the water enters the core through the paths and accumulates in the cells of the honeycomb structure. Mechanical deterioration of the honeycomb core such as delamination of the face-sheet from the core can be provoked by the accumulated water. This trouble can be detected measuring the nonuniform temperature distribution induced by the water in the cell of the honeycomb core. Using this principle and a Brillouin distributed fiber optic system the water accumulation monitoring of large-scaled structures was effectively detected [92].

In large AC power generators, fatigues on electrical insulation and mechanical components could cause deterioration of the stator coil tightness and provoke a failure. Additionally, during the installation of an AC power generator, the fiberglass ripple spring, which is located just below the stator wedge, must be sufficiently compressed to be flat or nearly flat to ensure the maximum tightness, to compensate for the stator coil ground-wall insulation creep that is the primary factor in stator coil looseness resulting in "slot pounding". Using a DPP-BOTDA this deformation is measured [93]. The sensing fiber is glued to a fiberglass flat plate, which is subjected to a periodic side force from a fiberglass ripple spring and is consequently characterized by a periodic deformation. The measurement of the longitudinal strain of the flat plate caused by the compression of the ripple spring is performed with the Brillouin technique. Then the shape of the flat plate is reconstructed according to the strain-displacement relation. Measurements of distributed lateral displacement of the flat plate with different periods of 3 and 3.25 cm are reported by the authors with a maximum displacement of 0.43 mm and a minimum measurable displacement of ~40 μm [93].

Steel corrosion has become a major problem worldwide, especially for structures exposed to aggressive environments. When corrosion happens, the volume of the steel rebar will increasingly expand due to the rust product accumulation on the surface of the steel rebar. Using this principle, a fiber BOTDA sensing technique was implemented to monitor

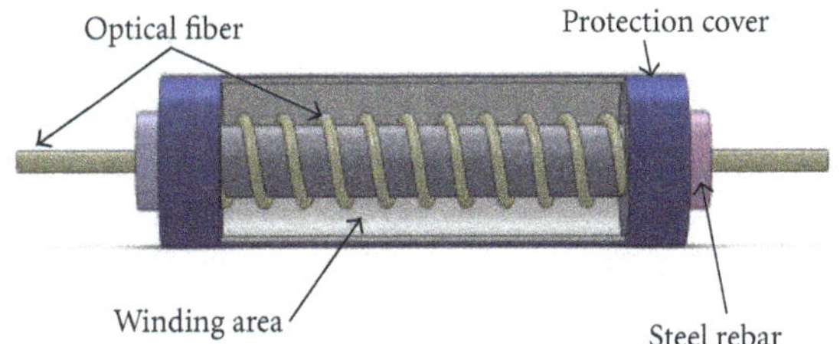

Figure 17: Basic packaging structure of the Brillouin corrosion expansion sensor.

corrosion expansion of steel rebars in steel reinforced concrete structures [94]. This kind of fiber optic coil winding Brillouin corrosion expansion sensors with different fiber optic coil winding packaging schemes was embedded into concrete specimens to monitor expansion strain caused by steel rebar corrosion. It was demonstrated by the authors that the expansion strain along the fiber optic coil winding area can be detected and measured using this kind of sensors during development of the corrosion. Three different types of these transducers were constructed by winding the authors around a polished steel rebar into a pretensioned fiber optic coil (Figure 17). The fiber optic will be stretched when the volume of the steel rebar expands due to corrosion and hence the tension strain change of the fiber optic coil in corrosion area can be monitored using a BOTDA analyzer.

The first kind of sensor offered a rather high sensitivity being able to detect signals of steel corrosion just at the beginning of the process. The second kind of sensor offered a relative low sensitivity compared with the first one; however, the larger monitoring range with a maximum expansion of 5690 $\mu\varepsilon$ compensates for this to make it suitable for application in the resinous corrosion situations. Finally, the third kind of sensor showed the lowest sensitivity but the largest monitoring range (6738 $\mu\varepsilon$) of the three ones. It can perform steel corrosion monitoring where there may exist an extremely corrosive environment.

5. Summary and Conclusions

Brillouin scattering can be successfully used to develop distributed fiber sensor systems. In these sensing approaches the optical fiber works, simultaneously, as transducer and as channel. Temperature and/or the strain Brillouin scattering dependence in conjunction with time, frequency, polarization, and others interrogation techniques, are employed to measure the change of specific parameters (measurands) along the entire fiber transducer with a given spatial resolution. Hence, the spatial resolution (minimum fiber length required to measure two consecutive perturbations or events) and the dynamic range (maximum fiber length of the transducer) are key factors for this kind of sensors.

Spontaneous and stimulated Brillouin scattering effects can be used for developing distributed fiber sensors. In the former scattering only an incident light is launched into the optical fiber and there is not any additional optical stimulus. Stimulated Brillouin sensors need additional stimulus for the phonon generation. Stimulated process can be obtained by an additional counter propagated light.

Brillouin distributed fiber sensors are useful for structural health monitoring systems because they can provide relevant information concerning the structural integrity of structures, even those that work in hazardous environments. Some of the most interesting applications in civil structures and in the smart materials have been briefly presented. Civil structures, bridges SHM (Götaälv Bridge, 1939 in Sweden, Streicker Bridge at Princeton University campus, among others), rails, geotechnical, pipelines structures and monitoring cases have been mentioned, as well as, distributed Brillouin sensor applications on composite materials such as carbon fiber reinforced plastic (CFRP) or honeycomb sandwich-based structures. Major problems especially for structures exposed to aggressive environments such as the measurement of fatigue on the electrical insulators, mechanical components and steel corrosion were briefly commented.

Acknowledgments

This work has been supported by the Spanish TEC2010-20224-C02-02 Project. Authors acknowledge Hany Shokry for his collaboration.

References

[1] J. M. Lopez-Higuera, Ed., *Handbook of Optical Fibre Sensing Technology*, John Wiley & Sons, New York, NY, USA, 2002.

[2] D. Leandro, A. Ullan, A. Loayssa, J. M. López-Higuera, and M. López-Amo, "Remote (155 km) fiber bragg grating interrogation technique combining Raman, Brillouin, and erbium gain in a fiber laser," *IEEE Photonics Technology Letters*, vol. 23, no. 10, pp. 621–623, 2011.

[3] C. Galindez, F. Madruga, A. Ullan, and J. M. Lopez-Higuera, "Técnica y dispositivo para conformar el espectro de ganancia de Brillouin en guías de onda ópticas," Patent, 2009.

[4] A. Rogers, "Distributed optical-fibre sensing," in *Handbook of Optical Fibre Sensing Technology*, J. M. Lopez-Higuera, Ed., chapter 14, John Wiley & Sons, New York, NY, USA, 2002.

[5] G. P. Agrawal, *Nonlinear Fiber Optics, Quantum Electronics-Principles and Applications*, Academic Press, Rochester, NY, USA, 1995.

[6] M. Niklès, L. Thévenaz, and P. A. Robert, "Brillouin gain spectrum characterization in single-mode optical fibers," *Journal of Lightwave Technology*, vol. 15, no. 10, pp. 1842–1851, 1997.

[7] R. W. Boyd, *Nonlinear Optics*, Academic Press, Rochester, NY, USA, 2nd edition, 2003.

[8] T. Horiguchi, K. Shimizu, T. Kurashima, M. Tateda, and Y. Koyamada, "Development of a distributed sensing technique using Brillouin scattering," *Journal of Lightwave Technology*, vol. 13, no. 7, pp. 1296–1302, 1995.

[9] T. Kurashima, T. Horiguchi, and M. Tateda, "Distributed-temperature sensing using stimulated Brillouin scattering in optical silica fibers," *Optics Letters*, vol. 15, no. 18, pp. 1038–1040, 1990.

[10] C. A. Galindez, F. J. Madruga, M. Lomer, A. Cobo, and J. M. Lopez-Higuera, "Effect of humidity on optical fiber distributed sensor based on Brillouin scattering," in *Proceedings of the 19th International Conference on Optical Fibre Sensors (OFS '08)*, vol. 70044W, pp. 70041–70044, April 2008.

[11] C. Galindez, F. J. Madruga, and J. M. Lopez-Higuera, "Influence of humidity on the measurement of Brillouin frequency shift," *IEEE Photonics Technology Letters*, vol. 20, no. 23, pp. 1959–1961, 2008.

[12] C. Galindez, F. J. Madruga, and J. M. Lopez-Higuera, "Brillouin frequency shift of standard optical fibers set in water vapor medium," *Optics Letters*, vol. 35, no. 1, pp. 28–30, 2010.

[13] D. Culverhouse, F. Farahi, C. N. Pannell, and D. A. Jackson, "Potential of stimulated Brillouin scattering as sensing mechanism for distributed temperature sensors," *Electronics Letters*, vol. 25, no. 14, pp. 913–915, 1989.

[14] M. N. Alahbabi, Y. T. Cho, and T. P. Newson, "150 km-range distributed temperature sensor based on coherent detection of spontaneous Brillouin backscatter and in-line Raman amplification," *Journal of the Optical Society of America B*, vol. 22, no. 6, pp. 1321–1324, 2005.

[15] M. Alahbabi, Y. T. Cho, and T. P. Newson, "Comparison of the methods for discriminating temperature and strain in spontaneous Brillouin-based distributed sensors," *Optics Letters*, vol. 29, no. 1, pp. 26–28, 2004.

[16] Y. T. Cho, M. Alahbabi, M. J. Gunning, and T. P. Newson, "50 km single-ended spontaneous-Brillouin-based distributed-temperature sensor exploiting pulsed Raman amplification," *Optics Letters*, vol. 28, no. 18, pp. 1651–1653, 2003.

[17] S. M. Maughan, H. H. Kee, and T. P. Newson, "57 km single-ended spontaneous Brillouin-based distributed fiber temperature sensor using microwave coherent detection," *Optics Letters*, vol. 26, no. 6, pp. 331–333, 2001.

[18] M. N. Alahbabi, Y. T. Cho, and T. P. Newson, "100 km distributed temperature sensor based on coherent detection of spontaneous Brillouin backscatter," *Measurement Science and Technology*, vol. 15, no. 8, pp. 1544–1547, 2004.

[19] T. Kurashima, T. Horiguchi, H. Hizumita, S. Furukawa, and Y. Koyamada, "Brillouin optical-fiber time domain reflectometry," in *Proceedings of the International Quantum Electronics Conference*, Dig Tech, Vienna, Austria, 1992.

[20] Y. Lu, H. Liang, X. Zhang, and F. Wang, "Brillouin optical time-domain reflectometry based on Hadamard sequence probe pulse," in *Proceedings of the 9th International Conference on Optical Communications and Networks (ICOCN '10)*, pp. 36–38, October 2010.

[21] P. C. Wait and T. P. Newson, "Landau Placzek ratio applied to distributed fibre sensing," *Optics Communications*, vol. 122, no. 4-6, pp. 141–146, 1996.

[22] Y. Mizuno, W. Zou, Z. He, and K. Hotate, "Proposal and experiment of BOCDR-Brillouin optical correlation-domain reflectometry," in *Proceedings of the International Society for Optical Engineering (SPIE '08)*, University of Western Australia, Optical and Biomedical Engineering Laboratory (OBEL), April 2008.

[23] Y. Mizuno, Z. He, and K. Hotate, "Measurement range enlargement in Brillouin optical correlation-domain reflectometry based on temporal gating scheme," *Optics Express*, vol. 17, no. 11, pp. 9040–9046, 2009.

[24] T. Horiguchi and M. Tateda, "BOTDA-nondestructive measurement of single-mode optical fiber attenuation characteristics using Brillouin interaction: theory," *Journal of Lightwave Technology*, vol. 7, no. 8, pp. 1170–1176, 1989.

[25] D. Iida and F. Ito, "Detection sensitivity of Brillouin scattering near Fresnel reflection in BOTDR measurement," *Journal of Lightwave Technology*, vol. 26, no. 4, pp. 417–424, 2008.

[26] C. A. Galindez and L. Thevenaz, "Effect of pulse chirp on distributed Brillouin fiber sensing," in *Proceedings of the 19th SPIE International Conference on Optical Fibre Sensors*

(OFS '08), vol. 7004, pp. 70041J–70044J, SPIE-The International Society for Optical Engineering, Perth, Australia, 2008.

[27] V. Lecoeuche, D. J. Webb, C. N. Pannell, and D. A. Jackson, "Transient response in high-resolution Brillouin-based distributed sensing using probe pulses shorter than the acoustic relaxation time," *Optics Letters*, vol. 25, no. 3, pp. 156–158, 2000.

[28] C. Galindez, F. J. Madruga, A. Cobo, O. Conde, and J. M. Lopez-Higuera, "Pulse shape effects on the measurement of Temperature using a Brillouin based Optical fiber sensor," in *Proceedings of the Enabling Photonics Technologies for Defense, Security, and Aerospace Applications III (EWOFS '07)*, vol. 65720, pp. 65721–65724, April 2007.

[29] H. Naruse and M. Tateda, "Optimum temporal pulse shape of launched light for optical time domain reflectometry type sensors using Brillouin backscattering," *Optical Review*, vol. 8, no. 2, pp. 126–132, 2001.

[30] K. Kishida and C. -H. Li, "Pulse pre-pump-BOTDA technology for new generation of distributed strain measuring system," in *Structural Health Monitoring and Intelligent Infrastructure*, pp. 471–477, Taylor & Francis, London, UK, 2006.

[31] C. A. Galindez, A. Quintela, M. A. Quintela, and J. M. Lopez-Higuera, "30cm of spatial resolution using pre-excitation pulse BOTDA technique," in *Proceedings of the 21st International Conference on Optical Fiber Sensors (OFS '11)*, vol. 77532, pp. 77531–77534, May 2011.

[32] C. A. Galindez, A. Quintela, M. A. Quintela, and J. M. Lopez-Higuera, "30cm of spatial resolution using pre-excitation pulse BOTDA technique," in *Proceedings of the 21st International Conference on Optical Fiber Sensors*, pp. 2344–2348, May 2011.

[33] L. Thevenaz and S. F. Mafang, "Distributed fiber sensing using Brillouin echoes," in *Proceedings of the 19th International Conference on Optical Fibre Sensors (SPIE '08)*, pp. 70043N–770044.

[34] W. Li, X. Bao, Y. Li, and L. Chen, "Differential pulse-width pair BOTDA for high spatial resolution sensing," *Optics Express*, vol. 16, no. 26, pp. 21616–21625, 2008.

[35] S. M. Foaleng, M. Tur, J. C. Beugnot, and L. Thévenaz, "High spatial and spectral resolution long-range sensing using Brillouin echoes," *Journal of Lightwave Technology*, vol. 28, no. 20, Article ID 5565369, pp. 2993–3003, 2010.

[36] A. W. Brown, B. G. Colpitts, and K. Brown, "Distributed sensor based on dark-pulse Brillouin scattering," *IEEE Photonics Technology Letters*, vol. 17, no. 7, pp. 1501–1503, 2005.

[37] Y. Dong, H. Zhang, L. Chen, and X. Bao, "2 cm spatial-resolution and 2 km range Brillouin optical fiber sensor using a transient differential pulse pair," *Applied Optics*, vol. 51, no. 9, pp. 1229–1235, 2012.

[38] T. Sperber, A. Eyal, M. Tur, and L. Thévenaz, "High spatial resolution distributed sensing in optical fibers by Brillouin gain-profile tracing," *Optics Express*, vol. 18, no. 8, pp. 8671–8679, 2010.

[39] K. Hotate, "Measurement of brillouin gain spectrum distribution along an optical fiber using a correlation-based technique-proposal, experiment and simulation-," *IEICE Transactions on Electronics*, vol. E83-C, no. 3, pp. 405–411, 2000.

[40] A. Minardo, R. Bernini, and L. Zeni, "Differential techniques for high-resolution BOTDA: an analytical approach," *IEEE Photonics Technology Letters*, vol. 24, no. 15, pp. 1295–1297, 2012.

[41] L. Thévenaz, S. M. Foaleng, K. Y. Song et al., "Advanced Brillouin-based distributed optical fibre sensors with submeter scale spatial resolution," in *Proceedings of the 36th European Conference and Exhibition on Optical Communication (ECOC '10)*, pp. 1–6, September 2010.

[42] J. C. Beugnot, M. Tur, S. F. Mafang, and L. Thévenaz, "Distributed Brillouin sensing with sub-meter spatial resolution: modeling and processing," *Optics Express*, vol. 19, no. 8, pp. 7381–7397, 2011.

[43] A. Minardo, R. Bernini, and L. Zeni, "Spatial resolution enhancement in preactivated BOTDA schemes by numerical processing," *IEEE Photonics Technology Letters*, vol. 24, no. 12, pp. 1003–1005, 2012.

[44] M. N. Alahbabi, Y. T. Cho, T. P. Newson, P. C. Wait, and A. H. Hartog, "Influence of modulation instability on distributed optical fiber sensors based on spontaneous Brillouin scattering," *Journal of the Optical Society of America B*, vol. 21, no. 6, pp. 1156–1160, 2004.

[45] Y. Dong, L. Chen, and X. Bao, "System optimization of a long-range Brillouin-loss-based distributed fiber sensor," *Applied Optics*, vol. 49, no. 27, pp. 5020–5025, 2010.

[46] M. A. Soto, G. Bolognini, and F. Di Pasquale, "Analysis of optical pulse coding in spontaneous Brillouin-based distributed temperature sensors," *Optics Express*, vol. 16, no. 23, pp. 19097–19111, 2008.

[47] R. Bernini, A. Minardo, and L. Zeni, "Long-range distributed Brillouin fiber sensors by use of an unbalanced double sideband probe," *Optics Express*, vol. 19, no. 24, pp. 23845–23856, 2011.

[48] M. Soto, M. Taki, G. Bolognini, and F. Di Pasquale, "Optimization of a DPP-BOTDA sensor with 25 cm spatial resolution over 60 km standard single-mode fiber using Simplex codes and optical pre-amplification," *Optics Express*, vol. 20, no. 7, pp. 6860–6869, 2012.

[49] M. A. Soto, G. Bolognini, and F. Di Pasquale, "Optimization of long-range BOTDA sensors with high resolution using first-order bi-directional Raman amplification," *Optics Express*, vol. 19, no. 5, pp. 4444–4457, 2011.

[50] A. Zornoza, R. A. Pérez-Herrera, C. Elosúa et al., "Long-range hybrid network with point and distributed Brillouin sensors using Raman amplification," *Optics Express*, vol. 18, no. 9, pp. 9531–9541, 2010.

[51] X. Bao and L. Chen, "Recent progress in distributed fiber optic sensors," *Sensors*, vol. 12, no. 7, pp. 8601–8639, 2012.

[52] Z. Liu, G. Ferrier, X. Bao, X. Zeng, Q. Yu, and A. Kim, "Brillouin scattering based distributed fiber optic temperature sensing for fire detection," in *Proceedings of The 7th International Symposium on Fire Safety Conference*, Worcester, Mass, USA, 2002.

[53] Y. Peled, A. Motil, L. Yaron, and M. Tur, "Slope-assisted fast distributed sensing in optical fibers with arbitrary Brillouin profile," *Optics Express*, vol. 19, no. 21, pp. 19845–519854, 2011.

[54] R. Bernini, A. Minardo, and L. Zeni, "Dynamic strain measurement in optical fibers by stimulated brillouin scattering," *Optics Letters*, vol. 34, no. 17, pp. 2613–2615, 2009.

[55] X. Bao, C. Zhang, W. Li, M. Eisa, S. El-Gamal, and B. Benmokrane, "Monitoring the distributed impact wave on a concrete slab due to the traffic based on polarization dependence on stimulated Brillouin scattering," *Smart Materials and Structures*, vol. 17, no. 1, Article ID 015003, 2008.

[56] P. Chaube, B. G. Colpitts, D. Jagannathan, and A. W. Brown, "Distributed fiber-optic sensor for dynamic strain

measurement," *IEEE Sensors Journal*, vol. 8, no. 7, pp. 1067–1072, 2008.

[57] C. A. Galindez, F. J. Madruga, and J. M. Lopez-Higuera, "Efficient dynamic events discrimination technique for fiber distributed Brillouin sensors," *Optics Express*, vol. 19, no. 20, pp. 7–5, 1891.

[58] Y. Peled, A. Motil, and M. Tur, "Fast Brillouin optical time domain analysis for dynamic sensing," *Optics Express*, vol. 20, no. 8, pp. 8584–8591, 2012.

[59] K. Y. Song, Z. He, and K. Hotate, "Distributed strain measurement with millimeter-order spatial resolution based on Brillouin optical correlation domain analysis," *Optics Letters*, vol. 31, no. 17, pp. 2526–2528, 2006.

[60] M. Belal and T. P. Newson, "Evaluation of a high spatial resolution temperature compensated distributed strain sensor using a temperature controlled strain rig," in *Proceedings of the 21st International Conference on Optical Fiber Sensors*, vol. 36, pp. 4728–4730, May 2011.

[61] D. Garus, K. Krebber, F. Schliep, and T. Gogolla, "Distributed sensing technique based on Brillouin optical-fiber frequency-domain analysis," *Optics Letters*, vol. 21, no. 17, pp. 1402–1404, 1996.

[62] D. Garus, T. Gogolla, K. Krebber, and F. Schliep, "Brillouin optical-fiber frequency-domain analysis for distributed temperature and strain measurements," *Journal of Lightwave Technology*, vol. 15, no. 4, pp. 654–662, 1997.

[63] R. Bernini, A. Minardo, and L. Zeni, "Accurate high-resolution fiber-optic distributed strain measurements for structural health monitoring," *Sensors and Actuators A*, vol. 134, no. 2, pp. 389–395, 2007.

[64] K. Y. Song, W. Zou, Z. He, and K. Hotate, "All-optical dynamic grating generation based on Brillouin scattering in polarization-maintaining fiber," *Optics Letters*, vol. 33, no. 9, pp. 926–928, 2008.

[65] K. Y. Song, "Operation of Brillouin dynamic grating in single-mode optical fibers," *Optics Letters*, vol. 36, no. 23, pp. 4686–4688, 2011.

[66] W. Zou, Z. He, K. Y. Song, and K. Hotate, "Correlation-based distributed measurement of a dynamic grating spectrum generated in stimulated Brillouin scattering in a polarization-maintaining optical fiber," *Optics Letters*, vol. 34, no. 7, pp. 1126–1128, 2009.

[67] W. W. Zou, Z. He, and K. Hotate, "Complete discrimination of strain and temperature using Brillouin frequency shift and birefringence in a polarization-maintaining fiber," *Optics Express*, vol. 17, no. 3, pp. 1248–1255, 2009.

[68] Y. Dong, L. Chen, and X. Bao, "Truly distributed birefringence measurement of polarization-maintaining fibers based on transient Brillouin grating," *Optics Letters*, vol. 35, no. 2, pp. 193–195, 2010.

[69] K. Y. Song, K. Lee, and S. B. Lee, "Tunable optical delays based on Brillouin dynamic grating in optical fibers," *Optics Express*, vol. 17, no. 12, pp. 10344–10349, 2009.

[70] K. Y. Song, W. Zou, Z. He, and K. Hotate, "Optical time-domain measurement of Brillouin dynamic grating spectrum in a polarization-maintaining fiber," *Optics Letters*, vol. 34, no. 9, pp. 1381–1383, 2009.

[71] J. M. Lopez-Higuera, L. Rodriguez Cobo, A. Quintela Incera, and A. Cobo, "Fiber optic sensors in structural health monitoring," *Journal of Lightwave Technology*, vol. 29, no. 4, pp. 587–608, 2011.

[72] F. Bastianini, A. Rizzo, N. Galati, U. Deza, and A. Nanni, "Discontinuous Brillouin strain monitoring of small concrete bridges: comparison between near-to-surface and "smart" FRP fiber installation techniques," in *Proceedings of the Smart Structures and Materials 2005: Sensors and Smart Structures Technologies for Civil, Mechanical, and Aerospace Systems (SPIE '05)*, pp. 612–623, March 2005.

[73] M. Enckell, B. Glisic, F. Myrvoll, and B. Bergstrand, "Evaluation of a large-scale bridge strain, temperature and crack monitoring with distributed fibre optic sensors," *Journal of Civil Structural Health Monitoring*, vol. 1, no. 1, pp. 37–46, 2011.

[74] B. Glisic, J. Chen, and D. Hubbell, "Streicker Bridge: a comparison between Bragg-grating long-gauge strain and temperature sensors and Brillouin scattering-based distributed strain and temperature sensors," in *Proceedings of the Sensors and Smart Structures Technologies for Civil, Mechanical, and Aerospace Systems*, pp. 79810–79812, March 2011.

[75] H.-J. Yoon, K.-Y. Song, H.-M. Kim, and J.-S. Kim, "Strain monitoring of composite steel girder bridge using distributed optical fibre sensor system," in *Procedia Engineering*, pp. 2544–2547, Elsevier, New York, NY, USA, 2011.

[76] I.-B. Kwon, M. Malekzadeh, Q. Ma et al., "Fiber optic sensor installation for monitoring of 4 span model bridge in UCF," in *Proceedings of the 29th IMAC, a Conference on Structural Dynamics*, pp. 383–388, February 2011.

[77] A. Minardo, R. Bernini, and L. Zeni, "Vectorial dislocation monitoring of pipelines by use of Brillouin-based fiber-optics sensors," *Smart Materials and Structures*, vol. 17, no. 1, Article ID 015006, 2008.

[78] H.-J. Yoon, K. Y. Song, J. S. Kim, and D. S. Kim, "Longitudinal strain monitoring of rail using a distributed fiber sensor based on Brillouin optical correlation domain analysis," *NDT and E International*, vol. 44, no. 7, pp. 637–644, 2011.

[79] M. Iten and A. M. Puzrin, "Monitoring of stress distribution along a ground anchor using BOTDA," in *Proceedings of the Sensors and Smart Structures Technologies for Civil, Mechanical, and Aerospace Systems*, vol. 76415, p. 76472, March 2010.

[80] M. Iten, D. Hauswirth, and A. M. Puzrin, "Distributed fiber optic sensor development, testing and evaluation for geotechnical monitoring applications," in *Proceedings of the Smart Sensor Phenomena, Technology, Networks, and Systems*, pp. 798207–798215, March 2011.

[81] N. Noether, A. Wosniok, K. Krebber, and E. Thiele, "Dike monitoring using fiber sensor-based geosynthetics," in *Proceedings of the III Eccomas Thematic Conference on Smart Structures and Materials*, 2007.

[82] N. Nöther, A. Wosniok, K. Krebber, and E. Thiele, "A distributed fiber optic sensor system for dike monitoring using Brillouin optical frequency domain analysis," in *Proceedings of the Smart Sensor Phenomena, Technology, Networks, and Systems*, vol. 6933, pp. 69330T-69331–69330T-69339, March 2008.

[83] N. Nöther, A. Wosniok, K. Krebber, and E. Thiele, "A distributed fiber optic sensor system for dike monitoring using Brillouin optical frequency domain analysis," in *Proceedings of the Smart Sensor Phenomena, Technology, Networks, and Systems*, p. 700303, March 2008.

[84] N. Noether, A. Wosniok, K. Krebber, and E. Thiele, "A distributed fiber-optic sensor system for monitoring of large geotechnical strutures," in *Proceedings of the 4th International Conference on Structural Health Monitoring on Intelligent Infrastructure (SHMII '09)*, 2009.

[85] A. Wosniok, N. Noether, K. Krebber, and E. Thiele, "Distributed monitoring of mechanical deformation in river dikes," in *Proceedings of the Eurosensors XXIII conference*, 2008.

[86] A. Klar and R. Linker, "Feasibility study of automated detection of tunnel excavation by Brillouin optical time domain reflectometry," *Tunnelling and Underground Space Technology*, vol. 25, no. 5, pp. 575–586, 2010.

[87] Z. Zhou, J. He, M. Huang, J. He, J. Ou, and G. Chen, "Casing pipe damage detection with optical fiber sensors: a case study in oil well constructions," in *Proceedings of the International Society for Optical Engineering (SPIE '10)*, pp. 764908–764911, March 2010.

[88] M. Nikles, "Long-distance fiber optic sensing solutions for pipeline leakage, intrusion and ground movement detection," in *Proceedings of the 6th Fiber Optic Sensors and Applications (SPIE '09)*, pp. 731602–731613, April 2009.

[89] B. Eisler, G. Lanan, M. Niklès, and L. Zuckerman, "Distributed fiber optic temperature sensing system for buried subsea arctic pipelines," in *Proceedings of the Deep Offshore Technology Conference 2008 (DOT '08)*, Houston, Tex, USA, 2008.

[90] H. Murayama, K. Kageyama, H. Naruse, A. Shimada, and K. Uzawa, "Application of fiber-optic distributed sensors to health monitoring for full-scale composite structures," *Journal of Intelligent Material Systems and Structures*, vol. 14, no. 1, pp. 3–13, 2003.

[91] S. Minakuchi, N. Takeda, S. I. Takeda, Y. Nagao, A. Franceschetti, and X. Liu, "Life cycle monitoring of large-scale CFRP VARTM structure by fiber-optic-based distributed sensing," *Composites A*, vol. 42, no. 6, pp. 669–676, 2011.

[92] S. Minakuchi, H. Tsukamoto, and N. Takeda, "Detecting water accumulation in honeycomb sandwich structures by optical-fiber-based distributed temperature measurement," *Journal of Intelligent Material Systems and Structures*, vol. 20, no. 18, pp. 2249–2255, 2009.

[93] D. Yongkang, B. Xiaoyi, and C. Liang, "Online monitoring of the distributed lateral displacement in large AC power generators using a high spatial resolution Brillouin optical fiber sensor," *Smart Materials and Structures*, vol. 20, no. 11, pp. 115001–115006, 2011.

[94] X. Zhao, P. Gong, G. Qiao, J. Lu, X. Lv, and J. Ou, "Brillouin corrosion expansion sensors for steel reinforced concrete structures using a fiber optic coil winding method," *Sensors*, vol. 11, no. 11, pp. 10798–10819, 2011.

11

High-Density Fiber Optical Sensor and Instrumentation for Gas Turbine Operation Condition Monitoring

Hua Xia,[1,2] Doug Byrd,[3] Sachin Dekate,[1] and Boon Lee[1]

[1] *Photonics Laboratory, Micro and Nano Structures Technologies, GE Global Research, 1 Research Circle, Niskayuna, NY 12309, USA*
[2] *College of Nanoscale Science and Engineering, State University of New York at Albany, NY 12222, USA*
[3] *Engineering Division, GE Energy, 300 Garlington Road, GTTC 200D, Greenville, SC 29615, USA*

Correspondence should be addressed to Hua Xia; hxia@albany.edu

Academic Editor: Joao Batista Rosolem

Gas turbine operation control is normally based on thermocouple-measured exhaust temperatures. Due to radiation shielding and bulky package, it is difficult to provide high spatial resolution for measuring can-to-can combustion temperature profile at the exhaust duct. This paper has demonstrated that wavelength-division-multiplexing-based fiber Bragg grating sensors could provide high spatial resolution steady and dynamic temperature measurements. A robust sensor package can be designed with either circumferential sensing cable or radial sensing rake for quasi-distributing multiple fiber sensors in the gas turbine environment. The field validations have demonstrated that quasi-distributed fiber sensors have not only demonstrated its temperature measurement accuracy compared to existing thermocouple sensors but also shown its unique dynamic response amplitude and power spectra that could be utilized for gas turbine transient operation condition monitoring and diagnostics.

1. Introduction

Accurate static and dynamic temperature detections are essential for safe and efficient operation and control in many industrial machinery systems, which include but are not limited to gas turbine, steam generator, boiler, combustor, compressor, gasifier, and so forth. In combustion control practice, an annular array of thermocouples is used to measure exhaust temperature profile, which is then used to ensure safe operation of the gas turbine. Whenever a fault temperature, either too cold or too hot, is detected, a gas turbine shutdown is initiated, which in many cases could be premature. Such a gas turbine operational control method requires accurate annular exhaust static and dynamic temperature measurement. Obviously, the purpose of the static and dynamic temperature detection is either for real-time industrial process monitoring and diagnostics or for operation control and optimization.

Since gas temperature is one of the critical measurement parameters for gas turbine operation, improvements of the temperature measurement accuracies can improve turbine efficiency. Direct combustor temperature measurement is highly desirable for robust turbine design and operation, but normally this requirement is beyond the capabilities of most thermocouple sensors. Therefore, most OEMs opt to use exhaust gas temperature measurement as a surrogate. Current exhaust temperature measurement using the annular array of thermocouples (TCs) provides limited discrete sensing points, and the sensing spatial resolution is not optimal. Accordingly, a conservative control and operation strategy of the gas turbine is warranted, which has a limiting effect on the gas turbine performance. A potential solution is to increase the number of thermocouples and improve their spatial arrangement. However, it is difficult to increase the number and location of the existing TCs from the current method due to their bulky packaging and excessive electrical wiring needs. Fiber Bragg grating (FBG) sensors are thought to be of great potential for temperature measurement in the harsh environments such as turbomachinery systems because of the advantages as having low mass, high sensitivity, multiplexing capabilities, multipoint distribution capabilities, multisensing functions, and electromagnetic interference immunity [1, 2].

A FBG is basically sensitive to both strain and temperature variations [3], but it can be packaged for static temperature measurement with so-called "loss package" or freestanding package method [4, 5]. However, a so-called "prestrain package" method could be used for measuring structural instability or dynamic signal. In the presurge condition for a gas turbine or compressor, such packaged fiber sensors could be used to measure temperature variation rates or thermal spikes either at the gas inlet or downstream exhaust. A sensing cable normally consists of a plurality of fiber sensors that have to be sealed in a metal, ceramic, or polymeric tube or capillary, depending upon the sensing environmental conditions. A sensing instrumentation is constructed by multiplexing different arrays or sensing cables either with multichannel optical sensing interrogator [e.g., Micron Optics 1 Hz sm125/1 kHz sm130, http://www.micronoptics.com/] or with an optical sensing interrogator plus an optical-switcher-based multiplexer [e.g., Micron Optics, sm041]. The static and dynamic signals from FBG sensors are analyzed either with time-division or with wavelength-division multiplexing method, or their combination.

This paper has presented recent progress in applying fiber sensors and instrumentation for measuring gas turbine exhaust temperature at both startup and steady operation conditions with wavelength-division multiplexing technology. For circumferential exhaust temperature detection from a gas turbine, FBG sensing elements are inscribed onto a single fiber with a spatial separation of 450 mm; three sensing fibers with 150 mm misplacement provide double spatial resolution for 9FB gas turbine exhaust temperature profile detection. For radial exhaust temperature detection, a precision exhaust temperature rake is used to support FBG sensors as an integrated radial sensing rake with 100 mm spatial resolution across exhaust duct wall to central barrier space. The obtained static wavelength data (1 data per second) are converted to temperatures with a precalibrated transfer function, while the dynamic wavelength data (1000 data per second) are analyzed to catch vibration modes and system displacement amplitude with fast Fourier transform algorithm.

2. High-Density Fiber Temperature Sensor

It is understood that a grating is a telecom filter device for wavelength multiplexing or demultiplexing applications; it is absolute not a temperature sensor despite its intrinsic natures of temperature and strain dependence. Conventional Type-I FBG has been used in various infrastructural health-monitoring applications [6, 7]. However, this type of FBG has poor thermal survivability and accuracy degradation at relative elevated temperature of greater than 300°C. Research has shown that some Type-IIA, Type-II, regenerated FBGs (fabricated either by ultraviolet lasers or femtosecond lasers), and chiral gratings have also shown survivability even at 1000–1100°C [8–13]. With thermal stable FBGs, they can be used as temperature sensor. In addition, a mechanical package has to be properly designed for maintaining FBG sensor's integrity, survivability, functionality, and durability [4, 5].

FBG elements used in this work have used high-power ultraviolet (UV) laser with a phase mask technology for inscribing Bragg grating in a single-mode fiber core [12, 13]. This forms a periodic mass density-modulated grating pattern that can be characterized by a Bragg resonant peak at

$$\lambda_B = 2 \cdot n_{\text{eff}} \cdot \Lambda(\rho), \tag{1}$$

where the effective refractive index n_{eff} in the fiber core includes the laser-inscription-process-produced refractive index modulation, n, and mass density modulation induced refractive index n_ρ. The thermal dependence of a Bragg grating can be described by

$$\Delta\lambda = \lambda_B \cdot \left[\kappa_\varepsilon \cdot \varepsilon + \left(\beta + \alpha + \frac{1}{n_{\text{eff}}}\frac{\partial n_\rho}{\partial T}\right) \cdot \Delta T + \left(\frac{1}{n_{\text{eff}}}\frac{\partial n_\rho}{\partial t} + \frac{1}{\Lambda}\frac{\partial \Lambda}{\partial t}\right)\Delta t\right], \tag{2}$$

where α is the coefficient of fiber thermal expansion, $\lambda_B \cdot \kappa_\varepsilon$ is the strain sensitivity (~1.2 pm/$\mu\varepsilon$), β is the thermooptic coefficient, and the Δt reflects the time-dependent fiber material microstructure relaxation process. For low-density Type-I FBGs, the temperature dependence of the FBG Bragg resonant wavelength can be simply expressed as

$$\begin{aligned}\Delta\lambda(t) &= \lambda_B \cdot [\kappa_\varepsilon \cdot \varepsilon + (\beta + \alpha) \cdot T(t) - T(0)] \\ &= a(0) + a(1) \cdot \Delta T(t),\end{aligned} \tag{3}$$

where $a(0)$ can be treated as an offset to initial wavelength shift, and $\kappa_T(0) = \lambda_B \cdot (\beta + \alpha)$ is defined as temperature sensitivity. For low-temperature application, the FBG has a linear response to external temperature by

$$\Delta\lambda(t) = \lambda(t) - \lambda_B(0) - a(0) = a(1) \cdot [T(t) - T(0)], \tag{4}$$

where $a(1) = \kappa_T(0) = \lambda_B \cdot (\beta + \alpha) \approx$ 10–12 pm/°C. High-density FBG is more or less related to compact fiber material with mass density and refractive index comodulation in the fiber core, induced by laser inscription process. Specifically, the operation at elevated temperature could turn an amorphous floppy microstructure of the fiber material to a nanocrystalline even microcrystalline morphology. With such a structural transition, the FBG thermal response function can be also described by

$$\begin{aligned}\Delta\lambda(t) &= \lambda_B \cdot \left[\kappa_\varepsilon \cdot \varepsilon + (\beta + \alpha + \beta_\rho) \cdot \Delta T + (1 + \xi) \cdot \left|\frac{\Delta\rho}{\rho}\right|\right] \\ &= b(0) + b(1) \cdot \Delta T,\end{aligned} \tag{5}$$

where $0.2 < \xi < 0.3$. This equation describes the fiber material which has experienced a structural transition

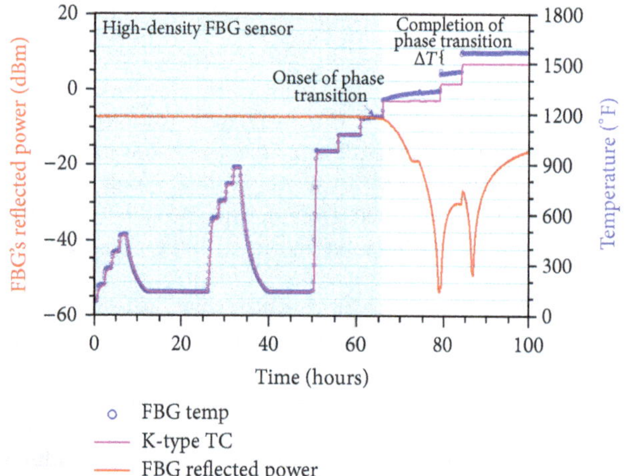

FIGURE 1: Fiber sensor response to external temperature change from ambient to 500°F, 900°F, 1200°F (shaded area), and beyond its maximum operation temperature of 1200°F.

with a baseline upshift determined by fiber material density change. Note that the temperature sensitivity has an additional term, where $\kappa_T = \lambda_B \cdot (\alpha + \beta + \beta_\rho)$. When $\beta_\rho > 0$, the temperature sensitivity κ_T at high-density phase will be greater than the low-density phase $\kappa_T(0) = \lambda_B \cdot (\alpha + \beta)$, as seen from most of regenerated and Type-II FBGs with $\kappa_T \sim$ 14–16 pm/°C. For femtosecond inscribed FBGs in pure silica or sapphire fibers, the temperature sensitivity κ_T is $\lambda_B \cdot (\alpha + \beta_\rho)$, where β_ρ depends mostly upon the used fiber material and laser power level or irradiation energy. Moreover, (5) could be rewritten as (4) and this would approximate the inherent nature of the fiber sensor response, determined by external temperature.

The FBG elements used in this paper were fabricated with high-power ultraviolet (UV) laser with a phase mask technology on a single-mode fiber material. The fiber material and inscription process can be found from [14], which has summarized detailed grating fabrication process. The obtained FBG elements will follow a postthermal annealing process to stabilize material structural morphology. Figure 1 has shown a typical high-density-FBG-based temperature sensor and its comparison with thermocouple measurement in three temperature ranges. First is from ambient to 500°F, second is from ambient to 900°F, and third is from ambient to ~1200°F (~650°C) and beyond. Since the fiber material microstructure has been stabilized previously annealing temperature at 1200◦F, a maximum allowed temperature, the wavelength shift beyond 1200°F (as in shaded area) will not correspond to real temperature because the fiber material microstructure or morphology is transforming into another thermal stabilized state. From 1200°F to 1500°F, accompanying the fiber material structural transition, the FBG's reflected power has experienced a dramatic variation, and the Bragg resonant wavelength eventually stabilized at 1500°F, which is defined as a new maximum allowed temperature. Obviously, this fiber material morphology transition has resulted in an offset of $\Delta T \sim 80$°F than real temperature, which is unacceptable for any machinery system control because of this material physical phenomenon if the FBG is operated up to 1500°F. It should be pointed out that the observed "dips" in the FBG's reflected power are more or less related to so-called grating regeneration process, similar to described in [2, 5, 13]. However, to guarantee a FBG sensor's accuracy and reliability, the maximum operation temperature is normally 10% less than its maximum allowed temperature. However, if a pulsed and short-term thermal shock occurs than the maximum allowed temperature, the deterioration of the fiber sensor's accuracy may be not noticeable. On the contrary, a new calibration is required for recovering FBG temperature accuracy by correcting the fiber material phase transition induced density and refractive index variations.

3. Fiber Temperature Sensor Calibration and Performance

Before its use as a temperature sensor, a FBG has to be calibrated to obtain a so-called transfer function for converting the relative measured Bragg resonant wavelength shift to relative temperature change, as shown in (4)-(5). On the other hand, a transfer function may be invalid if the fiber material is experiencing microstructural transition process, which occurs at $T > 570$°C for Type-I FBGs and $T > 870$°C for Type-II FBG. Whenever this situation occurs, a new calibration process will be required to ensure the measurement accuracy. That is because the wavelength shift of the FBG is no longer associated with temperature. Specifically, when the fiber material is experiencing a microstructural transition, its temperature sensing function is temporarily lost until a new fiber microstructure is thermally stabilized. Afterwards, the FBG can be again used as a temperature sensor with a new calibrated transfer function, as discussed in the previous section. Calibration process of the FBG versus standard temperature gauge can be done in a programmable furnace. FBGs and standard temperature gauge are installed inside the furnace chamber where the temperature profile is uniform. The maximum allowed temperature is defined as about 10% higher than the desired operation temperature. Between the lowest and maximum operation temperature range, at least 10 data points corresponding to isothermal values will be required for extracting transfer function. After such a calibration process, the obtained transfer function can be used directly to convert wavelength shift to temperature by embedding it into the software.

To calibrate the Bragg resonant wavelength shift of a FBG against a standard thermometer, it is normally to measure both signals in the temperature range of interested under an isothermal process. For high accurate temperature measurement, a resistance temperature detector (RTD) or precision resistance thermometer (PRT) could provide better than ±0.2°C accuracy while k-Type TC only provides ±1°C accuracy. RTD and PRT are popular because of their excellent stability and exhibit the most linear signal with respect to temperature of any electronic temperature sensor. They are generally more expensive than alternatives, however, because of the careful construction and use of platinum. They are also characterized by a slow response time, low sensitivity, and limited operation temperature (650°C). However, an RTD or PRT is commonly characterized by nonlinear quadric or cubic polynomial function of $R_T = R_0[1+a\cdot T+b\cdot T^2+c\cdot T^3]$ or $R_T = R_0\cdot[1+a\cdot T+b\cdot T^2]$, where R_T is resistance at temperature T, R_0 is nominal resistance, and a, b, and c are constants used to scale the RTD or PRT.

For fiber sensor calibration, on the other hand, a transfer function is not always linear as (4)-(5). For high-temperature sensing application, the thermooptic coefficient is a nonlinear function of temperature:

$$\beta_{\text{eff}} \equiv \beta + \beta_\rho = A + B \cdot \Delta T + C \cdot \Delta T^2, \quad (6)$$

where β_{eff} may be positive or negative depending upon fiber material and wavelength and A, B, and C are constants. The associated transfer function of the Bragg wavelength shift to temperature will have a similar form of cubic polynomial function for high-density FBG sensors:

$$\Delta\lambda(T) = a(0) + a(1)\cdot\Delta T + a(2)\cdot\Delta T^2 + a(3)\cdot\Delta T^3$$

$$a(0) = \lambda_B \cdot \left(\kappa_\varepsilon \cdot \varepsilon + (1+\xi)\cdot\left|\frac{\Delta\rho}{\rho}\right|\right),$$

$$a(1) = \lambda_B \cdot (A+\alpha), \qquad a(2) = \lambda_B \cdot B, \qquad a(3) = \lambda_B \cdot C. \quad (7)$$

For temperature $T < 200$°C, the FBG sensor response can be approximately expressed by $\Delta\lambda(T) = a(0) + a(1)\cdot\Delta T$, where $a(1) \approx \kappa_T$ is low-density FBG temperature sensitivity. The nonlinear equation (7), however, is regarded as a generic transfer function for accurate wavelength conversion for a broad temperature range measurement. It should be pointed out that $a(0)$ term in (7) reflects both strain effect and potential fiber material compactness or/and phase transition induced mass density contributions. Whenever an FBG sensor has experienced extreme temperature or suffered from package-related strain effects, the $a(0)$ could be changed, which leads to a temperature bias. In another word, (7) is only valid to be used for a constant strain thermal stabilized FBG, which either has not experienced any morphology change or the microstructural transitions have been fully completed.

The thermal response of FBG elements is measured with an optical sensing interrogator [MOI 1 Hz, sm125–500/sm041 channel multiplexer], in which the wavelength can be tuned from 1510 nm to 1590 nm with ±1 pm accuracy. A LabVIEW-based software automatically identifies FBG peak position and saves both peak wavelength and converted temperature values. Figure 2(a) has shown a typical FBG spectrum with its Bragg resonant wavelength at 1544.75 nm and peak power of −7.26 dBm at ambient. This wavelength is upshifted, following the furnace temperatures as seen in Figure 2(b). The isothermal values from FBG (blue lines) and from thermocouple (TC) (red lines) are plotted in Figure 2(c) and fitted to a cubic polynomial function. The obtained coefficients are used to convert FBG wavelength shift values to temperatures (in circles). Figure 2(d) has plotted temperatures, measured from both FBG (blue circles) and TC (red lines), during 150 hours calibration process.

With calibrated FBG sensors, the thermal stability or drifting trend has been studied under an isothermal condition. Figure 3 has shown the measured wavelength and FBG peak power stability at 650°C for degradation and lifetime estimations. The data (circles) in Figure 3(a) has shown a trend of the wavelength shift is ~5.4×10^{-6} nm/hr. at 650°C, and that of FBG peak power loss is about $-(5$ to $20) \times 10^{-5}$ dBm/hr. If ignoring any oven thermal drift arising from PID-controlled K-Type TC feedback control, one can attribute such a trend as FBG sensor long-term thermal trend. If using approximate temperature sensitivity of 0.015 nm/°C for $T > 600$°C, the thermal variation caused temperature error is about ±0.4°C per 1000 hours at 650°C. Note that these estimated thermal drifts are within K-Type TC accuracy range of ±0.4%, namely, ±2.5°C at 650°C. On the other hand, the measured FBG peak power loss trend could be used to estimate the lifetime of the FBG sensor at 650°C operation. If the FBG peak power degradation is at a rate of $-(5-20)\times 10^{-5}$ dBm/hr, it will lose 0.44–1.76 dBm per year. A common quality of an FBG sensor is of a typical peak power of 25 dBm; the corresponding time for losing 25 dBm peak power will take 10 more years.

However, the real lifetime of the FBG actually strongly depends upon external environment conditions and mechanical package. On the other hand, the linearly estimated lifetime could be underestimated if the degradation trend is nonlinear. In normal case, a lifetime of 3–6 years is required for industrial machinery system application.

4. Fiber Temperature Sensor Package and Installation

In order to install FBG sensor for high-density temperature profile measurement from a gas turbine engine or from a compressor interstage, the operation within such a harsh environment requires the fiber sensors to be packaged for reliability. Consequently, a fiber sensor package should be hermetically sealed for maintaining strong mechanical integrity against vibration, thermal cycles, and stress-corrosion-induced mechanical fatigues. This is not only due to fiber fragility itself but also due to the detrimental long-term effect of corrosive gases, moisture, and acidic and alkaline chemicals potentially attacking the sensors. In addition, a proper sensor package will not impede dynamic thermal measurement during gas turbine startup transient. Since each industrial system application may vary in temperature, pressure, flow rate, vibration, and corrosion, for example, the sensor packages may differ from one industrial system to another industrial system and survive different harsh environmental conditions.

FBG temperature sensors can be installed at precombustor flow path, turbine inlet, downstream of the combustor or exhaust, and interstage locations for measuring dynamic temperature anomalies. Presurge condition in a turbine engine causes abnormal temperature fluctuation in the gas phase. Key measurements for surge control are suction/discharge temperature and pressure and flow rate through the compressor. The FBGs can be positioned at

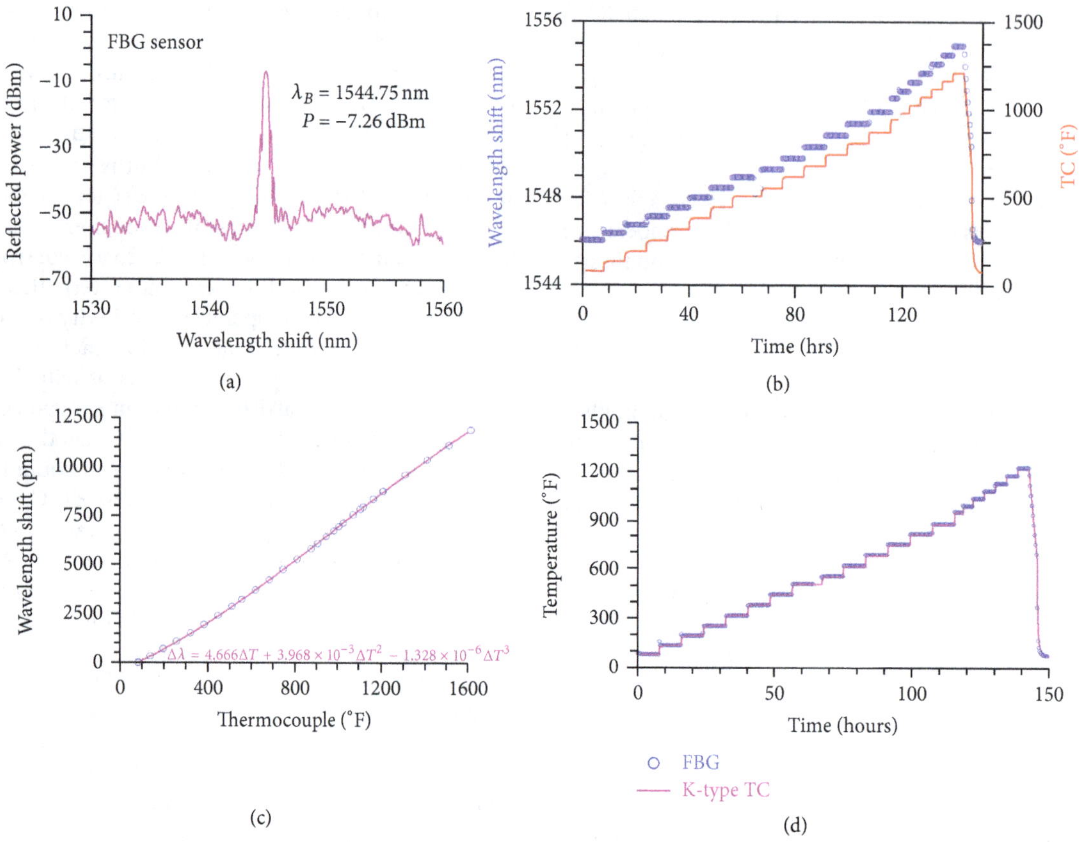

Figure 2: Fiber sensor calibration process for converting the measured Bragg resonant wavelength shift to temperature.

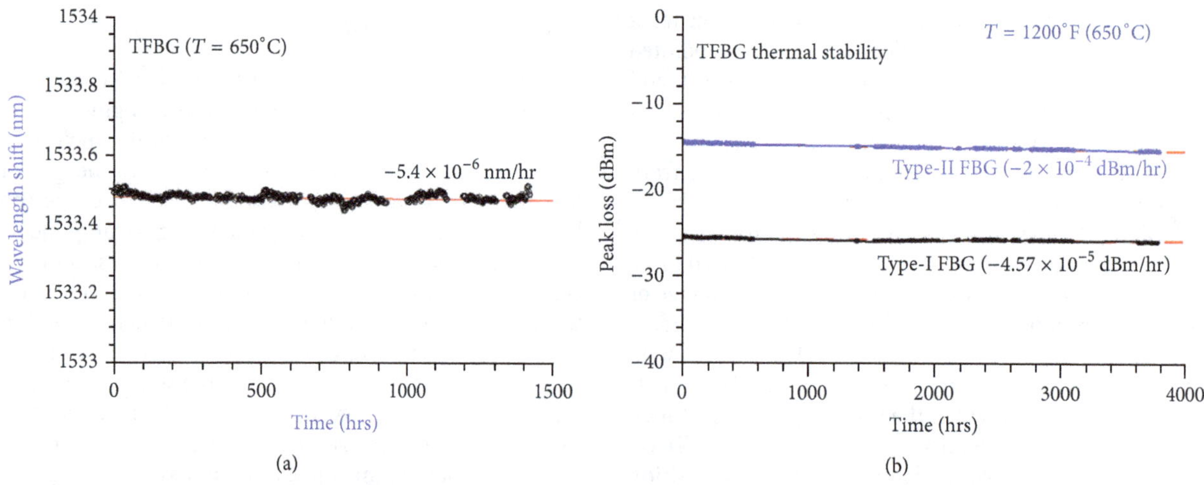

Figure 3: Short-term measured FBG sensor thermal stability (a) and FBG peak power loss at 650°C.

multiple locations in the compressor. For example, the FBG temperature sensor installed around the periphery of the turbine engine air inlet could detect presurge temperature spikes. In addition, the fiber sensing array may be installed in the pre-combustor flow path, downstream of the combustor, turbine engine inlet, and between the compressor stages. If the fiber sensors are installed in the turbine exhaust duct, it will enable the monitoring temperature variation that provides a means of detecting combustor health. This measured temperature combined with pressure enables better diagnostics and compressor surge protection.

Two fiber sensor packages have been designed for gas turbine exhaust temperature measurement. First one is a cable based for circumferential exhaust temperature detection, in which the FBG elements are inscribed onto three fibers and sealed inside an Inconel 625 tube, as described by [15]. The

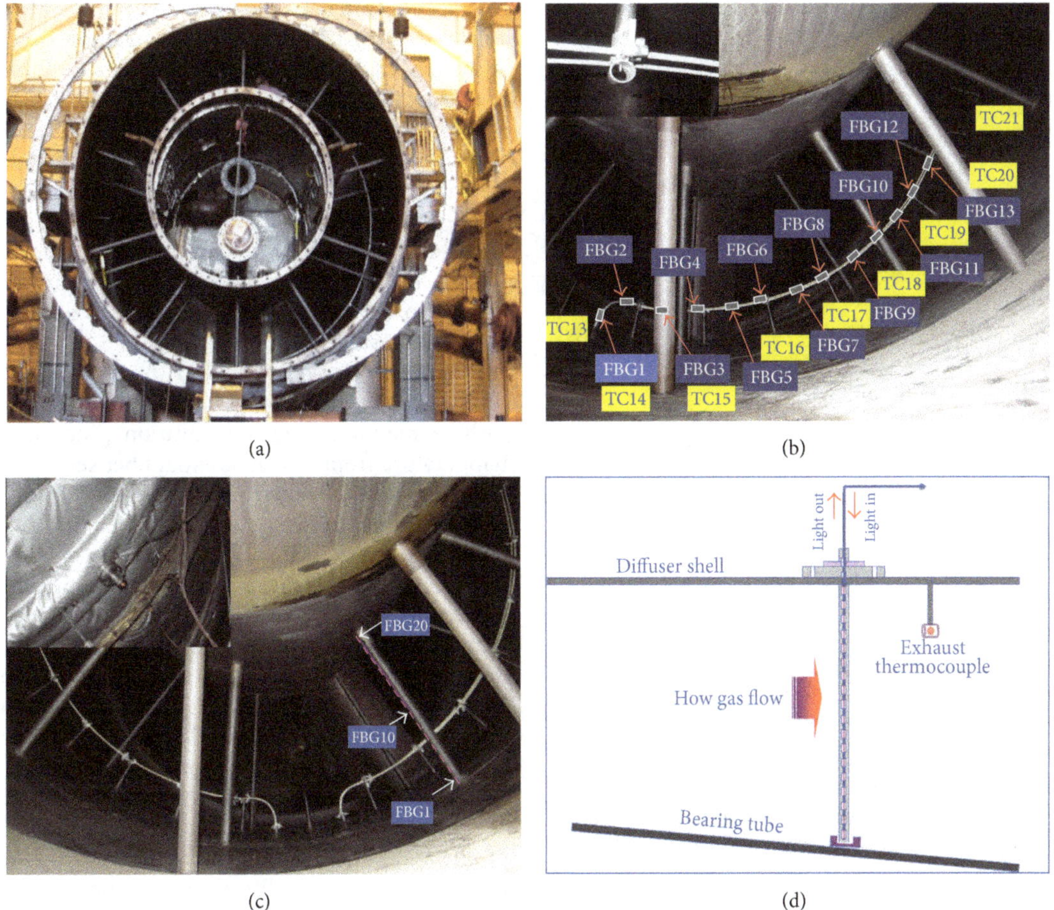

FIGURE 4: Fiber sensor installation in the gas turbine exhaust duct with circumferential fiber sensing cable in (a) and (b) and with fiber sensing rake in (c) and (d).

second one is based on a precision exhaust temperature rake with FBG sensors attached to its surface.

This integrated radial sensing rake has 100 mm spatial resolution across exhaust duct wall to central barrier space, described by [15]. Figure 4 shows two configurations for measuring both static and dynamic gas turbine exhaust temperature response. For addressing the entitlement of the FBG sensor performances in accuracy, reliability, and durability as a backup candidate technology of the existing thermocouple sensor, a fiber sensing cable is installed close to the existing exhaust thermocouples with a circumferential configuration, as shown in Figures 4(a) and 4(b). For evaluating dynamic performance in response time to transient operation condition variation, a fiber sensing rake is installed in a gas turbine that can measure radial exhaust temperature from diffuser to central bearing tube, as shown in Figures 4(c) and 4(d). A 1–5 Hz optical sensing interrogator with 4 channels capability is used to detect static signal, while a 1000 Hz (1 ms) optical sensing interrogator is used for dynamic signal detection.

For circumferential exhaust temperature detection, FBG elements are inscribed onto a single fiber with a spatial separation of 450 mm; three sensing fibers with 150 mm misplacement provide double spatial resolution for gas turbine exhaust temperature profile detection. The obtained wavelength data are converted to temperatures with a precalibrated transfer function that enables the measured FBG sensor static data to be compared with the thermocouple sensors. The fiber sensing cable is clamped to an existing radiation TC shield with entrances holes in between two thermocouples [15]. The down selection of a metal tube geometry and size has been determined with a simulation that has focused on thermal stress severity, mechanical strength, and natural vibration frequency of the metal tube. Since rotor blade rotation speed is 3600 rpm for a 60 Hz gas turbine and 3000 rpm for a 50 Hz gas turbine, its fundamental, first and second high-order vibration frequencies range from 50 to 300 Hz. To avoid an overlap between the natural vibration frequency of the metal tube packaged fiber sensing cable and gas turbine's vibration band, the modeling has shown that an Inconel tube could meet such a requirement with appropriate diameter and wall thickness.

For radial exhaust temperature detection, a precision exhaust temperature rake is used to support FBG sensors as an integrated radial sensing rake. There are 9 TCs that are embedded into the rake with a ⌀3 mm open hole to enable hot gas to pass through the hole for TC measuring hot gas temperature. However, the fiber sensing array is sealed in a fine-gage Inconel 625 tube, which is welded onto the rake surface with one-to-one correspondence between the FBG and TC location. Since the Inconel tube is attached to the rake surface and faces the hot gas flow, the combustion-dynamics-induced transient total temperature, $T_t(t)$, is proportional to flow velocity by

$$T_t(t) = T_S(t) + \frac{v(t)^2}{2c_p}, \quad (8)$$

where $T_S(t)$ is static temperature, c_p is gas-specific heat coefficient, and $v(t)$ is transient gas flow velocity. When FBG sensors are used to measure both static and total temperatures with (4), the gas flow velocity or combustion-induced flow dynamics can be directly measured by

$$v(t) = \frac{2c_p}{\kappa_T(0)}\left(\Delta\lambda\left[T_t(t)\right] - \Delta\lambda\left[T_S(t)\right]\right)^{1/2}$$

$$\text{Vortex shedding frequency} \propto \text{FFT} \quad (9)$$

$$\left\{\frac{2c_p}{\kappa_T(0)}\left(\Delta\lambda\left[T_t(t)\right] - \Delta\lambda\left[T_S(t)\right]\right)\right\}.$$

The static temperature corresponds to exhaust temperature with flow velocity close to zero, or near duct boundary layer. The vortex shedding frequency can be used to indirectly characterize the flow characteristics using Strouhal number, which is determined by frequency $*$ length/velocity. It is clear that the fast Fourier transform of the differentiation of static and dynamic signals could be used to correlate to different dynamic events and parameters, such as antisurge, turbine blade rotation speed variation, and combustion dynamics, and so forth. For example, when fuel quality changes, it leads to gas-specific heat (c_p) variation or vortex shading frequency variation. To measure dynamic FBG response to turbine operation condition, it normally requires high-band optical sensing interrogator that ranges from 1 kHz to 100 kHz.

5. Fiber Sensor Response to Static Exhaust Temperature

During a typical gas turbine factory test, different instruments will monitor various parameters, such as cranking motor output, turbine HP shaft speed, fuel stroke reference, IGV angle, averaged compressor inlet temperature, vibration amplitude, turbine exhaust static pressure, and air flow to fully characterize a gas turbine's performance before its delivery to customer. Corresponding to this baseline performance test, the gas turbine is under full speed without loading test. Figure 5 is a typical test from the fiber sensing rake with 20 FBG sensors. Since the fiber sensing rake is an integration of the existing exhaust precision temperature rake and FBGs, there are also nine TCs inside the rake as a reference. Since the first 3 FBGs are close to gas turbine diffuser wall, the boundary layer flow near the wall surface causes the measured temperature to be lower than the others. In Figure 5(b), these averaged TCs are plotted against additional 17 FBGs. To see details from both TC and FBG sensors, Figures 5(c) and 5(d) have plotted measured temperature during the steady-status operation condition.

Five of nine TCs data have been shown about 5°F variation during 100 min steady operation time while fiber sensors have shown about 9-10°F variation. This discrepancy is due to the fact that the TCs are located inside the rake with an open hole of ~ϕ3 mm, and the fiber sensing tube is welded to the rake surface across nine holes. Another potential origination could reflect the nature of the total temperature measured by the fiber sensors in contrast to the static temperature measured by the TCs. Meanwhile, the fiber sensors have shown comparable noise characteristics with respect to the TCs, which may relate to flow rate variation during state status operation.

Figure 6 has further shown the measured temperature profile from a full test of nearly 7 hours from a 7FB gas turbine manufacturing qualification process. The exhaust temperature from circumferential fiber sensing cable can be plotted in 3D as shown in Figure 6(a). During the startup stage of the gas turbine, performance monitoring is done via a polar diagram to highlight anomalies quickly. As shown in Figure 6(b), the measured exhaust temperature appears not to be uniform across 27 TCs. The FBG sensor at 170 locations also indicates the same anomalous variation. It is noticeable that the thermal profile is not symmetrical and it is very likely due to varied can-to-can combustion and hot gas flow variation. In addition, the dynamic process during startup provides variation as compared to steady base load operation which shows up on the polar diagram as temperature variation. It is clear that both 3D thermal profile and polar diagram measured with high-density fiber sensors could provide directly information on the gas turbine performance, which are comparable to the data measured from existing exhaust thermocouples.

6. Fiber Sensor Response to Dynamic Process

The fiber sensing rake and surface-attached fiber sensors have shown to be comparable to standard production TCs and also to be more sensitive to dynamic processes. This is illustrated by the fiber sensor response amplitude as shown in Figure 7. Since the first FBG is close to exhaust diffuser duct, its response amplitude is relatively small but with the increase of the radial distance towards the internal bearing tube, the dynamic amplitudes of the FBGs have shown corresponding increase. This increased amplitude mainly arises from the startup transient process with increased fuel gas flow, combustion dynamics, and pressure variation from compressor to gas turbine exhaust. However, the dynamic amplitude may be used for monitoring anomalous structural instability, presurge, and incomplete combustion events. For monitoring these dynamic events, the relative dynamic amplitude is more important than the absolute exhaust temperature measurement.

The other critical dynamic response feature of the fiber sensor is illustrated in Figure 8, where the FBG sensor's dynamic temperature response is plotted against the same location TC data as a function of time during first several minutes during the gas turbine startup transient. The small difference could be due to hot gas flow variation induced sensing rake oscillation or strain effect.

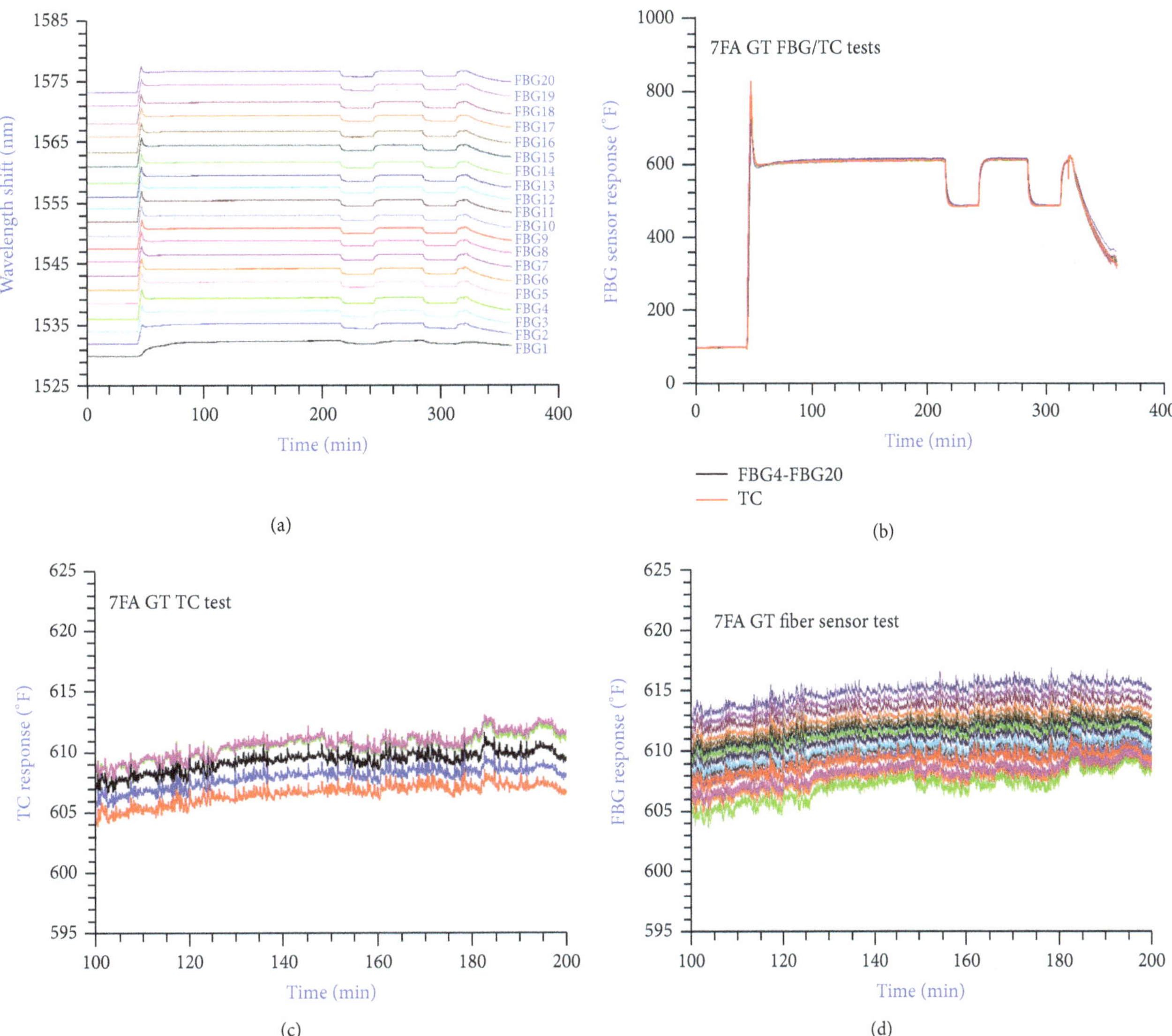

FIGURE 5: Fiber sensor and TC measured exhaust temperatures during a 7FA GT test with radial fiber sensing rake.

In addition, Figure 9 shows the dynamic response of the fiber sensors after leaving the transient startup condition. The bottom power spectra of the Figure 9 clearly show that the gas turbine cranking process has a high-level inharmonic oscillation. At gas turbine startup, the measured frequency is close to 60 Hz of the gas turbine blade rotation speed, but it gradually shifts to exhaust rake natural vibration frequency of ~40 Hz during steady-status operation. Such a dynamic response could be leveraged as a precursor for determining the health of the gas turbine. Thermal response of the FBG sensors can be used similar to annular array of thermocouples as a sensor input for fuel control to the combustion system. The dynamic temperature and power spectra can be utilized for antisurge, gas volumetric flow rate, combustion efficiency, and turbine-blade-rotation-related structural instability analyses.

It is also important to note that the converted power spectra and associated vibration mode profile can be used for gas turbine long-term performance degradation analysis. One of the features is its broad frequency band profile and its response amplitude. With increase vibration, mechanical issue becomes more of a concern, which may cause the gas turbine to degrade at an increased rate. These mechanical issues could come from the torque variation between the compressor and turbine coupling, rotor blade vibration and tip clearance change, bearing wearing, bucket parts loosening. Another valuable feature is fiber sensor's multifunctional nature that can be used for simultaneously measuring dynamic and static temperature, while its response amplitude can be associated with various dynamic and transient responses that could be related to various mechanical, thermal, and electric variations.

7. Conclusions

Fiber-Bragg-grating-based sensing technology has been further developed and validated in power generation system, such as gas turbine, for operation condition monitoring.

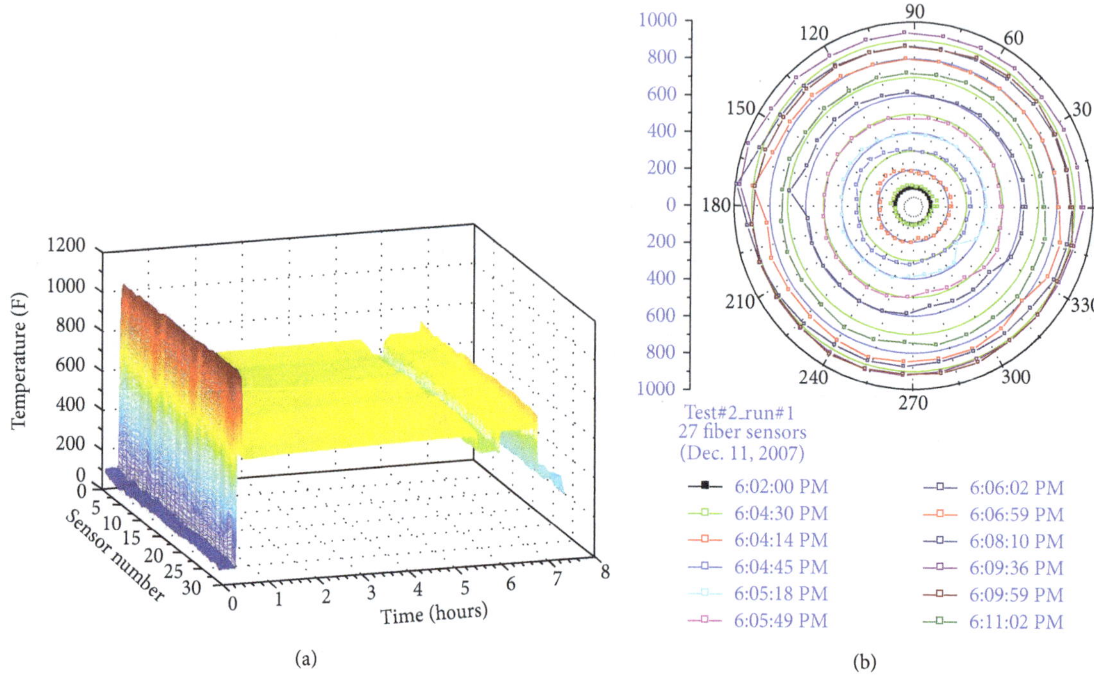

FIGURE 6: Exhaust thermal profile measured from a circumferential fiber sensing cable of 29 FBGs (a) and polar diagram of all the fiber sensors during a gas turbine startup process (b).

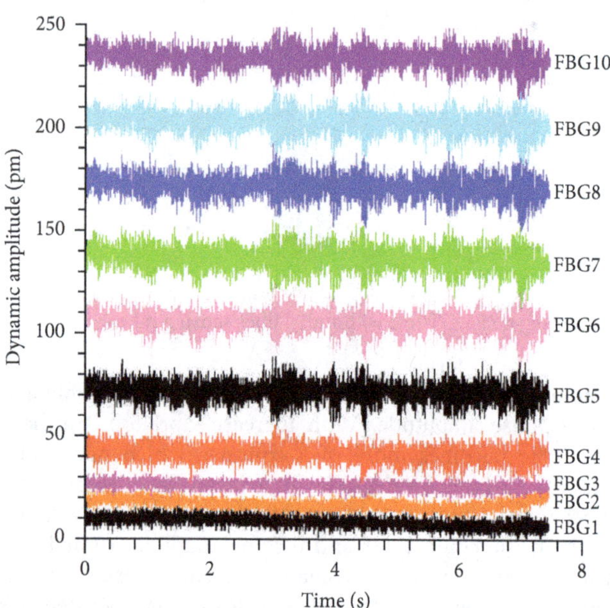

FIGURE 7: The measured dynamic response amplitude from FBG1 to FBG10 from a fiber sensing rake.

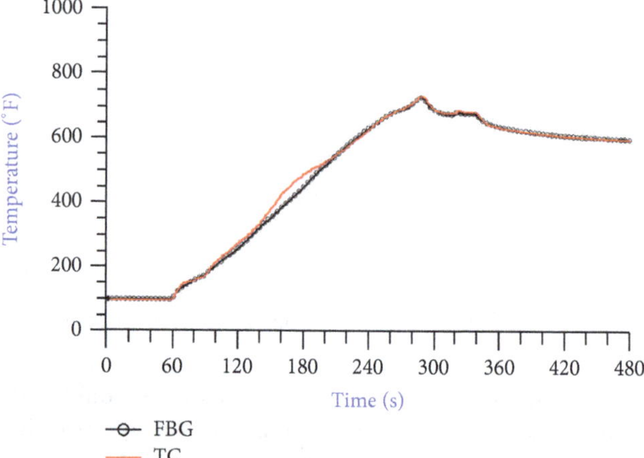

FIGURE 8: The transient response comparison between the fiber sensor and thermocouple from a 7FA gas turbine manufacturing qualification test.

The fiber sensor has demonstrated its unique advantage in providing high-density, multipoint, and multifunction capability in measuring static and dynamic responses from a gas turbine. To meet industrial harsh environmental sensing needs, one of the critical engineering work is the fiber sensor package, which varies from one machinery system to another system. The fiber sensor package is not only required to provide accurate measurement but also to prevent mechanical strength degradation from elevated temperature, massive hot gas flow, and blade-rotation-related structural vibration. The work presented in this paper is to demonstrate a new test/measurement technology that also can be used to measure multipoint temperature from a gas turbine with only one fiber sensing cable and one penetration. This fiber sensing method discussed herein is not limited to gas turbine exhaust temperature measurement. Its unique multifunction, static and dynamic responses, and distributed sensing capabilities make the high-density fiber sensor more valuable for

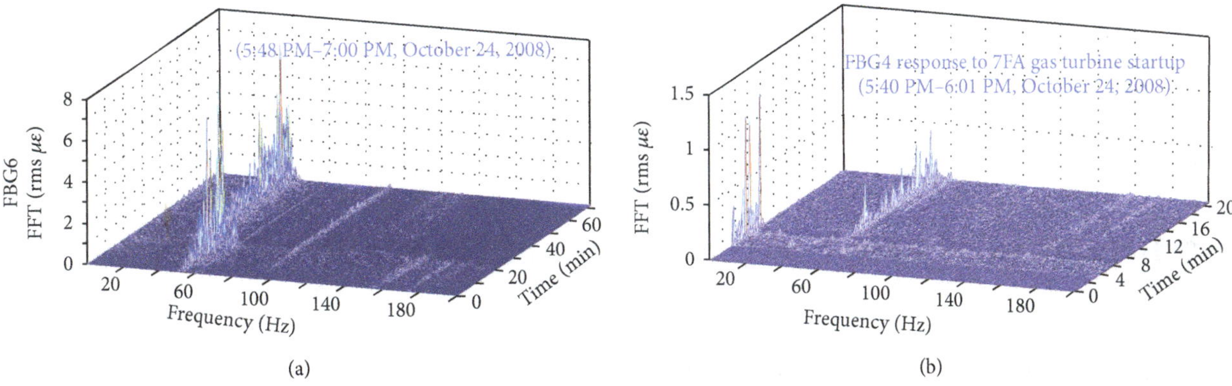

FIGURE 9: Fiber sensor measured dynamic signal and corresponding power spectra from 7FA gas turbine startup transient qualification test.

next-generation power generation turbomachinery system operation with respect to condition monitoring, diagnostics, control, and optimization.

Acknowledgments

During this development work, Kevin McCarthy, Juntao Wu, Renato Guida, Mohamed Sakami, Jerry Lopez, Alex Ross, David O'Connor, C. Wagner, James Nutt, Ronald Gilstrap, and Mike Krok have been involved in many fundamental research, laboratory test work, field sensor installations, and validations.

References

[1] K. O. Hill, Y. Fujii, D. C. Johnson, and B. S. Kawasaki, "Photosensitivity in optical fiber waveguides: application to reflection filter fabrication," *Applied Physics Letters*, vol. 32, no. 10, article 647, 3 pages, 1978.

[2] M. Fokine, "Underlying mechanisms, applications, and limitations of chemical composition gratings in silica based fibers," *Journal of Non-Crystalline Solids*, vol. 349, no. 1–3, pp. 98–104, 2004.

[3] S. Pal, T. Sun, K. T. V. Grattan et al., "Strain-independent temperature measurement using a type-I and type-IIA optical fiber Bragg grating combination," *Review of Scientific Instruments*, vol. 75, no. 5, pp. 1327–1331, 2004.

[4] H. Xia, K. McCarthy, and C. M. Young, "Harsh environment temperature sensing system and method," US Patent 7912334, 2011.

[5] D. Barrera, V. Finazzi, J. Villatoro, S. Sales, and V. Pruneri, "Packaged optical sensors based on regenerated fiber bragg gratings for high temperature applications," *IEEE Sensors Journal*, vol. 12, no. 1, pp. 107–112, 2012.

[6] H. Guo, G. Xiao, N. Mrad, and J. Yao, "Fiber optic sensors for structural health monitoring of air platforms," *Sensors*, vol. 11, no. 4, pp. 3687–3705, 2011.

[7] L. Hui and O. Jinping, "Structural health monitoring: from sensing technology stepping to health diagnosis," *Procedia Engineering*, vol. 14, pp. 753–760, 2011.

[8] B. Zhang, "High-temperature resistance fiber bragg grating temperature sensor fabrication," *IEEE Sensors Journal*, vol. 7, no. 4, pp. 586–591, 2007.

[9] J. Canning, S. Bandyopadhyay, M. Stevenson, and K. Cook, "Fiber Bragg grating sensor for high temperature application," in *Proceedings of the Opto-Electronics and Communications Conference and the Australian Conference On Optical Fibre Technology (OECC/ACOFT '08)*, July 2008.

[10] G. Brambilla, "High-temperature fibre Bragg grating thermometer," *Electronics Letters*, vol. 38, no. 17, pp. 954–956, 2002.

[11] H. Xia, "Advanced fiber optical sensor and instrumentation for power generation industrial monitoring and diagnostics," in *Fiber Optic Sensors and Applications*, vol. 8370 of *Proceedings of SPIE*, 2012.

[12] B. Zhang and M. Kahrizi, "High temperature resistance temperature sensor based on the hydrogen loaded Fiber Bragg grating," in *Proceedings of the IEEE Conference on Sensors*, pp. 624–627, November 2005.

[13] J. Li, D. Zhang, X. Wen, and L. Li, "Fabrication and temperature characteristics of thermal regenerated fiber bragg grating," in *Proceedings of the Symposium on Photonics and Optoelectronics (SOPO '12)*, pp. 1–4, 2012.

[14] H. Xia, "Fiber Bragg grating and fabrication method," US Patent 7574075, 2009.

[15] H. Xia, K. McCarthy, J. Nutt et al., "Bragg grating sensing package and system for gas turbine temperature measurement," US Patent 8306373, 2012.

12

Instrument for Label-Free Detection of Noncoding RNAs

Peter Noy,[1] Roger Steiner,[1] Joerg Voelkle,[1] Martin Hegner,[2] and Christof Fattinger[1]

[1] *F. Hoffmann-La Roche Ltd., Pharma Research and Early Development, Discovery Technologies, 4070 Basel, Switzerland*
[2] *CRANN—The Naughton Institute, School of Physics, Trinity College Dublin, Dublin 2, Ireland*

Correspondence should be addressed to Peter Noy, peter.noy@roche.com

Academic Editor: Maria Tenje

We set up a label-free direct binding assay for the detection of noncoding RNAs. The assay is based on nanomechanical cantilever arrays for the detection of surface stress induced by immobilized biomolecules and their interaction partners. We used various means to significantly reduce the drift of the cantilever readout that was a prominent feature in experiments with readout in stationary fluid before and after sample injection. Major improvements were achieved by focusing on a faster system equilibration (for instance temperature control and diffusion independence). Experimental protocols were improved to provide user-friendly and less time-consuming measurements. Further enhancements were achieved by, for example, using pre-gold-coated cantilever array wafers compared to individually prepared ones and a directly implemented data analysis tool as real-time feature of the measurement software. We have demonstrated picomolar specific biomarker target detection and can easily distinguish modified targets with single-nucleotide mismatches that hybridize with lower affinity.

1. Introduction

Nanomechanical sensing systems based on cantilever arrays are a basic research tool for exploring label-free assays. Investigators have shown several static mode applications for the detection of biological binding partners such as DNA hybridization [1–3] and receptor-ligand binding [4–7]. Our focus lies on the label-free detection of noncoding RNAs for medium throughput assays where half automated processes and less time-consuming protocols play an important role. Therefore our intention was to set up a stable and reliable device for this application in the field of genomics.

The detection of noncoding RNAs is of interest for monitoring miRNA or siRNA levels as biomarkers or for therapeutic approaches [8]. The present state of the art detection method for RNA is the branched DNA assay or DNA ELISA. As in an ELISA assay, an immobilized capture probe binds the target sequence. Afterwards the sandwich structure is completed with a detection probe (annotated as label extender). This label extender then binds the branched DNA with label probe. The labeled branches ensure a strong enough signal for detection [9]. The advantages of the DNA ELISA is that no amplification is necessary and no reverse transcription such as that in qPCR is needed. Measurements can be done directly on cell lysates. The fact that time-consuming assay protocols are inherent for this ELISA type assay is a disadvantage. Furthermore, there is one major limitation: to attach the label we need a certain amount of nucleotides from the target strand which are not available for recognition and to ensure specificity.

For comparative measurement we refer to a publication where the label-free detection of biomarker transcripts in human RNA with a nanomechanical cantilever setup was shown [10].

As proof of concept for the newly designed setup our goal was to detect a single-stranded 21mer oligonucleotide at 100 pM in a physiological buffer solution.

For the detection of successful hybridization experiments we measured the transduced surface stress which accumulated depending on the amount of specifically bound ssDNA biomolecules. We operated our device in static mode and measured in liquid. The induced bending of the cantilever (which lies in the nanometer range) is measured by reflecting a laser beam on the top of the cantilever and pointing it towards a position sensitive detector (PSD) as described in [11, 12]. Surface stress is induced by the interaction between immobilized biomolecules on the ssDNA biofunctionalized side of the cantilever bar and their interaction partners in an injected solution. Various forces such as intermolecular interactions, electrostatic forces, and changes in the electronic density of the cantilever surface lead to the resulting surface stress [13].

By subtracting the deflection signal of a nonspecific reference cantilever from the main signal, parasitic effects such as drift due to small temperature changes and nonspecific binding can be eliminated [13, 14].

Since measurable amount of signal drift is present in all known label-free detection methods we focused on its reduction by stabilizing the major external factors which affect drift in our nanomechanical setup. This was achieved by implementing a fast local temperature regulation system and measurement in continuous liquid flow. Our goal was to optimize the system towards semiautomatic device handling, which is essential for industrial applications.

To assist the interpretation of the recorded data we developed a real time analysis software which applies simple operations and plots the results concurrently with the measurement.

2. Instrumentation, Materials, and Methods

The cantilever deflection is measured by tracking a reflected laser spot on a position-sensitive detector (PSD) (1L10-10-A SU15, SiTek Electro Optics, Sweden). As laser source we chose pigtail laser diodes of 635 nm wavelength (HL6320G, Opnext Japan Inc., Japan) and operated them in constant power mode. By arranging eight laser coupled fibers in a linear array we achieved readout of the eight cantilevers through their sequential illumination.

Our setup is divided into three parts. The main part is a temperature-controlled box containing the cantilever instrument and the fluidic system (Figure 1). To keep the temperature at the cantilever array stable, we installed two controlled loops. (i) An external flow cycle thermostat (ministat 125, Peter Huber Kaltemaschinenbau GmbH, Germany) to stabilize the temperature inside the temperature-controlled box. (ii) The second temperature regulation module is a Peltier element mounted inside the measurement chamber at a distance of about 2 mm from the cantilevers. The Peltier element was regulated by a Peltier controller which is normally used for laser temperature stabilization (LDT-5525, ILX LIGHTWAVE, USA).

We performed all measurements in liquid phase. Two syringe pumps (neMESYS system, Cetoni GmbH, Germany) pull the system liquid and samples through the measurement chamber. To compensate for the pressure loss due to pulling we applied 80 mbar (nitrogen) overpressure on all sample vessels and the system buffer reservoir. Halar tubing (Ercatech AG, Switzerland) was used to reduce loss of probe molecules in the sample due to adsorption onto the tubing surface.

For measurements the cantilever can be installed either under dry conditions or in a prefilled system where the chamber and tubing are filled with buffer. In both cases we flush the system with CO_2 prior to filling with buffer. CO_2 dissolves 80 times better in water than nitrogen and leads to a gas bubble-free fluidic system.

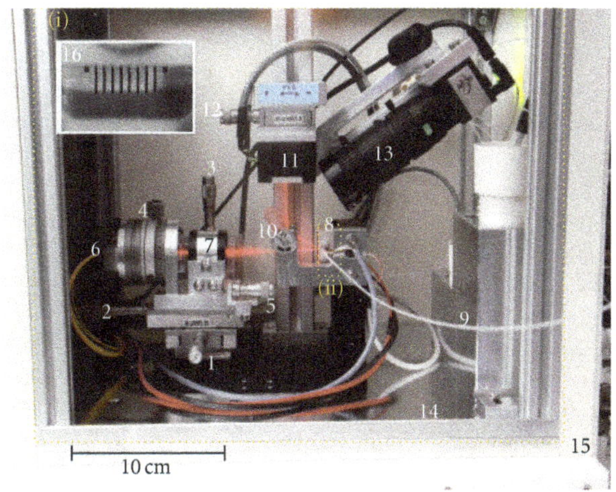

Figure 1: View inside temperature-controlled box containing the cantilever instrument. Laser ray path visible due to slight haze. (1, 2, 3) x, y, z positioning; (4) parallel alignment of fibers to cantilevers; (5) longitudinal focusing on cantilevers; (6) optical fibers (laser sources); (7) lens; (8) flow chamber (holds cantilever array chip); (9) tubing to syringe pump; (10) mirror with tilt function; (11) position-sensitive detector (PSD); (12) PSD alignment; (13) camera module; (14) ground plate connected to flow cycle thermostat; (15) thermal insulated box; (16) inset of cantilever array image mounted in flow chamber (8) taken with the camera module (13); (i) and (ii) illustrate the two temperature-controlled zones.

In addition to the temperature-controlled box the setup comprises a 19″ rack containing the laser controller and power supply for the PSD.

The setup is controlled by LabView (NI PCI-6221 interface and LabView software kit, National Instruments, Switzerland). All measured values are recorded and processed by LabView software. The data analysis is based on algorithms which were tested and previously applied for kinetic microarray signals [15].

We used cantilever arrays with eight cantilever sensors precoated with 2 nm titanium 20 nm gold (IBM Research GmbH, Switzerland). External dimensions of these sensors are as follows: 500 μm length, 100 μm width, and 0.5 μm thickness. To regenerate and clean the gold surface of environmental organics for subsequent ssDNA functionalisation, the arrays were treated with UV ozone for 60 minutes (radiation flux at 185 nm: ~4 W; ambient O_2) prior to use [16]. An oxygen plasma treatment to clean the gold surface is not recommended due to the widely distributed electron energy leading to radiation damages and a poor controllability [16].

All measurements were performed under continuous flow (10 μL/min for equilibration before and after the injections and 150 μL/min for the probe injection and wash step) using the above mentioned syringe pumps. Cantilever arrays were functionalized with thiol-modified ssDNA (Microsynth, Switzerland) for 60 minutes in acetic acid-triethylamine solution buffer in a home-built capillary device. Capillaries allow individual functionalisation of

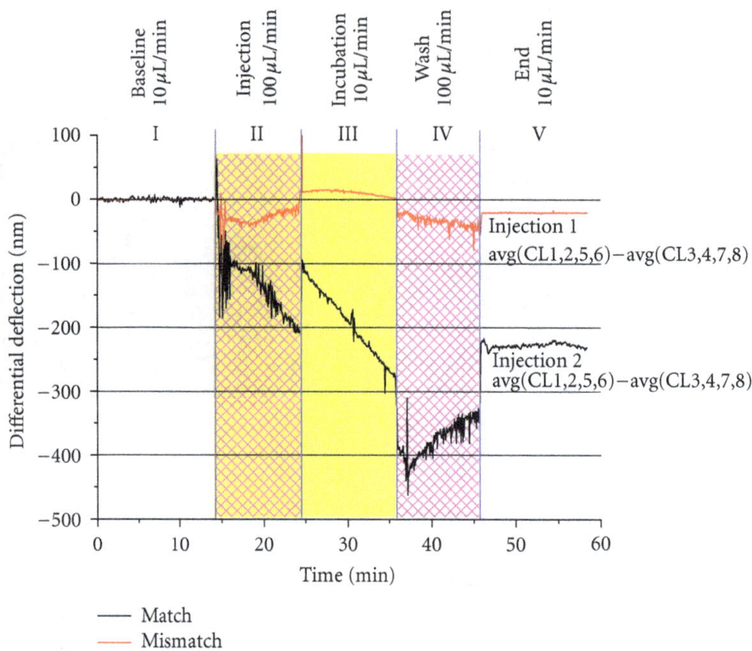

FIGURE 2: Overlay of two consecutive experiments to prove the detection of a 100 pM antisense strand. The graph shows the significant difference between an injection of 100 pM antisense match strand (black curve) and an injection of 100 pM antisense mismatch strand (red curve, the mismatch has two nonmatching base pairs in the centre of the target). Injecting the match sample induces approximately −200 nm differential deflection, where else the injection of the mismatch configuration leads to almost no differential signal. Phase (I) shows the recorded baseline at 10 μL/min buffer flow. (II) 1,000 μL sample injection at 100 μL/min. (III) incubation phase at 10 μL/min. (IV) flushing with buffer 100 μL/min. (V) resulting differential deflection after injection cycle is completed (10 μL/min buffer flow). Curves correspond to the differential deflection signal of positive minus reference cantilever (CL). Therefore the bending of the cantilevers is not absolute but differential deflections. The two injections were performed in series on the same cantilever array chip. A baseline correction, normalization, averaging, and differential signal calculation (probe minus reference) were done according to the literature [15]. Hatched area highlights the increased flow speed during injection and wash phase. Colored area indicates the presence of probe molecules in the flow chamber.

the various sensors. The DNA sequences chosen were AGAATAGGTATTTTTCCACAT for the biomarker target and AGAATAGGTATAATTCCACAT for the mismatch sequence. The chosen sequences do not tend to form hairpins and do not dimerize. In all experiments the following thiolated ssDNA oligonucleotides were used to functionalize the cantilever interface. (Sensor sequence: ATGTGGAAAAAT-ACCTATTCT-C6 linker-SH, Reference sequence: CTTACG-CTGAGTACTTTGA-C6 linker-SH). We used PBS (Invitrogen, Switzerland) as running and hybridization buffer.

3. Results and Discussion

With the described setup, we could detect a 100 pM antisense strand and differentiate between a perfect match and mismatch sequence. Figure 2 shows the overlay of two consecutive experiments. Before each injection a stable baseline was recorded to ensure that all cantilevers were equilibrated (phase (I)). Due to the automated injection program the timing for the following injection steps was the same for each experiment. This allowed the overlay of the two sequential experiments shown in Figure 2. The effect of switching the valve from running buffer reservoir to the probe container and changing the flow speed from 10 μL/min to 100 μL/min is visible at the beginning of the sample injection in phase (II). It takes about 3 minutes until the sample reaches the chamber with the cantilevers. This explains why the slope did not change significantly until mid phase (II). The heavy fluctuations can be explained by the change in refractive index, flow effects, and the exchange of molecules in the chamber before a new equilibration is set. After 10 min the sample (1,000 μL) is completely injected, the valve switches back to running buffer, and the flow speed is decreased to 10 μL/min (transition to phase (III)). In phase (III) the chamber is still filled with probe solution. A stable equilibrium is not reached during this incubation period. Several reactions leading to a cantilever deflection as described in [13] tend to occur. To remove the remaining probe solution and wash the chamber the buffer flow was increased (phase (IV)) to 100 μL/min. We flushed with 1,000 μL buffer. Here we see again that the delay before the probe solution in the system was fully replaced by buffer (change in slope). The peak at the changing point can be explained by the change in electrostatic conditions of the plain buffer solution compared to the buffer solution with probes. Fast effects such as valve switching and bulk buffer changes cannot be fully recorded due to the comparatively slow data acquisition (0.25 Hz), and therefore sequential injection traces are not completely identical. Finally the program switches back to the standby conditions

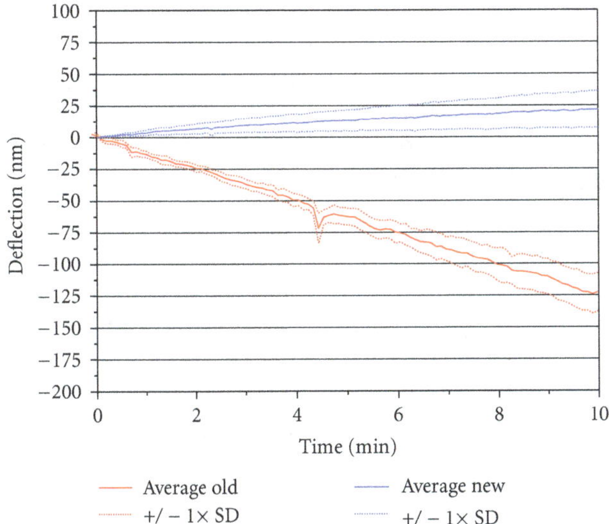

Figure 3: Typical drift before (red curve) and after implementation (blue curve) of means for drift reduction (continuous flow, temperature regulation, etc as described in this paper). Curves show the average of the raw data from 8 recorded cantilevers and the corresponding standard deviation. Curves have an offset at zero. Red curve measured under stationary conditions (flow: 0 μL/min). Blue curve measured in flow (10 μL/min).

(10 μL/min buffer flow), and the resulting deflection values are monitored. Compared to the end point of the deflection in phase (III), the start point of phase (V) is slightly higher (~50 nm) although we have the same flow speed in phase (III) and (V): 10 μL/min. A small amount of deflection is lost due to the dissolution of weakly bound strands (not fully hybridized) during the washing step. The two injections shown (red curve and black curve) were recorded sequentially. First, the negative probe (mismatch configuration) was injected and after a new equilibration the match injection was monitored. Finally the two starting points of the baselines were shifted to zero and the graphs plotted in an overlay. The resulting net deflection of ~200 nm for the 100 pM matching probe injection is repeatedly measured in our experiments. The resulting surface stress of about 9 mN m^{-1} is relatively high compared to previous experiments such as [10] (Young's modulus (Si): 130 GPa, Poisson ratio: 0.28). Reasons therefore could be due to longer cantilever functionalization times and due to different buffer properties which affect steric hindrance and ionic repulsion of the molecules.

By means of temperature stabilization and continuous flow measurements drift in the raw deflection signal was reduced from ~12 nm/min to ~2.5 nm/min as shown in Figure 3. The described setup and protocols represent a significant drift reduction by a factor 5 compared to previous experiments with readout in stationary fluid before and after sample injection. The gain in accuracy is especially of importance for the hybridization measurement with reaction times >1 min. The typical drift shown in Figure 3 was observed in all actual measurements.

In terms of electronic parts we used state of the art components. The amplifier has a noise level of approximately 1 μV, the PSD ~3 μV (BW = 100 Hz). The analogue digital converter NI PCI-6221 with ~122 μV noise level is therefore the main source of electric disturbance (values from datasheet stated in V_{RMS} to illustrate the critical components). Therefore we adjusted the full range scale to the maximum signal voltage and took the average over several measurement points (1,000 samples in 500 ms). This is possible due to the slow reaction time (>1 Hz) compared with the sampling rate characteristics of the electronic parts. Furthermore we optimized the settling time of the laser controller and adjusted the data processing to let the laser stabilize after switching. Before averaging, we discard the first half of the data points to be sure to have a stable laser signal. The remaining 500 samples are still enough for noise reduction by averaging. Due to the sequential readout a too long sampling time might lead to missing a reaction event.

By placing a temperature-controlling element close to the cantilever array we obtained a controlled loop with very short time constant for temperature equilibration. Time to regulate the temperature in the chamber from 21°C room temperature to 25°C setpoint is approximately 0.5 min. The much slower flow cycle thermostat regulation loop than that of the Peltier element leads to a stable temperature for all probe vessels, the buffer reservoir, and surrounding elements. In addition, a large (23 × 35 × 2.5 cm) aluminum ground plate provides a good heat exchange. To regulate the temperature from room temperature to setpoint by the flow cycle thermostat it takes ~50 min.

The two pulsation-free syringe pumps were embedded in our LabView control software. With two dosing modules an endless flow could be programmed, even for running measurements overnight. Besides electronic and temperature drifts the main portion of the overall drift visible in the deflection signal is drift due to diffusion effects (e.g., ionic exchanges between the cantilever surface and the surrounding liquid). The continuous flow led to a fast equilibration between the cantilever surface and the surrounding liquid which is diffusion independent. One feature which has to be taken into account when measuring in flow is the effect of the laminar flow on the cantilevers, as we see a deflection due to flow forces. In experiments with readout in stationary fluid before and after sample injection the liquid phase is moving during the injection process as well. This leads to significant flow induced deflections (see for instance in [10]). Depending on the position of the sensors relative to the liquid chamber channel the flow forces will be different for the eight cantilevers, inducing different additional bending that could potentially affect the measured deflection values. By measuring the baseline and the actual hybridization signal at equivalent buffer flow speeds, the comparability is given. The typical flow-induced bending by switching from stationary fluid to 10 μL/min is up to 7 nm. For the increased injection flow rate the induced bending is up to 400 nm (see, e.g., Figure 2). Stop flow read out with only a short "stop" phase to record the data points (much smaller time period than the drift kinetics) could add additional improvement.

Due to instrument design restrictions (flow path, lack of space, and temperature sensibility) we decided to set up the flow with a syringe pump in pulling mode. The disadvantage with this pulling method is the risk of sucking air into the flow path. Small air bubbles will stick to the cantilever array and lead to an abortion of the measurement. By compensating the pressure loss with a positive pressure on the probe side we avoided these problems. Additionally, Halar tubing was chosen to avoid gas diffusion into the system. The gas permeability value for oxygen for Halar is similar to PEEK and ~30 times less than Teflon (according to the specification guide from the provider). Moreover, Halar tubing is almost as flexible as Teflon tubing, in contrast to PEEK which would otherwise be a perfect material in terms of gas diffusion and low affinity for biomolecules.

Air bubbles tend to stick in small corners in the fluidic path and require a time-consuming procedure for their removal. CO_2 sparging allows fast fluidic system priming without any bubbles. The buffering characteristics of the solution and closing the CO_2 connection after priming ensure that the effect of the CO_2 on the acidity of the buffer is negligible.

4. Conclusion

Equilibration time and drift were significantly reduced by the fast temperature control system and continuous flow measurement. After installing the cantilever chip, it takes about 1.5 h until the system is ready to measure. The major time-consuming step is the cantilever functionalization although the protocol was simplified by using pre-gold-coated arrays and UV/O_3 activation. With CO_2 sparging, pressure compensation, and Halar tubing the formation of gas bubbles and their time-consuming removal was avoided. Further investigations into the effect of the continuous flow on the cantilevers will be carried out. The gain in drift reduction (approximately 10 nm/min) compared to the flow-induced bending (~7 nm) leads to the assumption that a measurement under continuous flow is an improvement. State-of-the-art electronic components and investigations into signal stability led to a stable and reliable device (fluctuations <5 nm for functionalized cantilever in liquid with a typical recording timescale of 0.25 Hz). Device control, measurement, and data analysis by LabView lead to a fast and straightforward workflow. The specific detection of a short oligonucleotide strand at 100 pM concentration in physiological buffer conditions demonstrated proof of concept of this setup.

Acknowledgments

We wish to thank Martin Hegner's group at the Trinity College in Dublin for guidance in setting up and testing the instrument and for providing valuable data for reference measurements. We acknowledge Ernst Meyer, University of Basel and Remo Hochstrasser, F. Hoffmann-La Roche, for their ideas and help for improvement of the instrumentation, Gregor Dernick, F. Hoffmann-La Roche, for his support in assay development, and Ulrich Certa, F. Hoffmann-La Roche, for support with the biological background.

References

[1] R. McKendry, J. Zhang, Y. Arntz et al., "Multiple label-free biodetection and quantitative DNA-binding assays on a nanomechanical cantilever array," *Proceedings of the National Academy of Sciences of the United States of America*, vol. 99, no. 15, pp. 9783–9788, 2002.

[2] F. Huber, N. Backmann, W. Grange, M. Hegner, C. Gerber, and H. P. Lang, "Analyzing gene expression using combined nanomechanical cantilever sensors," *Journal of Physics*, vol. 61, no. 1, article 090, pp. 450–453, 2007.

[3] J. Mertens, C. Rogero, M. Calleja et al., "Label-free detection of DNA hybridization based on hydration-induced tension in nucleic acid films," *Nature Nanotechnology*, vol. 3, no. 5, pp. 301–307, 2008.

[4] Y. Arntz, J. D. Seelig, H. P. Lang et al., "Label-free protein assay based on a nanomechanical cantilever array," *Nanotechnology*, vol. 14, no. 1, pp. 86–90, 2003.

[5] T. Braun, M. K. Ghatkesar, N. Backmann et al., "Quantitative time-resolved measurement of membrane protein-ligand interactions using microcantilever array sensors," *Nature Nanotechnology*, vol. 4, no. 3, pp. 179–185, 2009.

[6] T. Braun, N. Backmann, M. Vögtli et al., "Conformational change of bacteriorhodopsin quantitatively monitored by microcantilever sensors," *Biophysical Journal*, vol. 90, no. 8, pp. 2970–2977, 2006.

[7] N. Backmann, C. Zahnd, F. Huber et al., "A label-free immunosensor array using single-chain antibody fragments," *Proceedings of the National Academy of Sciences of the United States of America*, vol. 102, no. 41, pp. 14587–14592, 2005.

[8] R. W. Carthew and E. J. Sontheimer, "Origins and Mechanisms of miRNAs and siRNAs," *Cell*, vol. 136, no. 4, pp. 642–655, 2009.

[9] M. L. Collins, B. Irvine, D. Tyner et al., "A branched DNA signal amplification assay for quantification of nucleic acid targets below 100 molecules/ml," *Nucleic Acids Research*, vol. 25, no. 15, pp. 2979–2984, 1997.

[10] J. Zhang, H. P. Lang, F. Huber et al., "Rapid and label-free nanomechanical detection of biomarker transcripts in human RNA," *Nature Nanotechnology*, vol. 1, no. 3, pp. 214–220, 2006.

[11] H. P. Lang, R. Berger, C. Andreoli et al., "Sequential position readout from arrays of micromechanical cantilever sensors," *Applied Physics Letters*, vol. 72, no. 3, pp. 383–385, 1998.

[12] G. Meyer and N. M. Amer, "Simultaneous measurement of lateral and normal forces with an optical-beam-deflection atomic force microscope," *Applied Physics Letters*, vol. 57, no. 20, pp. 2089–2091, 1990.

[13] M. Godin, V. Tabard-Cossa, Y. Miyahara et al., "Cantilever-based sensing: the origin of surface stress and optimization strategies," *Nanotechnology*, vol. 21, no. 7, Article ID 075501, 2010.

[14] H. P. Lang, M. Hegner, E. Meyer, and C. Gerber, "Nanomechanics from atomic resolution to molecular recognition based on atomic force microscopy technology," *Nanotechnology*, vol. 13, no. 5, pp. R29–R36, 2002.

[15] T. Braun, F. Huber, M. K. Ghatkesar et al., "Processing of kinetic microarray signals," *Sensors and Actuators, B*, vol. 128, no. 1, pp. 75–82, 2007.

[16] W. Kern, *Handbook of Semiconductor Wafer Cleaning Technology—Science, Technology, and Applications*, Noyes Publications, 1993.

13

Measurement of Hepatitis B Surface Antigen Concentrations Using a Piezoelectric Microcantilever as a Mass Sensor

Sangkyu Lee,[1] Jongyun Cho,[1] Yeolho Lee,[2] Sangmin Jeon,[3] Hyung Joon Cha,[3] and Wonkyu Moon[1]

[1] *Department of Mechanical Engineering, Pohang University of Science and Technology, Pohang 790-784, Republic of Korea*
[2] *Corporate Technology Operations SAIT, Samsung Electronics Co., Ltd., Yongin 446-712, Republic of Korea*
[3] *Department of Chemical Engineering, Pohang University of Science and Technology, Pohang 790-784, Republic of Korea*

Correspondence should be addressed to Wonkyu Moon, wkmoon@postech.ac.kr

Academic Editor: Martin Hegner

Hepatitis B surface antigen (HBsAg) concentrations were measured using a piezoelectric microcantilever sensor (PEMS) developed by the authors. The developed PEMS is label-free and detects the sensing signal electrically. It was designed to measure the mass of biomolecules attached to it using an accurate mass-microbalancing technique; its probe area is confined to the end of the cantilever, and its equivalent spring constant is relatively high to minimize the effect of changes in the surface stress when the biomolecules are attached to it. The "dip- and-dry" technique was used to enable the probe area of the sensor to react with reagents in controlled environmental conditions. HBsAg was detected by an immunoreaction whereas the reaction time, antibody density, and its area on the probe were kept at a constant level. The mass of the detected HBsAg was measured in the range of 0.1–100 ng/mL.

1. Introduction

Hepatitis B virus (HBV) infection causes the disease hepatitis B and may also lead to cirrhosis and hepatocellular carcinoma [1]. An estimated two billion people worldwide have been infected by HBV, and of these, 350 million are chronically infected. Hepatitis B surface antigen (HBsAg) forms part of the surface of the virus and is used as a biomarker for the HBV infection [1]. We need to detect HBsAg in very low concentrations to accurately diagnose HBV. In hospitals, chemiluminescence immunoassay is widely used to detect HBsAg, and its detection limit is approximately 0.05 ng/mL [2].

A piezoelectric microcantilever sensor (PEMS) offers many advantages as a biosensor and is suitable for the detection of HBsAg. A PEMS is a highly sensitive label-free sensor that is sufficiently small to be developed as a portable device; multiplexed detection and electrical readout are also available [3]. Many studies have been conducted using a PEMS to detect various biomarkers [4].

The principle of detection in a PEMS is based on changes in the resonant frequency of the PEMS before and after a target protein is attached to it; target proteins are captured on the probe area of the PEMS by an immunoreaction. The frequency changes depend on changes in the surface stress and mass loading due to the attached biomolecules. The influence of the surface stress on the resonant frequency decreases as the effective stiffness (spring constant) of the PEMS increases [5, 6]. If we use a PEMS as a mass sensor based on a mass-microbalancing technique [7], the effective stiffness should be sufficiently large, and then, the resonant frequency will vary only in response to the mass loading effects.

The experimental setup consists of a part that measures the resonant frequency and a part that enables the probe area of the sensor to react with reagents. An impedance analyzer is usually used to measure the resonant frequency of the PEMS by detecting the peak point of the phase angle [8], dielectric loss [6], and so forth. The "dip- and-dry" technique is widely used in detection experiments [9]. The PEMS is dipped into reagents to either immobilize the antibody or bind the antigen, and the resonance frequencies are detected in air. The quality factor of the PEMS should be large because it is related to the accuracy of the detection of the resonant frequency change. A number of studies in the literature have reported the detection of biomolecules with a PEMS that is operated in a liquid, but the quality factor is greatly reduced in liquids [10]. Environmental conditions such as humidity and temperature must be controlled at a constant level because the resonance frequency of a PEMS can be affected by these conditions.

In this study, we demonstrate the measurement of HBsAg concentrations with a PEMS we developed that functions as a mass sensor. In order for the PEMS to function as a mass sensor, it was designed to have relatively large effective stiffness and the probe area was confined to the end of the device. The "dip- and-dry" technique was used, and the masses of detected HBsAg were measured in different concentrations and with different reaction times. Moreover, a control test using other proteins was performed.

2. Piezoelectric Microcantilever as Mass Sensor

The developed PEMS was previously designed by the authors to have sufficient sensitivity and reliability as a mass sensor [7, 11]; the geometrical shape and dimensions of the sensor are shown in Figure 1(a). The piezoelectric layer is composed of lead zirconate titanate (PZT), whose composition is $Pb(Zr_{52}Ti_{48})O_3$, and the cantilever structure is made of silicon. Gold is patterned on the end of the cantilever in the probe area.

A mechanical lumped parameter model, which is shown in Figure 1(b), is used to understand the mechanical characteristics of the PEMS. The lumped parameters are calculated from the modal analysis results of the PEMS using a commercial finite element analysis tool, COMSOL Multiphysics software. The material properties and the calculated value of the lumped parameters are listed in Table 1.

The mass sensitivity of the developed PEMS is over a million times higher than that of a commercial quartz crystal microbalance (QCM). The mass sensitivity of the PEMS is approximately 175 Hz/pg, which is calculated from an eigenvalue analysis before and after the addition of mass on the probe area (Figure 1(a)), and the mass sensitivity of the widely used QCM with a 5 MHz crystal is approximately 79 Hz/μg [13]. On the other hand, both sensors are comparable in the mass sensitivity per area because the probe area of the QCM ($\sim$1.267 cm^2) is over a million times larger than that of the PEMS ($\sim$50 μm^2). The mass sensitivity per area of the PEMS is 87.5 Hz/($\mu g/cm^2$) and that of the QCM is 100 Hz/($\mu g/cm^2$).

Because the surface stress effect on the resonant frequency of the developed PEMS is quite small, the PEMS can be used as a mass sensor. If we assume that the initial surface stress s is zero and that the probe area covers the entire area on one side of the rectangular-shaped microcantilever, the surface stress sensitivity is shown as follows [5]:

$$\frac{1}{f_0}\frac{df}{ds} = \frac{3}{16}\frac{1}{K_{\text{eff}}}. \quad (1)$$

The surface stress sensitivity is inversely proportional to the effective stiffness. If the probe area is reduced to one-tenth

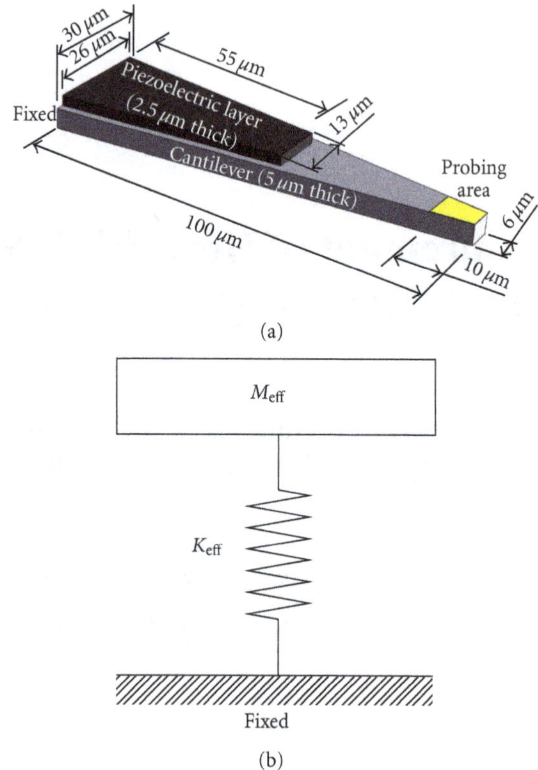

Figure 1: (a) Schematic diagram and (b) mechanical lumped parameter model of piezoelectric microcantilever sensor (PEMS).

of its original area and that area is located at the end of the cantilever, the surface stress sensitivity is reduced to approximately one quarter of its original value. The change in surface stress that is induced by the attached biomolecules is typically in the range of 0.01–0.08 N/m [6], and therefore, the change in the resonant frequency is calculated to be approximately in the range of 2.7–21.6 Hz. The frequency changes due to surface stress are less than or equal to the frequency resolution of our PEMS.

The developed PEMS was fabricated using a standard micromachining technique, the details of which are described in our previous paper [14]. Figure 2(a) shows a cross-sectional diagram, and Figure 2(b) shows an optical image of the developed PEMS. The deep trench that can be seen in Figure 2(b) was created to improve the electrical properties of the PEMS.

3. Experiments

3.1. Materials. Recombinant HBsAg, monoclonal anti-HBsAg, and alpha-fetoprotein (AFP) were purchased from HBI (South Korea). Phosphate-buffered saline (PBS, pH 7.4) and bovine serum albumin (BSA) were purchased from Affymetrix. Thiolated protein A/G (Protein A/G—SH) was prepared in the Magic Laboratory (POSTECH, South Korea) using Protein A/G (BioVision) and Traut's reagent (2-iminothiolane hydrochloride) according to the instructions for the Traut's reagent (number 26101, Thermo Scientific).

TABLE 1: Material properties and parameters used in mechanical lumped parameter model of piezoelectric microcantilever sensor (PEMS) [12].

Quantity	Unit	Value	Expression	Remark
Silicon				
E	GPa	170	Young's modulus	
ρ	kg/m^3	2329	Density	
ν		0.28	Poisson's ratio	
Piezoelectric layer				
E	GPa	Anisotropic		PZT5H
ρ	kg/m^3	7500		
Resonance frequency (f_0)	MHz	1.345479	$f_0 = (1/2\pi)\sqrt{K_{\text{eff}}/M_{\text{eff}}}$ In fundamental mode	Theoretical data
Effective stiffness (K_{eff})	N/m	234	$K_{\text{eff}} = 2E_{\text{strain}}/\delta_{\max}^2$, where E_{strain} is strain energy and $\delta_{\max}$ is the maximum displacement of the free end of the PEMS in fundamental mode	Theoretical data
Effective mass (M_{eff})	ng	3.272	$M_{\text{eff}} = k_{\text{eff}}/4\pi^2 f_0^2$	Theoretical data

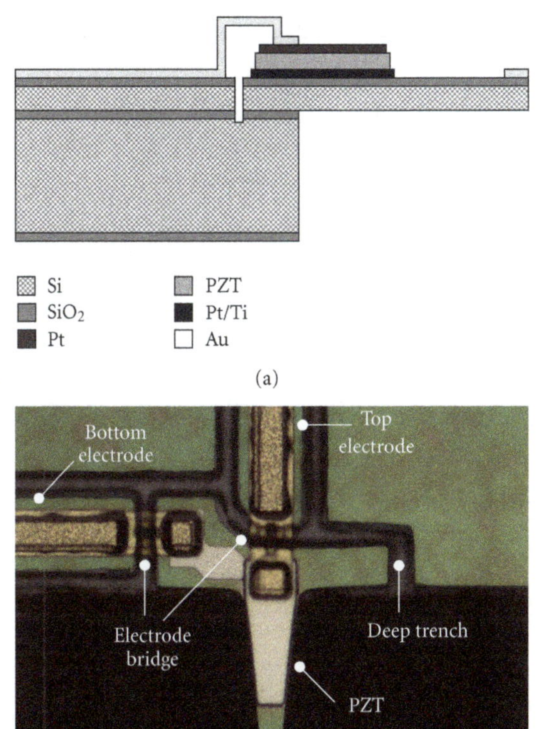

FIGURE 2: (a) Schematic cross-sectional diagram and (b) optical image of developed PEMS.

2-Propanol(isopropyl alcohol), ethanol, and methanol were purchased from Avantor Materials.

3.2. Experimental Setups. Figure 3 shows the experimental setups for the detection of HBsAg using the PEMS. The probe area of the PEMS was dipped into a droplet (<10 μL) of biochemical solution, and the dipping depth was controlled using a linear stage for monitoring through a CCD camera. The resonant frequency of the PEMS was measured using a PXI system, which is a computer-based measurement device. The peak point of the conductance spectra of the PEMS is the mechanical resonant frequency [15], and the frequency at the peak point was calculated using a program developed using LabVIEW, a graphical program language. A thermo-hygrostat was used to maintain a constant humidity and temperature during the experiment.

3.3. Detection Procedure. The detection procedure used for HBsAg is shown in Figure 4, and its detailed steps are listed in Table 2. First, the sensor was subjected to a cleaning process that employed ultraviolet (UV) light (254 nm, UV-5D Short-Wave Lamp, Spectronics), and wet cleaning was also applied. Second, the gold surface on the probe area was reacted with the thiolated protein A/G, which was used to bind the antibody. Third, the sensor was passivated with BSA to prevent nonspecific binding. Fourth, anti-HBsAg was immobilized on the reaction site of the sensor. Fifth, a control test using PBS and AFP was performed for comparison with the detection results of HBsAg. Finally, the HBsAg was detected at a specific concentration while controlling the reaction time. PBS rinsing and DI water rinsing were carried out at every step, and the resonant frequency of the PEMS was also measured at every step after drying with nitrogen (N_2) gas. The mass of the detected HBsAg was calculated using the difference in the resonant frequency before and after detection of the HBsAg and the mass sensitivity of the PEMS. In the process of immobilizing the anti-HBsAg, the concentration and reaction time at every step was determined experimentally, as listed in Table 2. The biochemical solution in the form of droplets was replaced several times to compensate for the effect of evaporation on the concentration.

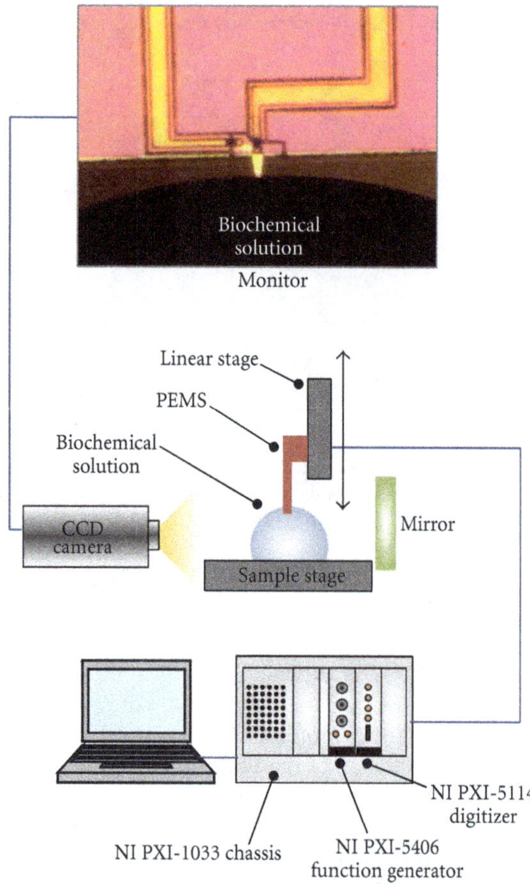

FIGURE 3: Experimental setups for detection of HBsAg with PEMS using "dip- and-dry" technique.

TABLE 2: Procedures and time requirements in detection process of HBsAg.

	Procedure	Time (min)
Sensor cleaning	UV cleaning	30
	2-Propanol cleaning	10
	Ethanol cleaning	30
	Methanol cleaning	10
	DI water rinsing	5
Protein A/G-SH	Protein A/G-SH (5 μg/mL)	40
	PBS rinsing	5
	DI rinsing	5
	N_2 gas dry	1
BSA	BSA (0.01%)	30
	PBS rinsing	5
	DI rinsing	5
	N_2 gas dry	1
Anti-HBsAg	Anti-HBsAg (25 μg/mL)	30
	PBS rinsing	5
	DI rinsing	5
	N_2 gas dry	1
Control	PBS	10
	DI rinsing	5
	N_2 gas dry	1
	AFP	10
	PBS rinsing	5
	DI rinsing	5
	N_2 gas dry	1
HBsAg	HBsAg	10
	PBS rinsing	5
	DI rinsing	5
	N_2 gas dry	1

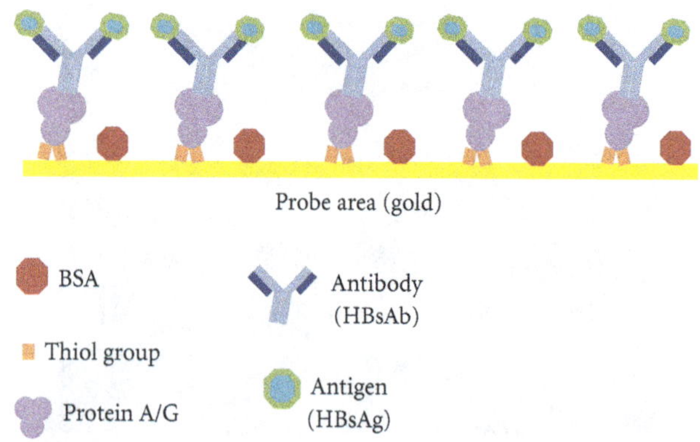

FIGURE 4: Procedure for detection of HBsAg with PEMS.

4. Results and Discussion

The resonant frequency of the PEMS decreases after biomolecules are attached to it. Figure 5 shows the conductance spectra of the PEMS at every step in the process of immobilization of the anti-HBsAg on the probe area. The resonant frequency decreases progressively at each step. It takes less than 20 s to obtain a conductance spectrum, and the quality factor of the PEMS is approximately 200.

The mass amount of the detected HBsAg depends on the concentration of the target solution, the reaction time, the density of the immobilized anti-HBsAg, and the area it occupies on the probe. The reaction time with the target was maintained at a constant level for 10 min. Furthermore, the anti-HBsAg density and its area on the probe were almost maintained at a constant level because the probe area was defined by identical patterned gold and employing the same processes used for immobilization of the antibody. In this way, the mass amount of the HBsAg depends only on the concentration; that is, the PEMS could measure the concentration of HBsAg in the target solution by measuring the mass of the detected HBsAg.

Figure 6 shows the mass amount of detected HBsAg in the range of 0.1–100 ng/mL. As the concentration increases, the binding speed of the HBsAg also increases. Therefore, the mass amount of detected HBsAg during the reaction time of 10 min increased as expected. Figure 6 also shows the results of the measured mass that were obtained from control tests using PBS and AFP. The results for PBS are related to the minimum detectable mass, and the results for AFP are related to the binding selectivity of the detection test of HBsAg performed with the PEMS.

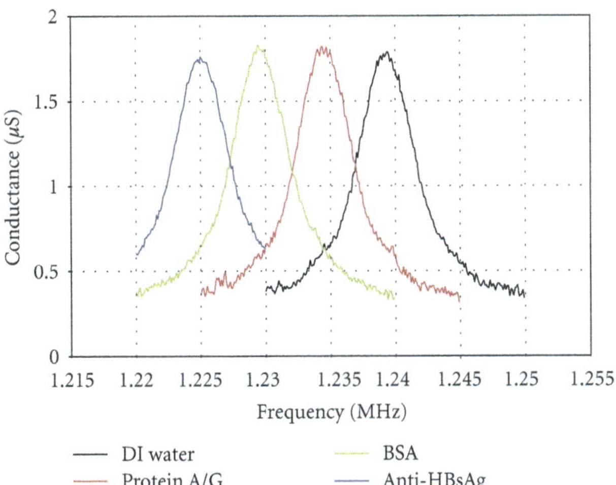

FIGURE 5: Conductance spectra of PEMS in the process of immobilizing anti-HBsAg on probe area.

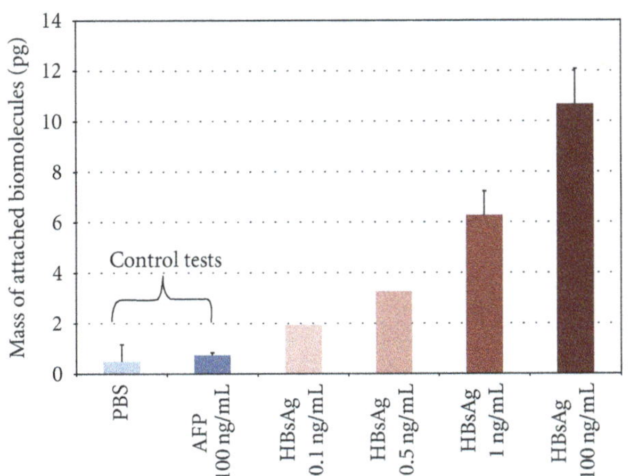

FIGURE 6: Mass of detected HBsAg in several concentrations.

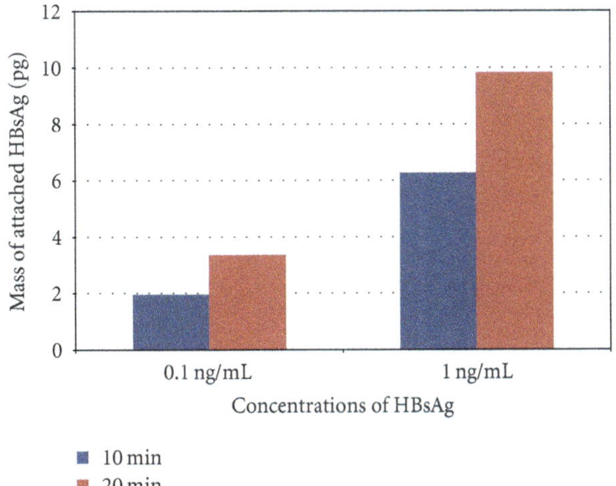

FIGURE 7: Time dependence of immunoreaction for different concentrations of HBsAg.

Figure 7 shows the time dependence of the immunoreaction for different concentrations. The mass increased by approximately 73% at a concentration of 0.1 ng/mL and 57% at a concentration of 1 ng/mL, as the reaction time increased twofold. The binding speed of the HBsAg may decrease as the reaction time increases because the binding speed will be zero at the saturation point; therefore, the increasing ratios of the mass of detected HBsAg are less than 100%. In addition, the increasing ratio has a lower value at a higher concentration of HBsAg because the binding speed of the HBsAg increases and then the binding reaction quickly reaches the saturation point. We need to maintain a constant reaction time while measuring the concentration of HBsAg, but we need to adjust the reaction time as the detection region of the concentration. That is, a shorter reaction time is needed in the lower concentration region and a longer reaction time is needed in the higher concentration region.

In the immunoassay, flow injection analysis (FIA) is also widely used; it allows tracking the kinetics of the immunoreaction [9] and it is relatively useful for multiplexed detection or automated system. However, FIA needs more complicated systems compared to the analysis using "dip-and-dry" technique. Furthermore, a PEMS operating in liquid has lower quality factor and lower resonant frequency due to viscous damping effect and added mass effect of the liquid [4, 7]; thus, FIA could deteriorate the reliability and sensitivity of the PEMS. Therefore, "dip- and-dry" approach is appropriate for measuring the concentration of HBsAg with the developed PEMS.

5. Conclusions

In this paper, HBsAg concentrations were measured in the range of 0.1–100 ng/mL using a PEMS that was developed for use as a mass sensor. The "dip- and-dry" technique was used, and the concentration was measured by measuring the mass of the detected HBsAg while keeping the reaction time for the target solution, the anti-HBsAg density, and its area on the probe constant. From the results obtained, we expect that the piezoelectric microcantilever mass sensor can be utilized for the measurement of the concentration of HBsAg and can also be used for sensitive diagnostic testing for HBV infection.

Acknowledgments

This work was supported by the "Development of Bio Robot Technology for POCT (point-of-care testing) (10024720)" program under the Industrial Source Technology Development Programs of the Ministry of Knowledge Economy (MKE) of Korea and was partly supported by a National Research Foundation of Korea (NRF) grant funded by the Korea government (MEST) (no. 2010-0019292).

References

[1] S. P. S. Monga, Ed., *Molecular Pathology of Liver Diseases (Molecular Pathology Library 5)*, Springer, New York, NY, USA, 2011.

[2] S. C. Lou, S. K. Pearce, T. X. Lukaszewska, R. E. Taylor, G. T. Williams, and T. P. Leary, "An improved Abbott ARCHITECT® assay for the detection of hepatitis B virus surface antigen (HBsAg)," *Journal of Clinical Virology*, vol. 51, no. 1, pp. 59–63, 2011.

[3] K. S. Hwang, S. M. Lee, S. K. Kim, J. H. Lee, and T. S. Kim, "Micro- and nanocantilever devices and systems for biomolecule detection," *Annual Review of Analytical Chemistry*, vol. 2, pp. 77–98, 2009.

[4] S. Xu and R. Mutharasan, "Cantilever biosensors in drug discovery," *Expert Opinion on Drug Discovery*, vol. 4, no. 12, pp. 1237–1251, 2009.

[5] Q. Ren and Y. P. Zhao, "Influence of surface stress on frequency of microcantilever-based biosensors," *Microsystem Technologies*, vol. 10, no. 4, pp. 307–314, 2004.

[6] S. Shin, J. P. Kim, S. J. Sim, and J. Lee, "A multisized piezoelectric microcantilever biosensor array for the quantitative analysis of mass and surface stress," *Applied Physics Letters*, vol. 93, no. 10, Article ID 102902, 2008.

[7] Y. Lee, G. Lim, and W. Moon, "A piezoelectric micro-cantilever bio-sensor using the mass-micro-balancing technique with self-excitation," *Microsystem Technologies*, vol. 13, no. 5-6, pp. 563–567, 2007.

[8] D. W. Chun, K. S. Hwang, K. Eom et al., "Detection of the Au thin-layer in the Hz per picogram regime based on the microcantilevers," *Sensors and Actuators A: Physical*, vol. 135, no. 2, pp. 857–862, 2007.

[9] L. Nicu, M. Guirardel, F. Chambosse et al., "Resonating piezoelectric membranes for microelectromechanically based bioassay: detection of streptavidin-gold nanoparticles interaction with biotinylated DNA," *Sensors and Actuators B: Chemical*, vol. 110, no. 1, pp. 125–136, 2005.

[10] T. Kwon, K. Eom, J. Park, D. S. Yoon, H. L. Lee, and T. S. Kim, "Micromechanical observation of the kinetics of biomolecular interactions," *Applied Physics Letters*, vol. 93, no. 17, Article ID 173901, 2008.

[11] Y. Lee, *Mass detection technique using piezoelectric micro cantilever and its application to bio sensor*, Ph.D. thesis, Pohang University of Science and Technology, Pohang, Republic of Korea, 2008.

[12] S. Lee et al., "Improvements in electrical properties of piezoelectric microcantilever sensors by reducing parasitic effects," *Journal of Micromechanics and Microengineering*, vol. 21, Article ID 085015, 2011.

[13] D. Lee, M. Yoo, H. Seo et al., "Enhanced mass sensitivity of ZnO nanorod-grown quartz crystal microbalances," *Sensors and Actuators B: Chemical*, vol. 135, no. 2, pp. 444–448, 2009.

[14] Y. Lee, S. Lee, H. Seo, S. Jeon, and W. Moon, "Label-free detection of a biomarker with piezoelectric micro cantilever based on mass micro balancing," *Journal of the Association for Laboratory Automation*, vol. 13, no. 5, pp. 259–264, 2008.

[15] IEEE Standard, "177-1966—IEEE Standard Definitions and Methods of Measurements for Piezoelectric Vibrators," 1966.

14

Cell Proliferation Tracking Using Graphene Sensor Arrays

Ronan Daly,[1] Shishir Kumar,[1] Gyongyi Lukacs,[2] Kangho Lee,[1] Anne Weidlich,[1] Martin Hegner,[2] and Georg S. Duesberg[1]

[1] CRANN and School of Chemistry, Trinity College Dublin, Dublin 2, Ireland
[2] CRANN and School of Physics, Trinity College Dublin, Dublin 2, Ireland

Correspondence should be addressed to Ronan Daly, dalyr1@tcd.ie and Martin Hegner, martin.hegner@tcd.ie

Academic Editor: Maria Tenje

The development of a novel label-free graphene sensor array is presented. Detection is based on modification of graphene FET devices and specifically monitoring the change in composition of the nutritive components in culturing medium. Microdispensing of *Escherichia coli* in medium shows feasibility of accurate positioning over each sensor while still allowing cell proliferation. Graphene FET device fabrication, sample dosing, and initial electrical characterisation have been completed and show a promising approach to reducing the sample size and lead time for diagnostic and drug development protocols through a label-free and reusable sensor array fabricated with standard and scalable microfabrication technologies.

1. Introduction

Controlled monitoring of bacterial growth has long been essential both as a diagnostic tool and as a standard drug development testing procedure. Common laboratory techniques involve bacterial proliferation on a Petri dish or in solution, providing an excess supply of nutrition and a controlled environment while sampling regularly for parallel tests by optical techniques. The need for higher throughput testing, more rapid diagnoses, and a more efficient use of samples has led to the implementation of miniaturised well-plate techniques. However, the drive for continuous improvement along with the concurrent growth in nanotechnology has led to a paradigm shift in sensing of biological activity. Significant advances in the coupling of proliferation to microcantilever [1–3] or quartz crystal microbalance (QCM) [4, 5] measurements have shown the potential for sensitivity to ultrasmall quantities of cells. In this work we are focused on the incorporation of graphene into label-free field effect transistor (FET) sensors to offer an alternative path to monitoring cell growth. In the approaches mentioned, it is the bacterium, the least abundant component, that acts as the analyte. Here we present initial results for the development of a novel label-free sensor for biological activity and specifically cell proliferation that relies upon measuring the change in the components of the bulk nutritive liquid. We propose the use of a scalable graphene FET microfabrication technology to (i) grow graphene films by chemical vapour deposition, (ii) transfer them to functional substrates and (iii) microstructure and contact graphene devices. These graphene FETs are functionalised by direct microdispensing of biological materials. We show initial evidence for cell proliferation on the microfabricated devices and the change in graphene charge transport responses with concentration changes of the lysogeny broth (LB) medium. This provides the basis for a scalable system allowing *in situ* tracking over the culture lifecycles in a range of parallel devices without the need for repeated sampling.

For diagnostics and drug development, one of the key drivers in sensor development is the reduction of the required sample volume. In a similar way that Moore's law drives the trend in decreasing transistor size for optimised device speed, there is a consistent decrease in sensor dimensions used for detecting proliferating bacteria. It has been noted in the literature that with a decrease in sample volumes there is an expected decrease in testing time. This is due to a number of factors, including the more rapid diffusion of nutrients because of the exponentially smaller system dimensions and the increased sensitivity requiring fewer lifecycles before detection occurs. This has been exploited

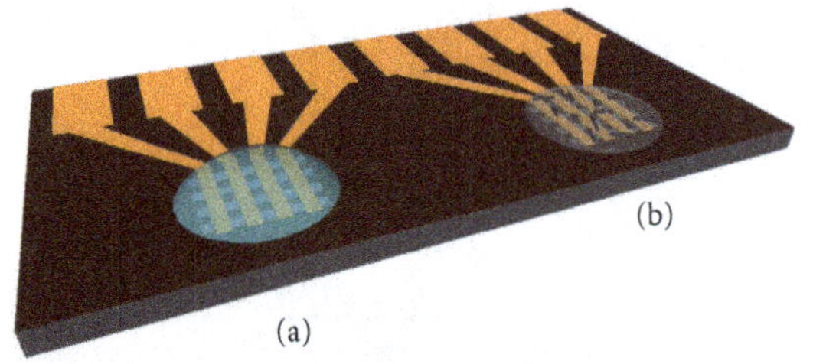

Figure 1: Metal electrodes contacting an underlying graphene layer acting as a sensor device with (a) bacteria proliferation occurring within a dosed volume of LB medium and (b) bacteria proliferation occurring on a thin agar layer, filled with a nutritive medium.

previously using micromechanical approaches, for example, by Gfeller et al. [1] where bacteria grew on an agar layer of a microcantilever array. However, these methods still have some drawbacks, namely, multistep fabrication techniques with limited device reusability and intricate surrounding measurement apparatus. For the devices proposed in this work, as shown in Figure 1, the sensor area is in complete contact with the sample and, through rapid diffusion and convection in such small sample volumes, is expected to be highly sensitive to changes. Large arrays of these sensors enable multiple parallel testing and improvement of the statistical confidence while still decreasing the batch time and conserving the low sample volume requirements. The robust nature of the graphene FETs allows repeated cleaning and reuse while the output is a simple electrical resistance measurement in the kΩ range. In standard laboratory techniques and the microfabricated devices shown in Figure 1, LB is used as a feedstock to promote the binary fission process and bacterial growth on an agar layer or in solution. The nutrition contained within the aqueous broth includes vitamins, minerals, and organic compounds such as amino acids all of which are essential to the proliferation and growth of *Escherichia coli (E. coli)*, the bacteria examined in this report.

The growth of cells is most often monitored by optical density (OD) measurements, where light absorption is used to identify the presence of bacteria in suspension. There are bulk-scale techniques to monitor the change in the LB content as a means to understanding the growth rate of bacteria. These track solution conductivity, pH, or fluorescence [6], but there are to-date limited attempts to scale this approach down to microscale arrays and to our knowledge no attempts to incorporate two-dimensional carbon sensors for this purpose. The unique electrical and mechanical properties of graphene lend themselves to incorporation into FET devices in this case. The relative freedom from catalytic impurities, the low levels of noise, flexibility, robustness, and ease of microstructuring have all been noted [7] as benefits to using this material as an ultrasensitive recognition element in biosensor devices. Such devices have proven effective in air and liquids for the sensing of individual gas molecules [8], proteins [9], and bacteria [10] when direct graphene-analyte interactions occur. The direct contact is believed to lead to charge transfer and hence a change in the electrical response of the graphene sheet.

2. Experimental Details

2.1. Sensor Fabrication. Sensor devices are manufactured on 15 × 15 mm pieces of p-doped (Boron) silicon (100) with a 300 nm layer of SiO_2 from Si-Mat Silicon Materials, Germany, and cut using the Disco DAD 3220 wafer dicer. Samples are cleaned prior to microfabrication using ultrasonication in HPLC grade acetone, ultrasonication and rinse in HPLC grade propan-2-ol and subsequent drying in a rapid flow of filtered, dry nitrogen. An oxygen plasma treatment is also carried out to remove organic contamination using the Diener PICO barrel asher. Masks for UV lithography were designed in-house and created using the Heidelberg DWL 66FS direct writing system. UV lithography was carried out with the OAI Mask Aligner using Microposit S1813 positive photo resist and MF319 developer (both from Rohm and Haas Electronic Materials). Metal sputter deposition was carried out using the Gatan 682 Precision Etching Coating System at a rate of 0.1 Å s^{-1}. After standard polymer lift-off procedures, residual polymer was removed by oxygen plasma treatment except when graphene was present, when solvent cleaning alone was used.

2.2. Graphene Transfer and Etching. Graphene, produced by chemical vapour deposition (CVD) as described in Results, is transferred from metal foil to the substrate as follows. A layer of poly(methyl methacrylate) (PMMA), (Mr-I 35 K PMMA from Microresist Technology GmBH) was spin coated on top of graphene film/copper foil pieces. Thermal-release tape was adhered on top of this PMMA support film, and the copper was then etched by floating the sample in etchant (0.25 M $FeCl_3$ + 0.2 M HCl). The resulting layered film of thermal-release tape/PMMA/graphene was cleaned with DI water, dried, and placed onto the substrate (as shown in Figure 3(b)). Because the graphene follows the contours of the PMMA/thermal-release tape layer, a uniform pressure was applied to the film to ensure close contact and conformation to the substrate. A range of pressures were used successfully ranging from 10 to 25 bar approximately. Heating the substrate from below promoted release of the upper tape layer. The remaining PMMA layer was removed by an initial soak in HPLC-grade acetone followed by an overnight soak in HPLC grade chloroform. The process can be carried out without thermal-release tape to avoid some contamination. In this case, PMMA-supported graphene is dredged from DI water onto the substrate. After the sample dries, PMMA can be removed as before.

2.3. Chemicals and Bacterial Culture. Chemicals and culturing medium were purchased from Sigma Aldrich (Arklow, Ireland) unless otherwise stated. *E. coli* CIP 53.126 was obtained from Collection de l'Institut Pasteur (Paris, France). Overnight cultures were prepared (200 rpm, 35°C, 15–18 h) in LB (1% NaCl, 1% tryptone, 0.5% yeast extract) from single colonies of *E. coli*. 1 mL of the overnight cultures were transferred into 30 mL of 50% LB, and 25% glycerol, 25% DI water and cultured (200 rpm, 35°C) for 110 min in order to reach a logarithmic growth rate. Glycerol was added

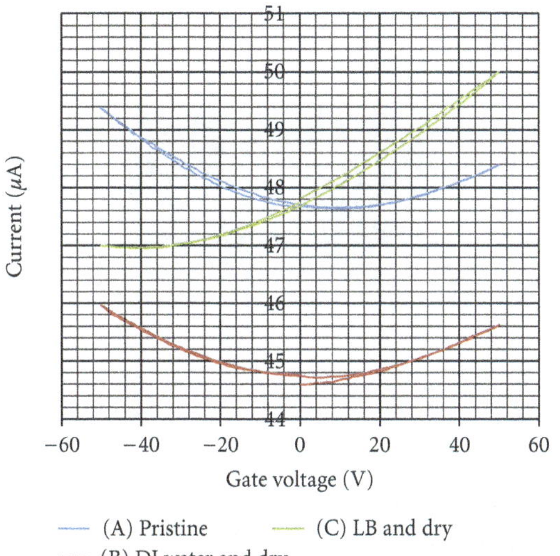

	Pristine	DI water	LB 1%$_{w/w}$	LB 10%$_{w/w}$	LB
Dirac point	+10 V	+4 V	−17 V	−34 V	−43 V

FIGURE 2: Performing gate sweep using back-gated graphene FET device shows that with exposure to LB medium there is a shift in the Dirac curve towards the negative bias voltage.

to ensure droplets did not evaporate prior to measurement. Observations confirmed evaporation was inhibited. Two additional diluted solutions were prepared from the LB stock. The first solution consisted of 500 μL LB, 250 μL glycerol, and 250 μL DI water. The second, more dilute solution consisted of 31 μL LB, 250 μL glycerol, and 469 μL deionised water (twice dilution and thirty-two times dilution, resp.). These are referred to as 2LB and 32LB in the text.

2.4. Microdispensing and Electrical Measurement. LB medium both with and without cells was dosed during this work using an Autodrop microdispensing system from Microdrop Technologies and a nozzle with a diameter of 50 μm. Subsequent electrical measurements on the graphene FET devices were carried out using a Keithley 2400 Sourcemeter attached to a Karl Suss probe station. Substrates were transferred between the dosing and measurement devices within a Petri dish containing a pad saturated with water to maintain humidity and inhibit evaporation of the dosed droplets during transport.

2.5. Additional Analysis. Raman spectroscopy was carried out using a Horriba Jobin Yvon LabRam HR system and a line of 632.8 nm. Scanning electron microscopy (SEM) was carried out using the Zeiss ULTRA Plus in the Advanced Microscopy Laboratory, CRANN, Trinity College Dublin. Prior to SEM imaging, bacteria were fixed by soaking in 5% v/v glutaraldehyde solution in 0.05 M phosphate buffer (pH7) and incubated at room conditions with gentle agitation for 3-4 h. Glutaraldehyde was then removed by 6 successive washes in fresh 0.05 M phosphate buffer, each of 10 minutes duration. Samples were subsequently dehydrated with a sequence of 10-minute rinses in 10, 30, 50, 70, 90, 100, and 100% v/v ethanol.

3. Results and Discussion

The fundamental premise of LB components affecting the conductance of graphene was confirmed using high-quality graphene flakes grown on Ni by chemical vapour deposition (CVD) and contacted with *e*-beam lithography. The graphene preparation and contacting process is described elsewhere [11]. The crucial step in this case is that the graphene has been cleaved by the Scotch tape to leave a clean surface. By dosing (i) 18 MΩ deionised water and (ii) LB medium onto a graphene FET device and comparing the electrical response of the sensor upon solvent evaporation, we see the precipitated materials from the LB medium lead to a slight increase in the measured resistance of the graphene strips and a clear shift in the Dirac point, as indicated in Figure 2. Graphene has linear dispersion in both valence and conduction bands. The degenerate point where these bands meet is known as the Dirac point. The Fermi level of graphene can move across the Dirac point under a bias, changing the concentration of charge carriers and therefore the resistance of samples. Thus, the minimum conductance (or maximum resistance) point observed in I–V characteristics displayed in Figure 2 corresponds to the Dirac point. A negative shift, as is observed for the samples exposed to LB, is equivalent to *n*-doping of the graphene.

With the development of a bulk graphene manufacturing technique, namely CVD, the incorporation of this remarkable material into scalable production of devices is now feasible [12]. The graphene used in this sensing application was grown by CVD, and all patterning was carried out by another scalable production technique, that of optical lithography. The CVD growth was carried out on 15 × 15 mm samples of copper foil in a tube furnace, as indicated in Figure 3(a) and described in detail in a separate publication [13]. For this work two techniques are detailed in the experimental section

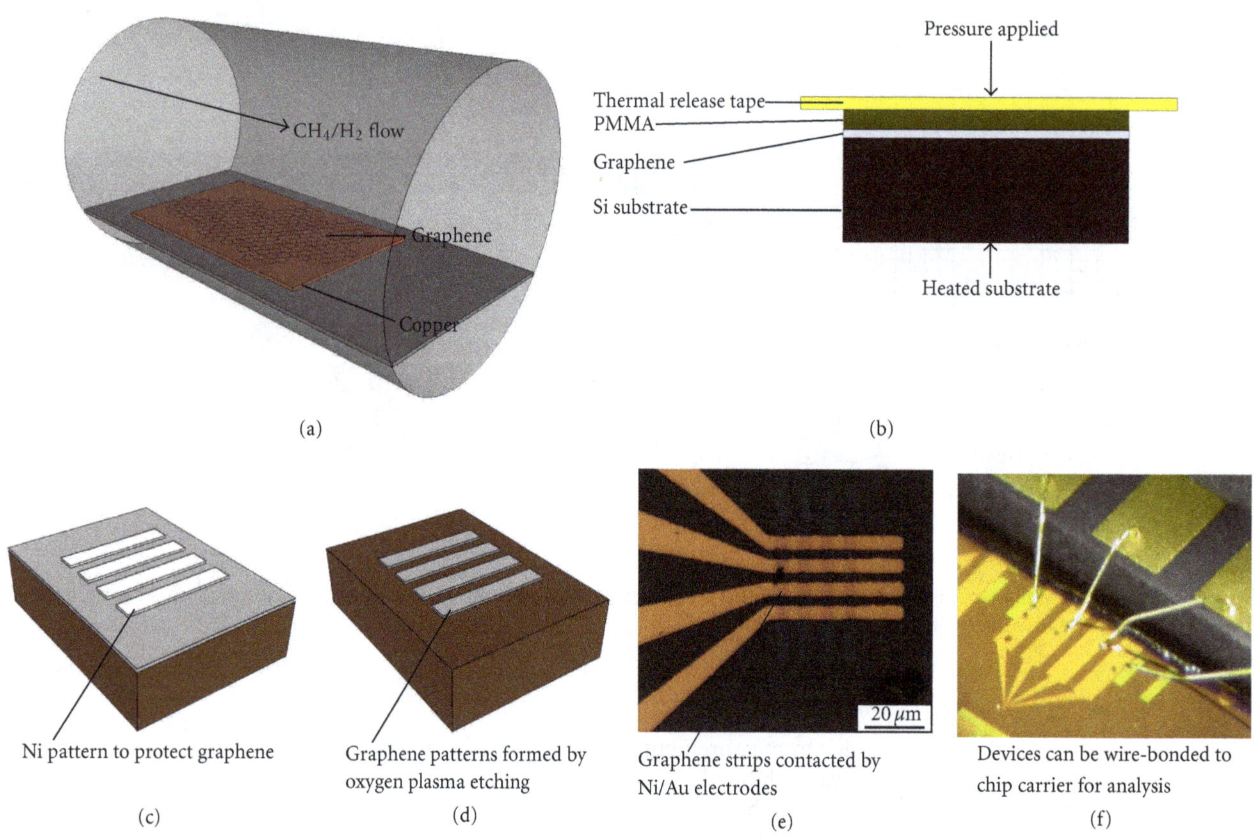

Figure 3: Production and microfabrication of graphene. (a) CVD growth of graphene on copper foil in a tube furnace, (b) graphene is attached to SiO_2/Si by attaching to a support layer of PMMA and thermal-release tape and transferring by a combination of heat and pressure, (c) after transfer to SiO_2/Si substrate, a Ni protection layer is patterned on the graphene, (d) oxygen plasma removes the unprotected graphene and the Nickel is subsequently removed by HCl etching. (e) shows the contacted graphene strips and (f) shows integration of the sensor into a chip carrier.

attempting to optimise the transfer of approximately 15 × 15 mm films of graphene to the SiO_2/Si substrates. In summary, after coating the graphene film/copper foil pieces with a PMMA support layer and a further layer of thermal release tape, the copper can be removed by etching with $FeCl_3$. Transfer of graphene to SiO_2/Si substrates is completed by applying pressure through the tape and PMMA support layers, pressing the graphene surface onto the substrate as indicated in Figure 3(b). Heat applied through the substrate allows easy release of the thermal-release tape leaving behind the PMMA/graphene layers with the graphene adhering to the substrate very strongly by the van der Waals forces [14]. The PMMA can be removed with solvent cleaning. Due to concerns regarding contamination from the thermal-release tape and the fracturing effects of the mechanical transfer method, a second approach was developed. The PMMA support layer is still applied to the graphene film/copper foil pieces and the etching occurs as before at the liquid-air interface, leaving a graphene/PMMA layer floating on the surface. This is carefully transferred to the substrate surface through dip coating, and the same solvent cleaning steps occur to remove PMMA. The substrates have been prestructured by UV lithography with distinct, chromium alignment marks. These were included to enable a sequence of UV lithography patterning steps to occur that lead to metal-contacted graphene strips with good adhesion to the substrate, using a technique described by Kumar et al., [13]. As shown in Figures 3(c)–3(f), a sacrificial masking pattern of nickel is formed to protect the areas required for the devices and the uncovered graphene is removed by an oxygen plasma. The nickel protection layers are then completely removed by an acid etch with 1 M HCl, and the remaining graphene strips are contacted by four Ni/Au electrodes (4 μm/48 μm) using a final UV lithography step. The contacted samples can then be probed directly using a needle prober or wire-bonded to a chip carrier for electrical measurements. This technique was modified from previous work to include a range of alignment marks for accurate positioning of all layers and a design that can be directly incorporated into an inkjet dosing system.

It is observed that the gate voltage behaviour and the scale of resistances recorded for graphene prepared with this technique have changed. This is partly due to the known issue of contamination during the incorporation of graphene into functional devices using multistep lithography processes. Graphene is notoriously difficult to maintain free of contamination and defects, and novel cleaning techniques will become essential for large-scale manufacture of graphene

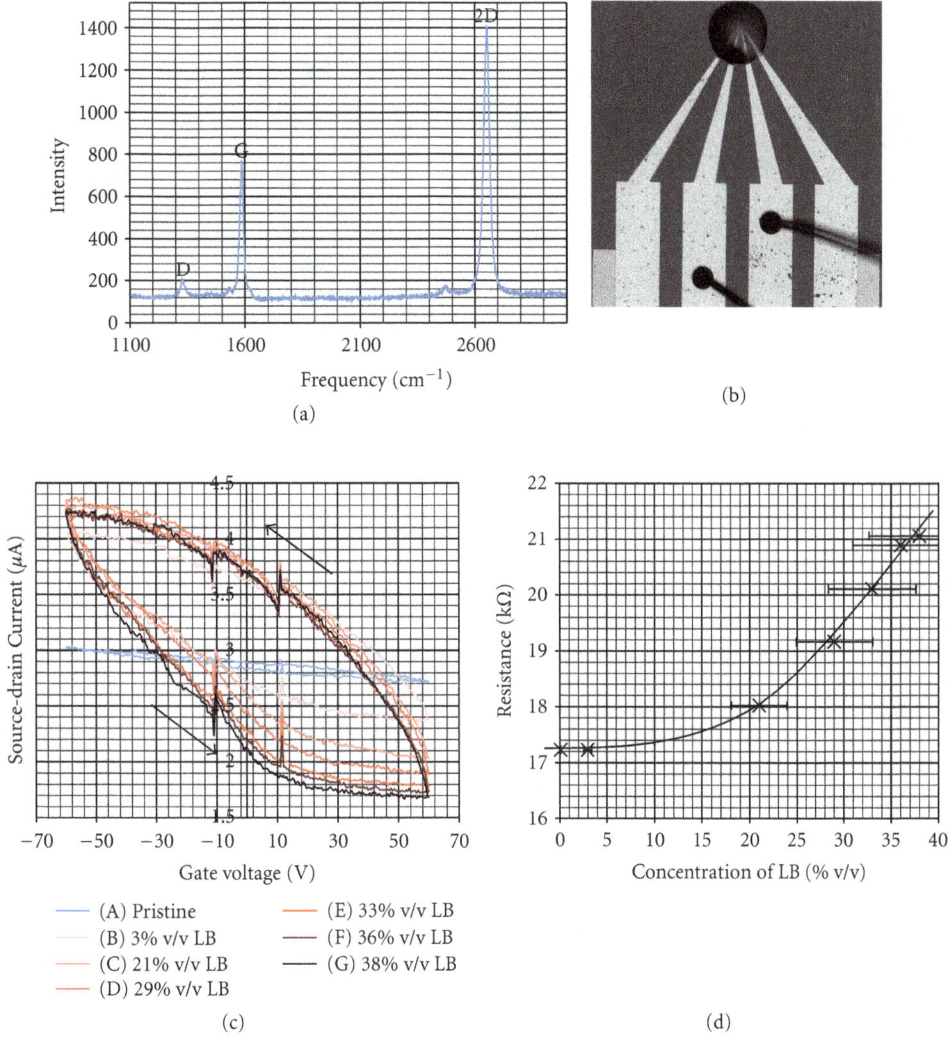

FIGURE 4: (a) Example of the Raman spectrum of CVD-grown graphene after transfer. (b) Controlled deposition of small volumes of LB/glycerol medium is shown to be accurate by optical microscopy. (c) The *p*-type behaviour of the pristine graphene is observed, and a clear increase in hysteresis is noted upon measurement of LB. A shift in gate dependency with LB concentration is also noted. The glycerol is added to reduce droplet evaporation and is maintained at the same concentration in each case. (d) A change in resistance with LB concentration is noted using two-probe measurements on the graphene FET. Error bars are calculated based on droplet repeatability findings by Lukacs et al. [19].

[15]. An example of the Raman spectrum of CVD growth of graphene on copper that was transferred to SiO_2 is shown in Figure 4(a). The G and 2D bands are clearly visible. The small bandwidth and the high 2D/G ratio are indicative of single-layer graphene [16]. A small D-band is also observed around $1350\,cm^{-1}$ indicating some defects/disorder present in our samples. Also, the unintentional doping of graphene due to the local environment or the substrate is a known phenomenon [17] and an observed *p*-type gate dependence of graphene is often attributed to this environmental factor [18]. This *p*-type behaviour is indicated in our results for pristine graphene shown in Figure 4(c), while no distinct Dirac point is found in the given gate voltage range. To understand how this modified graphene behaviour translates to a liquid sensing environment, a 32 times diluted LB medium (as described in Section 2) was dosed onto a sensor device using a microdispensing inkjet tool, as shown in Figure 4(b). To increase the LB concentration in this environment, additional drops were subsequently added containing a more concentrated solution (twice diluted LB). After each step change in concentration, the samples were transported to a probe station for electrical measurement as noted in Section 2.

A set of results using this method is presented in Figure 4(d). The increase in resistance with LB concentration is again noted. While intuitively, the inclusion of ions in solution would lead to a decrease in solution resistance, the observed increase in resistance with solution concentration is tentatively assigned to a charge transfer of negative charge from the LB solution to the graphene, counterbalancing its pristine *p*-type behaviour. This is consistent with the behaviour noted earlier for the dried LB scenario where there

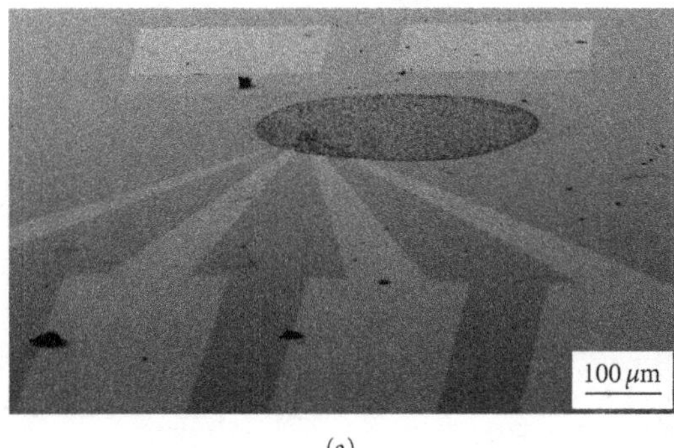

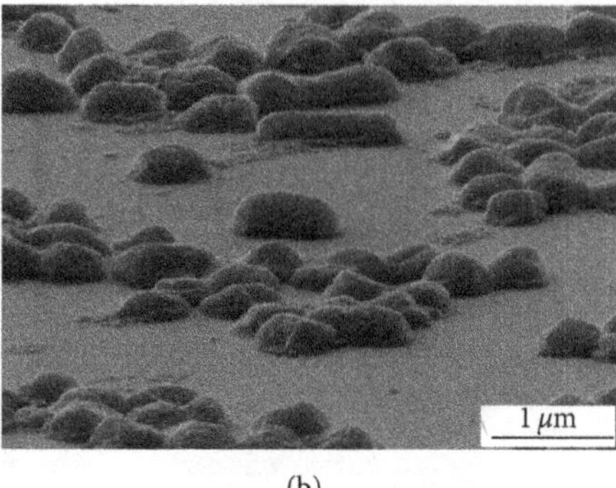

(a) (b)

Figure 5: Microdispensing of 50% LB/25% glycerol medium containing *E. coli* shows it is feasible to guide proliferation to the graphene sensor region. (a) An SEM image shows the graphene sensor and contacts at 80° tilt with a dense circular pattern of adhered bacteria fixed using a dehydration protocol and shown in (b) at a higher magnification.

is a shift in the Dirac point towards a negative gate voltage (Figure 2). As the result of this balance of charges, the system moves closer to neutrality and exhibits lower conductivity. Unlike single-molecule detection studies, the complexity of the medium does not allow a detailed interpretation. Each salt and biomolecule will contribute to the charge transfer in a different way, as will the balance struck between molecules converted to bacterial biomass and those excreted during proliferation. However, this initial approach allows insight into the potential sensitivity to changes in medium concentration.

With this proven ability to sense changes in a complex liquid system on top of graphene FET arrays, it was essential to show the capability to deliver *in situ* and localised bacteria proliferation. Through experiments microdispensing *E. coli* to sensor devices and comparing a device in an ambient atmosphere where the droplet is allowed to dry and a device maintained in an incubator for 1 day we show (i) survival of the *E. coli* through the dispensing protocol, (ii) no obvious ill effects of the substrate or possible contaminants from fabrication processes, and (iii) bacteria proliferation around the sensor area. Figure 5(a) shows an SEM image showing the dense drying pattern surrounding a graphene FET device made up of the *E. coli* shown in more detail by the second SEM image, see Figure 5(b). Imaging of the control and incubated samples shows an increase in cells/unit area by a factor of 8.6 over the course of 1 day. Fixation and drying steps, described in Methods, have been introduced to allow high-resolution images of cells in incubated samples.

4. Future Work and Conclusions

An optimised system for *in situ* electrical measurements is currently being developed. This has led to a system where both control and bioactive samples are dispensed within a single array and electrically analysed over the course of the experiment. Ongoing work is focusing on minimising background signals, optimising the device fabrication for high humidity environments and defining necessary sample concentrations to ensure a change within the detectable limits. As noted earlier, the rapid diffusion through such small volumes and the sensitivity of graphene-based sensors are expected to ensure short batch times. Future work will include cell growth on a nanoscale agar-coating layer directly over the graphene device. This will reduce the sample volume significantly and is expected to ensure a rapid measurement of bacteria proliferation. More fundamental work is required to isolate the influences of each component in the LB medium to quantify the contributing factors to the measurements as the nutrition is converted to biomass. During the initial rapid-growth stage of the bacterial proliferation process that we are targeting, the cells grow at the maximum rate for a given medium using the nutrition to form bacterial biomass. While in this initial work we assume this to be the main influence on changes to the solution properties, it is known that a range of metabolites are also produced and excreted in this stage. It is well understood from the fermentation industry that once the proliferation slows, the production of metabolites also changes leading to a variation in pH [20]. In fact the biomass can be estimated by the progressive change in pH this process causes [21]. Planned work also includes examination of additional influences from biomolecules excreted from the proliferating bacteria.

We have shown the feasibility of a label-free micron-scale graphene sensor array, fabricated with standard and scalable technologies, for monitoring the change in concentration of the nutritive medium used to promote bacteria proliferation. The introduction of different concentrations of nutritive medium could immediately be analysed with this novel electrochemical sensing method, and clear shifts in conductivity were detected in the liquid environment. The device has been tested in liquid and a surrounding humid environment and shows minimal drift that can in future be accounted for with additional control sensors. It is believed that the ability to microfabricate the sensors towards the length scale of individual bacteria will in future allow the targeting of just a few of the organisms, thus providing very specific and quantitative data. We have developed a fabrication, microdispensing, and analysis protocol for this novel sensing approach and demonstrated that the indirect measurement technique will be suitable for inexpensive, re-usable, and rapid diagnostic tools.

Acknowledgments

This work was supported by the Science Foundation Ireland under the CSET scheme SFI08/CE/I1432. The SEM imaging was enabled by the Advanced Microscopy Laboratory, Trinity Technology and Enterprise Campus, Dublin 2, Ireland under the framework of the INSPIRE programme, funded by the Irish Government's Programme for Research in Third Level Institutions, Cycle 4, National Development Plan 2007–2013. GSD acknowledges SFI for the PICA grant and SK the Embark Initiative for an IRCSET scholarship.

References

[1] K. Y. Gfeller, N. Nugaeva, and M. Hegner, "Rapid biosensor for detection of antibiotic-selective growth of Escherichia coli," *Applied and Environmental Microbiology*, vol. 71, no. 5, pp. 2626–2631, 2005.

[2] D. Ramos, J. Tamayo, J. Mertens, M. Calleja, and A. Zaballos, "Origin of the response of nanomechanical resonators to bacteria adsorption," *Journal of Applied Physics*, vol. 100, no. 10, Article ID 106105, pp. 106105-1–106105-3, 2006.

[3] N. Nugaeva, K. Y. Gfeller, N. Backmann, H. P. Lang, H. J. Güntherodt, and M. Hegner, "An antibody-sensitized microfabricated cantilever for the growth detection of Aspergillus niger spores," *Microscopy and Microanalysis*, vol. 13, no. 1, pp. 13–17, 2007.

[4] N. Kim and I. S. Park, "Application of a flow-type antibody sensor to the detection of Escherichia coli in various foods," *Biosensors and Bioelectronics*, vol. 18, no. 9, pp. 1101–1107, 2003.

[5] X. L. Su and Y. Li, "A QCM immunosensor for Salmonella detection with simultaneous measurements of resonant frequency and motional resistance," *Biosensors and Bioelectronics*, vol. 21, no. 6, pp. 840–848, 2005.

[6] C. Faber, "Assessment of the inhibitory potency by MRI," in *NMR Spectroscopy in Pharmaceutical Analysis*, U. Holzgrabe et al., Ed., Elsevier, 2008.

[7] W. Yang, K. R. Ratinac, S. R. Ringer, P. Thordarson, J. J. Gooding, and F. Braet, "Carbon nanomaterials in biosensors: Should you use nanotubes or graphene," *Angewandte Chemie—International Edition*, vol. 49, no. 12, pp. 2114–2138, 2010.

[8] F. Schedin, A. K. Geim, S. V. Morozov et al., "Detection of individual gas molecules adsorbed on graphene," *Nature Materials*, vol. 6, no. 9, pp. 652–655, 2007.

[9] Y. Ohno, K. Maehashi, Y. Yamashiro, and K. Matsumoto, "Electrolyte-gated graphene field-effect transistors for detecting ph and protein adsorption," *Nano Letters*, vol. 9, no. 9, pp. 3318–3322, 2009.

[10] N. Mohanty and V. Berry, "Graphene-based single-bacterium resolution biodevice and DNA transistor: Interfacing graphene derivatives with nanoscale and microscale biocomponents," *Nano Letters*, vol. 8, no. 12, pp. 4469–4476, 2008.

[11] P. N. Nirmalraj, T. Lutz, S. Kumar, G. S. Duesberg, and J. J. Boland, "Nanoscale mapping of electrical resistivity and connectivity in graphene strips and networks," *NanoLetters*, vol. 11, pp. 16–22, 2011.

[12] X. Li, W. Cai, J. An et al., "Large-area synthesis of high-quality and uniform graphene films on copper foils," *Science*, vol. 324, no. 5932, pp. 1312–1314, 2009.

[13] S. Kumar, N. Peltekis, K. Lee, H. Kim, and G. S. Duesberg, "Reliable processing of graphene using metal etchmasks," *Nanoscale Research Letters*, vol. 6, no. 1, p. 390, 2011.

[14] J. S. Bunch, S. S. Verbridge, J. S. Alden et al., "Impermeable atomic membranes from graphene sheets," *Nano Letters*, vol. 8, no. 8, pp. 2458–2462, 2008.

[15] N. Peltekis, S. Kumar, N. McEvoy, K. Lee, A. Weidlich, and G. S. Duesberg, "The effect of downstream plasma treatments on graphene surfaces," *Carbon*, vol. 50, no. 2, pp. 395–403, 2012.

[16] A. C. Ferrari, J. C. Meyer, V. Scardaci et al., "Raman spectrum of graphene and graphene layers," *Physical Review Letters*, vol. 97, no. 18, Article ID 187401, 2006.

[17] Y. H. Wu, T. Yu, and Z. X. Shen, "Two-dimensional carbon nanostructures: fundamental properties, synthesis, characterization, and potential applications," *Journal of Applied Physics*, vol. 108, no. 7, Article ID 071301, 2010.

[18] H. E. Romero, N. Shen, P. Joshi et al., "n-type behavior of graphene supported on Si/SiO2 substrates," *ACS Nano*, vol. 2, no. 10, pp. 2037–2044, 2008.

[19] G. Lukacs, N. Maloney, and M. Hegner, "Ink-jet printing: perfect tool for cantilever array sensor preparation for microbial growth detection," *Journal of Sensors*. In press.

[20] M. Scheidle, B. Dittrich, J. Klinger, H. Ikeda, D. Klee, and J. Buchs, "Controlling pH in shake flasks using polymer-based controlled-release discs with pre-determined release kinetics," *BMC Biotechnology*, vol. 11, p. 25, 2011.

[21] M. L. Christensen and N. T. Eriksen, "Growth and proton exchange in recombinant Escherichia coli BL21," *Enzyme and Microbial Technology*, vol. 31, no. 4, pp. 566–574, 2002.

15

Plastic Optical Fibre Sensing of Fuel Leakage in Soil

Masayuki Morisawa and Shinzo Muto

Department of Information and Communication System, Graduate School of Medicine and Engineering, University of Yamanashi, 4-3-11 Takeda, Kofu 400-8511, Japan

Correspondence should be addressed to Masayuki Morisawa, morisawa@yamanashi.ac.jp

Academic Editor: Kevin Chen

A basic operation of the very simple optical sensing system of fuel leakage in uniform sandy and clayey soils, which is consisting of a plastic optical fibre (POF) transmission line, the POF-type sensor heads, and a single LED photodiode pair, has been studied theoretically and experimentally. Its sensing principle is based on the POF structure change in the sensor head caused by fuels such as petrol. A scale-downed model prepared in the experimental room showed a possibility of optical detection of fuel leakage points in uniform soil. As this system operates without receiving the influence of water containing in fuels and soils, its application to fuel leak monitor around a filling station and oil tank can be expected.

1. Introduction

At present, various fuels such as petrol, kerosene, and light oil are used in many fields to make our life comfortable. However, leakage of fuels produces many serious problems such as pollution in soil or underwater, explosion, and combustion in a filling station or oil tank. To prevent such serious problems, quick and safety detection of fuel leakage point is strongly required. Of course, it goes without saying that an optical sensing is the safest method [1, 2], and, hitherto, several optical sensing systems such as an optical time domain reflectometry (OTDR) and a multipoint sensor system have been studied [3, 4]. However, considering the practical application around a filling station and oil tank, more simple and low-cost ones are desired. From this background, we made an attempt to develop a very simple and low-cost optical sensing system to detect leakage of fuel in soil. Therefore, we mainly used the POF because it has many advantages such as low cost, easiness in handling, and immunity to electromagnetic interference [5]. Its system consists of the POF transmission line, several POF-type sensor heads, and a single LED photodiode pair. According to the preliminary study, certain kinds of polymer such as polyisoprene (PIP) cause swelling by attachment of fuel molecules and then change its refractive index remarkably [6]. Based on this phenomenon, the POF-type sensor head to detect petrol was fabricated and tested. Specially, a scale-downed model of the sensing system for detecting fuel leakage in uniform sandy and clayey soils has been studied theoretically and experimentally. This paper reports on its basic property towards development of the practical system.

2. Application Model and Its Operation Principle

The model considered here is shown schematically in Figure 1. As can be seen from this figure, the proposed optical sensing system consists of the POF transmission line, several POF-type sensor heads, and a single LED photodiode pair. In addition, its system assumes that the POF is buried underground but is not subjected to such a large pressure that would deform it considerably by breaking and crushing it to the extent that its original shape and state cannot be restored. This model seems to be suitable for the practical case around a filling station or oil tank. The sensing principle in the POF-type sensor head is based on the swell phenomenon in its cladding layer. When is exposed to vapour phase alkane or petrol, the refractive index of the PIP film with a thickness of about 5 μm decreases from 1.52 to 1.48 as shown in Figure 2. The solid line in Figure 2 shows

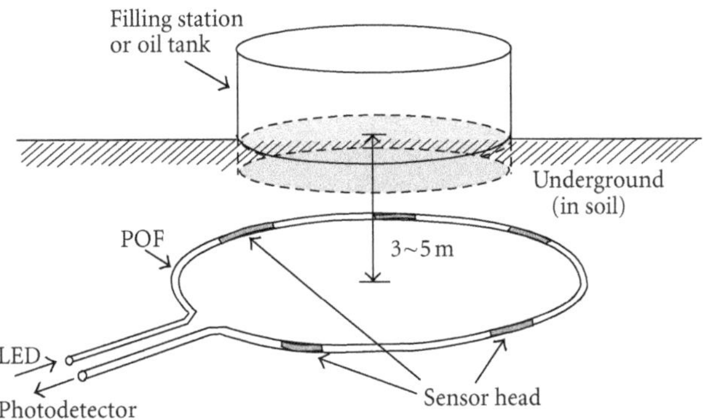

FIGURE 1: Application model of the POF-type sensing system.

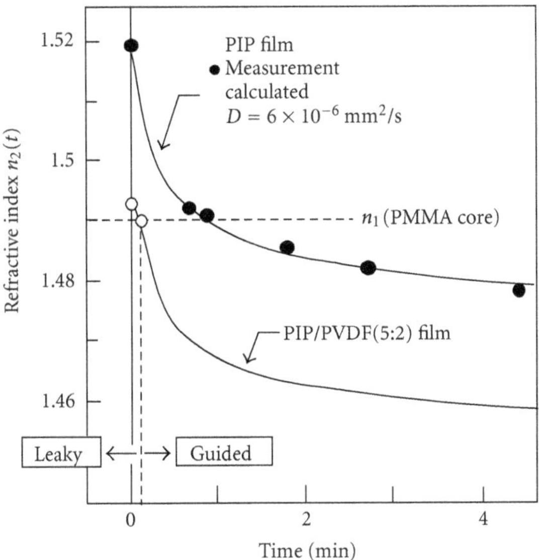

FIGURE 2: Refractive index changes in PIP and PIP/PVDF (5:2) films after exposing to hexane vapour.

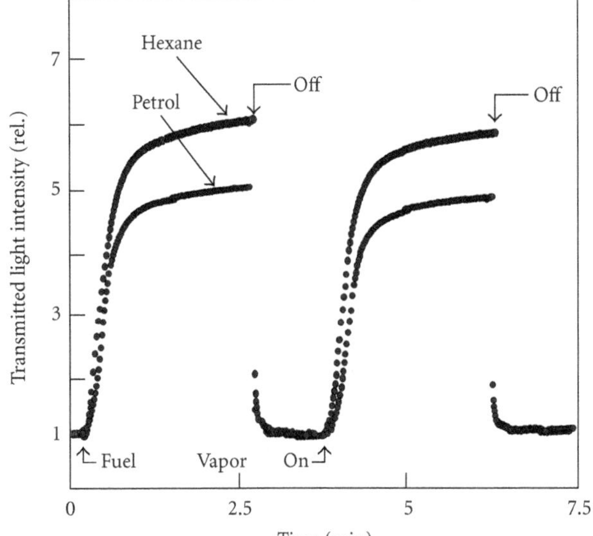

FIGURE 3: Response of POF-type sensor head with the PIP/PVDF (5:2) cladding layer to hexane and petrol vapours.

the theoretical change in the refractive index $n(t)$, which is calculated using a following one-dimensional diffusion equation for hexane concentration $N(t)$ in the PIP film with a diffusion coefficient of $D = 6 \times 10^{-6}$ mm/s and a relation between $N(t)$ and $n(t)$ [7, 8]:

$$\frac{\partial N(x,t)}{\partial t} = D\frac{\partial^2 N(x,t)}{\partial x^2}. \tag{1}$$

Assuming the initial condition of $N(x,0) = 0$ and $N(0,t) = N_0$, the solution of above one dimensional diffusion equation is easily obtained as follows:

$$N(x,t) = N_0\left(1 - \frac{2}{\sqrt{\pi}}\int_0^{x/(2\sqrt{Dt})} e^{-\alpha^2}\,d\alpha\right). \tag{2}$$

And then, as the density in the swelling film decreases with increasing $N(x,t)$, $n(x,t)$ can be expressed by the following equation:

$$n(x,t) = \frac{n_0 n_\infty(N_0)}{(n_0 - n_\infty(N_0))N(x,t)/N_0 + n_\infty(N_0)}, \tag{3}$$

where n_0 is the refractive index at $N(x,0)$ and $n(N_0)$ is the final one at $N(x,\infty) = N_0$.

Considering the POF structure change in the sensor head consisting of the swell polymer clad and polymethylmethacrylate (PMMA) core with a refractive index of $n_1 = 1.490$, the initial refractive index (n_2) in the cladding layer must be set at a slightly lager value than n_1 in the fibre core [8]. Therefore, the mixture of the PIP and polyvinylidenfluoride (PVDF) with a ratio of 5 to 2 was considered as a sensitive cladding layer. The bare PMMA core was obtained by removing the cladding layer from the commercialized POF without a jacket (Mitsubushi-Rayon, ESKA), by mean of organic solvent such as 1,4-dioxane. About 15 minutes immersion at 20°C and wiping off the dissolved cladding polymer using a soft tissue made us easily obtain the bare PMMA core with a smooth surface. The PIP/PVDF (5:2) cladding polymer was dissolved in dimethylsulfoxide (DMSO) and was dip-coated on the bare PMMA core with a diameter of about 1 mm. Its thickness is about 5 μm. According to these conditions, the repeatability of the sensor

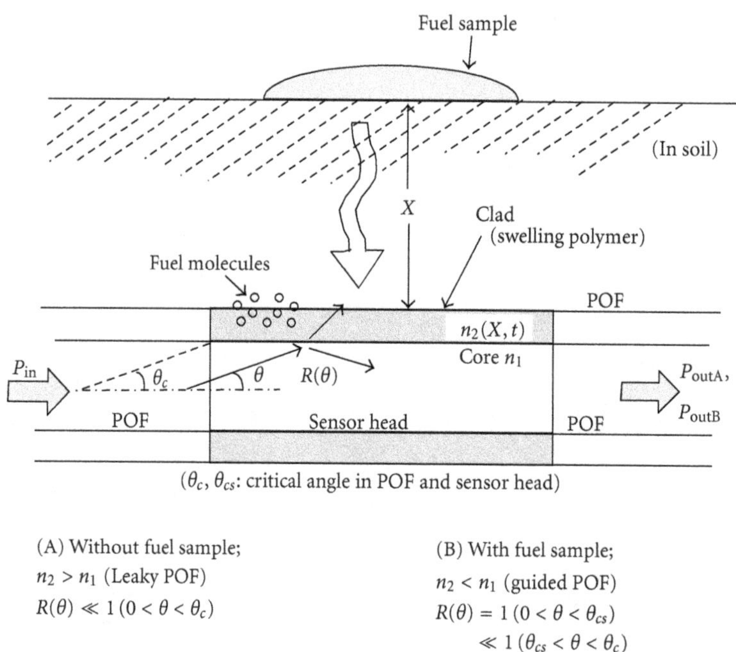

FIGURE 4: POF sensing model to detect fuel leakage in uniform soil. P_{outA} is the output light intensity, and P_{outB} is that one with fuel sample.

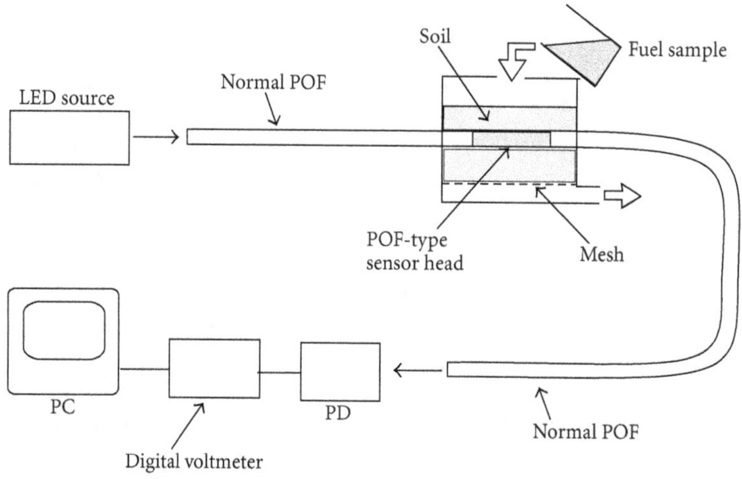

FIGURE 5: Experimental setup.

fabrication becomes very high. When was exposed to hexane vapour, its refractive index in cladding layer changed from 1.492 to 1.455, as also shown in Figure 2. From this result, it is clear that the POF-type sensor head changes quickly its structure from a leaky type to a guided one.

First, the sensor head with a length of 50 mm was tested to confirm its operation even in the wavelength region with a large propagation loss. So a blue LED was used as a light source. Figure 3 shows the results for vapour phase hexane and petrol. As can be seen from this figure, a large change in transmitted light intensity based on the POF structure change from leaky type to guided one was observed experimentally. In addition, the proposed sensor structure gave high sensitivity and good reproducibility for both petrol and hexane. Referring to these results, the sensor head with a length of 10 mm was connected to the normal POFs at the input and output edges and was set at the position $x = 55$ mm in uniform sandy soil or clayey soil as shown in Figure 4. Here, the light reflection at the clad-soil boundary was neglected to simplify the analysis. In this figure, for the case of (a) without fuel sample, the output light intensity P_{outA} is small because the power reflection coefficient $R(\theta)$ at the core-clad boundary becomes a very small value [8]. On the other hand, for the case of (b) with fuel sample, the output light intensity P_{outB} increases remarkably because of its change to the guided structure. Then, the value of P_{outB}/P_{outA} gives the change in transmitted light intensity, that is, the sensitivity. Its property

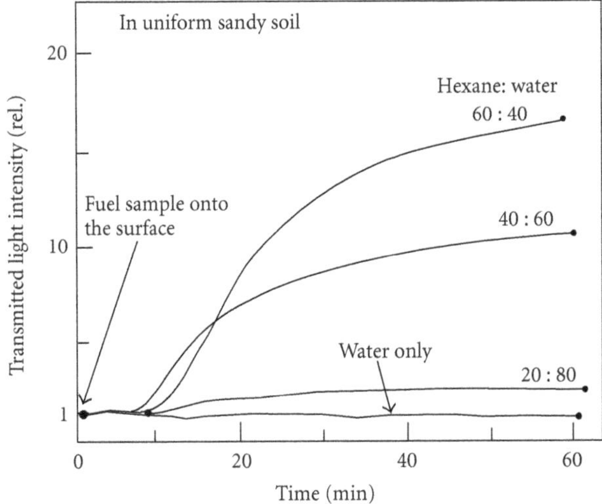

FIGURE 6: Sensor response observed in uniform sandy soil.

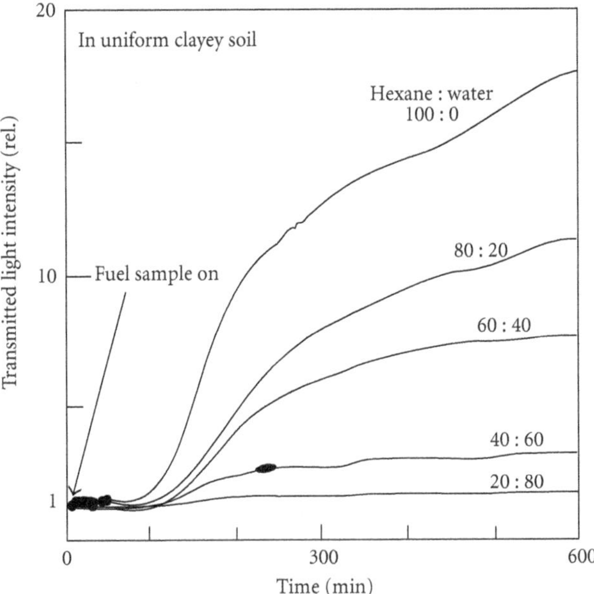

FIGURE 7: Sensor response observed in uniform clayey soil.

was measured using a photodetector, and its electrical signal was fed to a digital voltmeter connected to a computer for real-time data processing, as shown in Figure 5. In this experiment, to check an influence of water, the mixture of hexane and water was used as the fuel sample [7]. The obtained sensor responses are shown in Figures 6 and 7, respectively. From these figures, it is clear that the change in transmitted light intensity increases depending on the fuel concentration without receiving the influence of water and that diffusion of the fuel sample in clayey soil takes a much longer time. Furthermore, the diffusion coefficient of fuel sample in uniform soil was estimated from the start time of increase in the transmitted light intensity. For example, its value in uniform sandy soil used here was decided to be 0.7 mm^2/s. Using this value and a ray tracing method

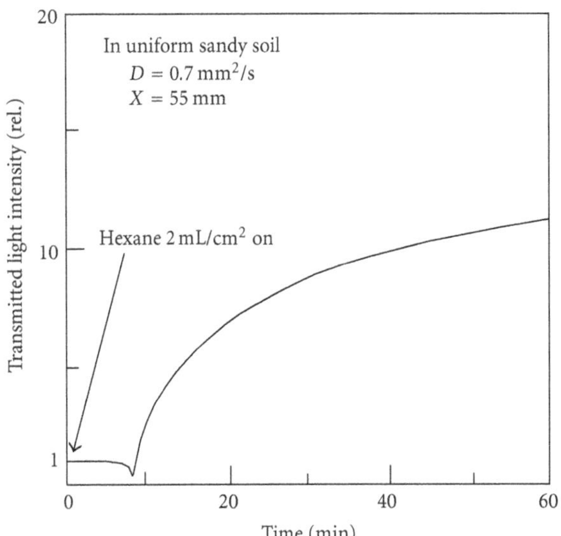

FIGURE 8: Calculated sensor response in uniform sandy soil.

[8], the change in transmitted light intensity through the sensor head can also be calculated theoretically. Its result is shown in Figure 8. As is clear from the figure, excepting the small dip at the critical point caused by the neglect of light reflection at the clad-soil boundary, the calculated sensor response coincided qualitatively with the experimental ones shown in Figure 6. In addition, as the similar operation was confirmed in the case with a long sensor head of 120 mm, a series connection of several sensor heads with a total length below 120 mm seems to be possible.

3. Detection Method of Fuel Leakage Point

Considering a practical application of the above sensor, detection of the leakage point is needed. Therefore, the POF sensing system with multisensor heads was considered. Its system is shown schematically in Figure 9, in which, the upper configuration is type (a) with equal sensor head lengths of 10 mm and the lower one is type (b) with different sensor head lengths of 30 mm, 20 mm, and 10 mm, respectively. This model also means a scale-downed sensor model of the fuel leakage point around a filling station or oil tank shown in Figure 1. However, to shorten the experiment time, these sensor heads were buried in uniform light sandy soil with a diffusion coefficient of $D_1 = 80\ mm^2/s$ and at 120 mm in depth. The value of D_1 was estimated experimentally using the same method described in Section 2. Furthermore, considering the directional diffusion of petrol in uniform light sandy soil, these values of D_2 and D_3 were also estimated from the value of D_1 to be 30 mm^2/s and 10 mm^2/s, respectively. Under these conditions, a certain amount of petrol was poured onto the surface points marked with the closed circles of A, B, and C, respectively. To simplify the analysis, we assumed that fuel leakages from each point do not occur at the same time. Then, we can consider the above system as a cascade connection of three sensor heads. The calculated transmitted light intensity changes for each

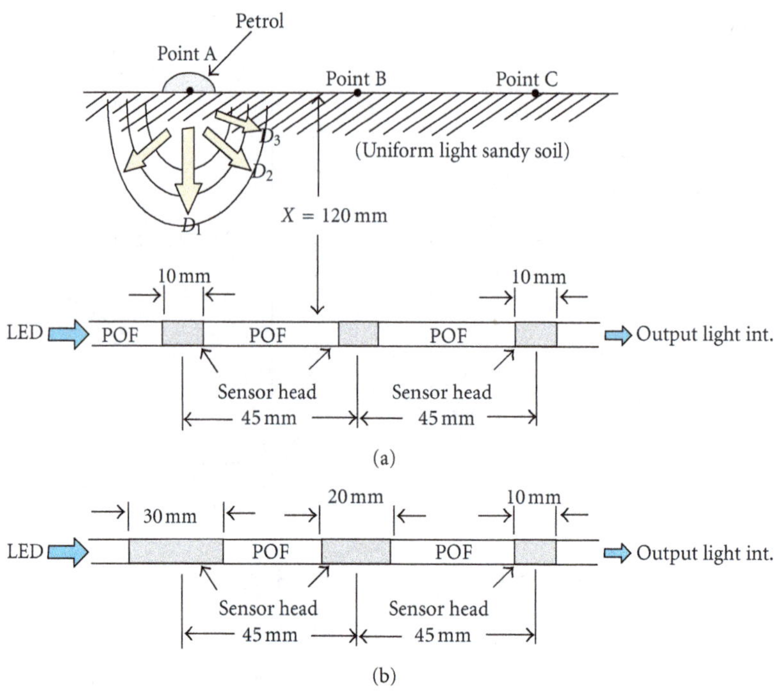

Figure 9: Three sensor heads system to detect petrol leakage point in uniform light sandy soil. (a) With equal sensor head lengths of 10 mm and (b) with sensor head lengths of 30 mm, 20 mm, and 10 mm.

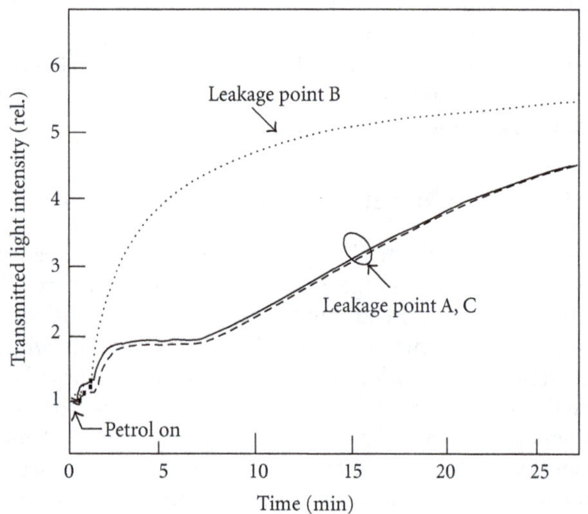

Figure 10: Calculated output light intensity change in type (a) for each leakage point.

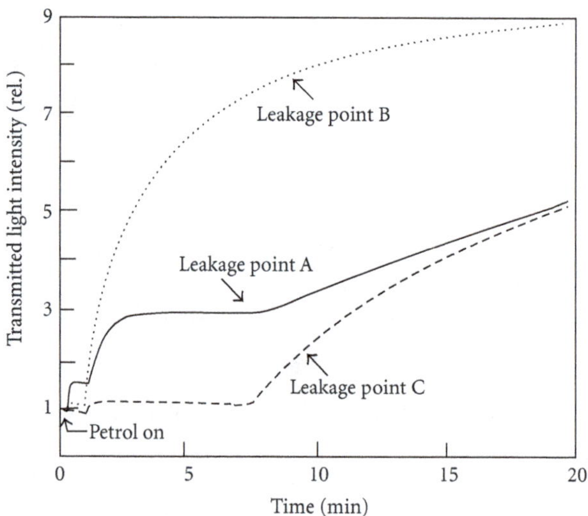

Figure 11: Calculated output light intensity change in type (b) for each leakage point.

leakage point A, B, and C are shown in Figures 10 and 11, respectively. From these figures, it is found that the change in light intensity for the leakage point A or C becomes small in the early stage but that one for the fuel leakage point B increases remarkably. The later property is due to that, a few minute after the petrol reaching to the second sensor head, the first and the third sensor heads operate as a guided POF at almost same time. In addition, the difference of the sensor head length between the first sensor head and third one made clear the change in the light intensity level in the early stage. Judging from these results, the system of type (b) showed a useful property to detect fuel leakage point.

To confirm the above characteristics experimentally, the same configuration of the three sensor heads and measurement setup were used. In this experiment, the transmitted light intensity changes in only early stage were measured because the continuous supply of petrol was difficult. The results obtained are shown in Figures 12 and 13, respectively. As is clear from these figures, the experimental properties in

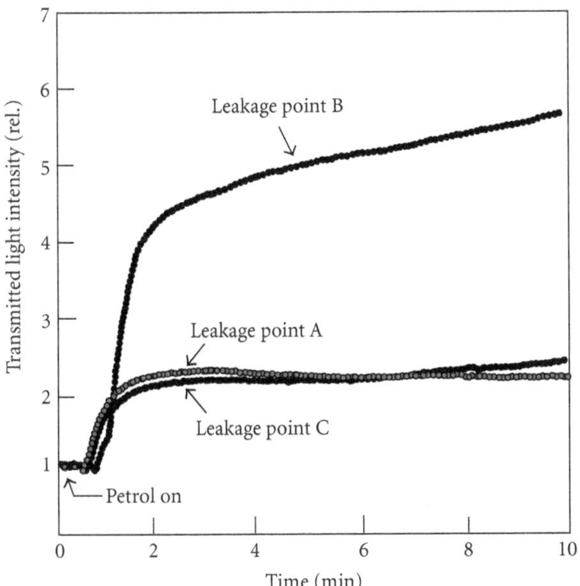

FIGURE 12: Observed light intensity change in early stage for type (a).

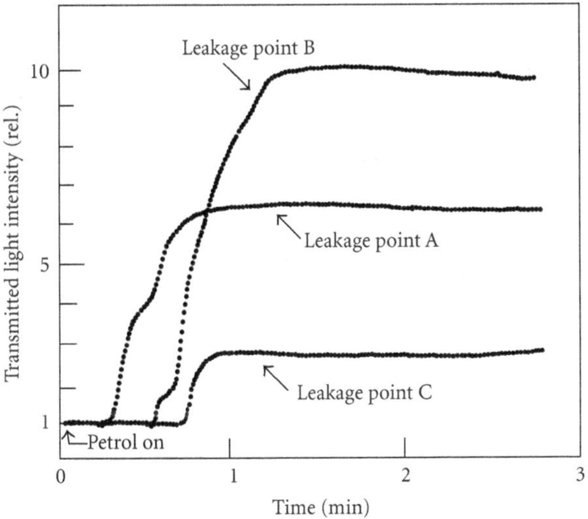

FIGURE 13: Observed light intensity change in early stage for type (b).

early stage of fuel leakage coincided qualitatively with the calculated ones. Thus, a very simple system consisting of the POF transmission line and the POF-type sensor head showed a possibility on optical detection of the fuel leakage point in uniform soil around a filling station and/or oil tank. In addition, this system will be used over whole visible region because it operated even in the blue wavelength. If the fuel leakages from some points occur simultaneously, a complicated change in the transmitted light intensity will be observed. Such a response also gives an important information showing the simultaneous fuel leakage from some points. Furthermore, when considering the practical uses of this sensor, it should be noted that this sensor system depends significantly on the diffusion coefficient of soil into gasoline. Therefore, for practical uses of this sensor, calibration should be performed by collecting a soil sample from the site where this system is set up and measuring the diffusion coefficient of the soil sample. However, when the diffusion coefficient is not uniform, the sensor response may become complicated. Therefore, we wish to treat them theoretically and experimentally in the near future.

4. Conclusions

The very simple POF-type optical sensing system of fuel leakage in uniform soils, which is based on the POF structure change in the sensor head produced by a swell phenomenon in the cladding layer, has been studied using a scale-downed model. Its theoretical and experimental results showed that a quick and safety detection of petrol leakage in uniform soil becomes possible without receiving an influence of water included in fuel and soil. Although this system is not suitable for the distributed leakage detection over a long distance, it seems to be used for a few point detection around a filling station and/or oil tank. Referring to these results, we are now studying the improved system which will be able to use for detecting fuel leakage from many points. Its result we will be reported in the near future.

References

[1] J. Darkin and B. Calshaw, *Optical Fiber Sensors*, vol. 4, ch 17, Artech, Boston, Mass, USA, 1997.

[2] M. Archenault, H. Gagnaire, J. P. Goure, and N. Jaffrezic-Renault, "A simple intrinsic optical-fibre chemical sensor," *Sensors and Actuators B*, vol. 8, no. 2, pp. 161–166, 1992.

[3] K. I. Aoyama, K. Nakagawa, and T. Itoh, "Optical time domain reflectometry in a single-mode fiber," *IEEE Journal of Quantum Electronics*, vol. 17, no. 6, pp. 862–868, 1981.

[4] P. A. Wallace, N. Elliott, M. Uttamlal, A. S. Holmes-Smith, and M. Campbell, "Development of a quasi-distributed optical fibre pH sensor," in *Proceedings of the 14th International Conference on Optical Fiber Sensors (OFS '00)*, pp. 456–459, October 2000.

[5] A. Weinert, *Plastic Optical Fibers*, ch 2, Publics MCD, Munich, Germany, 1999.

[6] S. Muto, K. Uchiyama, G. Vishnoi et al., "Plastic optical fiber sensors for detecting leakage of alkane gases and gasoline vapors," in *Photopolymer Device Physics, Chemistry, and Applications IV*, vol. 3417 of *Proceedings of SPIE*, pp. 61–69, July 1998.

[7] D. Matsuyama, M. Morisawa, C. X. Liang, and S. Muto, "Plastic Optical Fiber Sensing of Leakage of Fuel in Soil," in *Technical Digest of 16th International Conference on Optical Fiber Sensors (OFS-16)*, Proceedings of SPIE, p. 618, 2003.

[8] M. Morisawa, Y. Amemiya, H. Kohzu, C. X. Liang, and S. Muto, "Plastic optical fibre sensor for detecting vapour phase alcohol," *Measurement Science and Technology*, vol. 12, no. 7, pp. 877–881, 2001.

16

Supervised Expert System for Wearable MEMS Accelerometer-Based Fall Detector

Gabriele Rescio, Alessandro Leone, and Pietro Siciliano

Institute for Microelectronics and Microsystems, Italian National Research Council (CNR), Via Monteroni, c/o Campus Università del Salento, Palazzina A3, 73100 Lecce, Italy

Correspondence should be addressed to Gabriele Rescio; gabriele.rescio@le.imm.cnr.it

Academic Editor: Andrea Cusano

Falling is one of the main causes of trauma, disability, and death among older people. Inertial sensors-based devices are able to detect falls in controlled environments. Often this kind of solution presents poor performances in real conditions. The aim of this work is the development of a computationally low-cost algorithm for feature extraction and the implementation of a machine-learning scheme for people fall detection, by using a triaxial MEMS wearable wireless accelerometer. The proposed approach allows to generalize the detection of fall events in several practical conditions. It appears invariant to the age, weight, height of people, and to the relative positioning area (even in the upper part of the waist), overcoming the drawbacks of well-known threshold-based approaches in which several parameters need to be manually estimated according to the specific features of the end user. In order to limit the workload, the specific study on posture analysis has been avoided, and a polynomial kernel function is used while maintaining high performances in terms of specificity and sensitivity. The supervised clustering step is achieved by implementing an one-class support vector machine classifier in a stand-alone PC.

1. Introduction

The problem of falls in the elderly has become a health care priority due to the related high social and economic costs [1]. In fact the European population aged 65 years or more, which may be in need of assistance is increasing. This trend asks care-holders institutions to employ more efficient and optimized methods in order to be able to grant the required service at lower costs. The consequences of falls in the elderly may lead to psychological trauma, physical injuries, hospitalization, and even death in the worst scenario [2–5]. The main reason that pushed for the development of the presented system is to allow noncompletely self-sufficient people (e.g., older people) to live safely in their own houses as long as possible. This is important not only for aspects of health regarding assisted people, but also for the consequent social advantages. The European community issued and funded various projects and consortia. The mission focuses on several purposes, all addressed to older people, varying from the assistance in case of need, to the prevention of dangerous or unhealthy situations. The purpose of the work described in this paper is to focus on people fall detection.

Many solutions have been proposed in the detection and prevention of falls, and some excellent review studies were presented [1, 6]. Basically, fall-detection solutions can be classified in three main classes: wearable devices, ambient devices, and camera-based devices. The first approach requires that the elderly holds some kind of devices (e.g., an assistive cane) or wears sensors like accelerometers and/or gyroscopes to detect the motion of the body. In particular, recent miniaturization and cost reduction of MEMS accelerometers and the availability of reliable wireless communication technologies enabled the realization of affordable wearable monitoring systems that can be worn by people performing their normal daily activities [7–10]. For these reasons, in the last few years, the use of portable devices in the health monitoring of chronic patients has increased considerably. However, these devices have some drawbacks: they are prone to be forgotten, worn in a wrong body position, or accidentally damaged. Regarding fall detection, with respect to vision or acoustic sensors, the accelerometer module has the advantage of not having to be set up and installed in all rooms of the "smart home," as it is required for instance for 3D video trackers or acoustic scene analyzers. On the other hand,

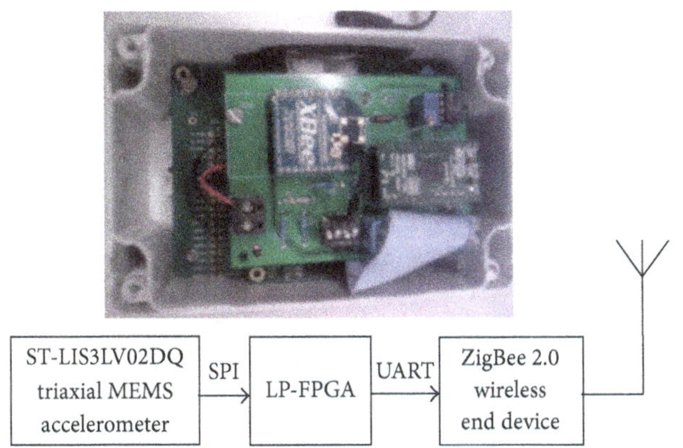

FIGURE 1: Picture and block diagram of the wireless accelerometer-based device.

the camera-based approach is a nonintrusive solution, since the sensor is a camera installed on a wall and is able to detect falls, switching off when the monitored person is no longer alone [11]. On the other hand, camera-based monitoring solutions suffer when large occluding objects (i.e., furniture) obstruct camera's viewing. Another significant drawback of any camera-based solution is the violation of monitored person's privacy.

In this paper a fall detector through a wearable triaxial MEMS accelerometer is presented. The proposed solution overcomes the limitation of well-known threshold-based approaches [12–17] for which the accelerometer-based device need, to be tuned in an appropriate way for each installation (i.e., the parameters setup may be different for different people). For these reasons, a machine-learning scheme [18] is used, showing high generalization capabilities in the fall detection discrimination process. The expert system uses robust features extracted taking into account important constraints and requirements of mobile solutions (workload). The extracted features are (quasi-)invariant both to specific characteristics of the mounting setup (device on chest, on waist, and on abdomen) and specific characteristics of the end users in terms of age, weight, height, and gender.

2. Materials and Methods

2.1. Triaxial Accelerometer Sensor System. The hardware used is a wearable device composed by commercial discrete circuits, according to the design proposed in [19]. The picture and the logical block diagram are shown in Figure 1. The system integrates an ST LIS3LV02DL tri-axial MEMS accelerometer with digital output, an FPGA for computing functionalities, and a ZigBee module for wireless communication up to 30 m. The power consumption is about 190 mW in streaming mode and 9 mW in idle.

The wearable device can operate in streaming (raw data are sent via ZigBee to an external computing platform for data analysis with a 10 Hz frequency) or in standalone modality, by running the threshold-based fall detection implementation on the on-board FPGA (the power consumption is limited since the ZigBee module is activated just when a fall event occurs). The LIS3LV02DL MEMS accelerometer is DC coupled, and it responds up to 0 Hz, with 16-bits resolution and a full scale in the range $\pm 2g$. Data can be transmitted in hexadecimal format. The sensor measures both static and dynamic accelerations along the 3 axes and allows one to receive information on the 3D spatial relative position (compared to the Earth gravity vector) of the person who wears it.

In stationary conditions, assuming a particular axis, the component of the acceleration (amplitude A, rif. (1)) is defined according to the value of the sine of the angle α between the considered axis and the horizontal plane, which is perpendicular to the Earth gravity component (g):

$$A = g \sin(\alpha). \quad (1)$$

In this way, if the accelerometer relative orientation is known, the resulting data can be used to determinate the angle of the user posture respect to the vertical direction.

2.2. Simulated Falls and ADL Tasks. In order to analyse the waveforms along each axis in the presence of falls and other kind of events (Activities of Daily Living, ADLs), a data collection has been defined in controlled (simulated) conditions by involving 11 healthy male and 2 healthy female volunteers. The simulated falls were performed by using a crash mat (height 2 cm) and knee/elbow pad protectors, meeting safety and ethical requirements. The range of actors age was 39.3 ± 12.3 years, weighing 73.7 ± 13.4 kg, and a height of 1.76 ± 0.1 m. 450 actions were simulated in which 250 were falls compliant to the specifications proposed by Noury et al. [20]. The following falls have been simulated for study:

(1) backward falls ending in the lying position,

(2) backward falls with recovery,

(3) forward falls ending in lying flat,

(4) forward falls with recovery,

(5) lateral falls.

During the data collection the wearable device was placed with an elastic band in a different positioning area on the upper part of the torso (on the chest, on the waist, and on the abdomen). Some ADLs were simulated in order to evaluate the ability of fall-detection algorithms to discriminate falls from ADLs. The simulated ADL tasks belong to the following categories:

(1) walking,

(2) sitting down on a chair and then standing up,

(3) lying down on a mat (height 33 cm) and then standing up,

(4) lying down on a mat (height 2 cm) and then standing up,

(5) kneeling on a mat (height 2 cm) and then standing up.

Each actor performed more than 15 simulated ADLs for a total of 205.

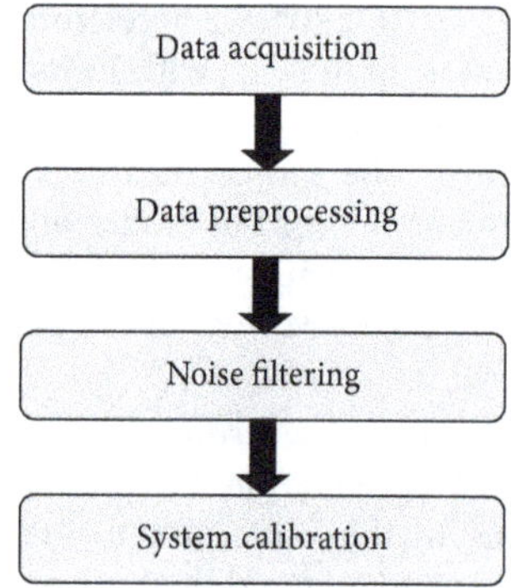

FIGURE 2: Logical framework overview.

2.3. Preprocessing and System Calibration. The first four computational steps (Figure 2) deal with data acquisition, data preprocessing, noise filtering, and system calibration.

The acceleration data on three axes (A_x, A_y, and A_z) is read out from the device worn by a user during the data collection. Data was stored into a portable computer and converted into gravitational units to represent acceleration data in the range $\pm 2g$, in order to make it possible to extract the angle α (described previously) and for not having orders of magnitude too different in the features (during the training of the classifier used to detect the fall events). The samples coming from the device are filtered out by a low pass 8 order, 8 Hz cut-off FIR (Finite Impulse Response) filter to reduce the noise due to electronic components, environment, and human tremor.

In order to correctly handle preprocessed data, a calibration procedure was accomplished by recovering the initial conditions after the device mounting. During this step the correct placing of hardware is verified by checking if two acceleration axes are orthogonal to the Earth gravity g (Figure 3): the acceleration values measured on the two orthogonal components must be close to zero.

The calibration procedure is composed by the following steps:

(i) the user wears the device in a standing position for 10 seconds;

(ii) the calibration routine calculates the average of the acceleration on each axis over this period. These are the initial acceleration values A_{x0}, A_{y0}, and A_{z0};

(iii) if A_{x0}, A_{y0}, and A_{z0} will be close to those expected, they will be considered as references in the fall detector algorithm, otherwise a routine to compensate the sensed misplacement will be enabled.

With respect to Figure 3, the acceleration along x-component will be close to -1 and almost zero for the others according to (1). The expected values of acceleration on three axes A_{x0}, A_{y0}, and A_{z0} are also reported in Table 1.

In practice, it is difficult to locate the device exactly in the right position, and the measured acceleration values will differ slightly from those expected. The calibration procedure is concluded if the following conditions are satisfied at the same time (and the values A_{x0}, A_{y0}, and A_{z0} will be recorded and used as references in the feature extraction phase):

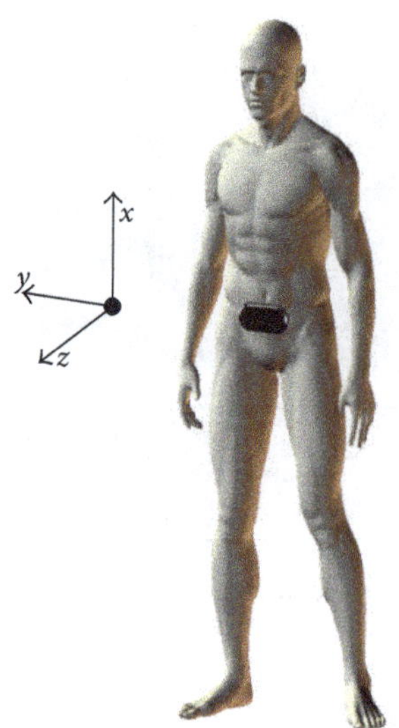

FIGURE 3: Mounting position and calibration.

TABLE 1: Acceleration values along the three axes during the initial position of Figure 3.

	A_{x0}^{ref}	A_{y0}^{ref}	A_{z0}^{ref}
Initial position	−1	0	0

$$\left|A_{x0} - A_{x0}^{\text{ref}}\right| < 0.3,$$
$$\left|A_{y0} - A_{y0}^{\text{ref}}\right| < 0.3, \quad (2)$$
$$\left|A_{z0} - A_{z0}^{\text{ref}}\right| < 0.3.$$

Since the values (A_{x0}^{ref}, A_{y0}^{ref}, and A_{z0}^{ref}) are (−1, 0, and 0);

$$\left|A_{x0} + 1\right| < 0.3, \qquad \left|A_{y0}\right| < 0.3, \qquad \left|A_{z0}\right| < 0.3. \quad (3)$$

Otherwise a routine to compensate the sensed misplacement is enabled, and the angles displacements of the sensor axes (αA_{x0}, αA_{y0}, and αA_{z0}) are calculated using the following trigonometric equation:

$$\alpha A_{x0} = \arctan\left(\frac{A_{x0}}{\sqrt{(A_{y0})^2 + (A_{z0})^2}}\right),$$
$$\alpha A_{y0} = \arctan\left(\frac{A_{y0}}{\sqrt{(A_{x0})^2 + (A_{z0})^2}}\right), \quad (4)$$
$$\alpha A_{z0} = \arctan\left(\frac{A_{z0}}{\sqrt{(A_{x0})^2 + (A_{y0})^2}}\right).$$

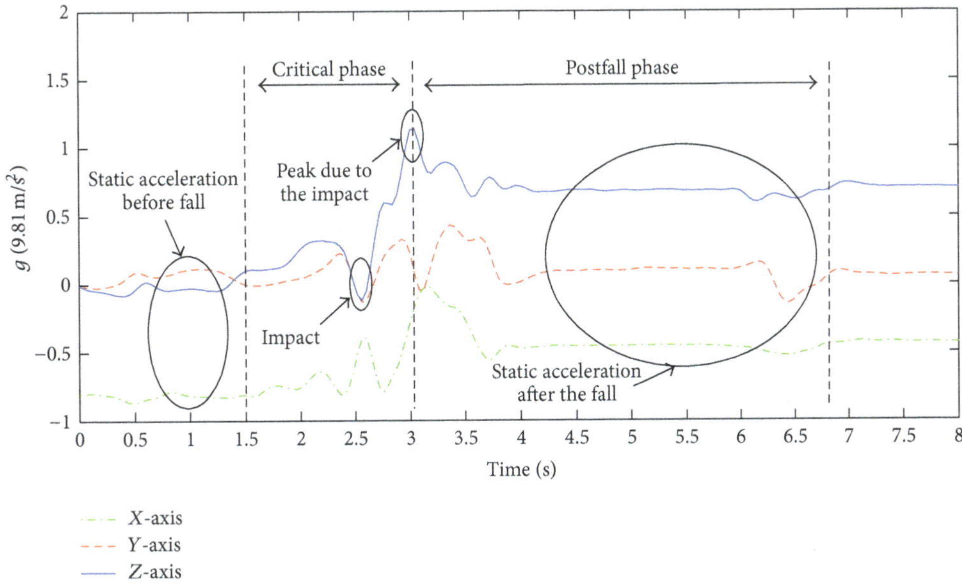

FIGURE 4: Typical acceleration waveform along each axis for a forward fall.

These values are stored and will be used for the correction of the misplacement during the feature extraction phase, described below.

2.4. Feature Extraction. Robust features are extracted in the time domain in a 5 sec sliding window by considering both quick and relevant acceleration changing along each axis (due to the fall) and by the change in position registered after the fall. The aim is to produce robust features taking into account all the information for the distinction of falls from other events. It is also important that such features have a low dependence on both the position of the sensor (whether it is placed on the waist, on the chest, or on the abdomen) and the human body characteristics of the user. Moreover, the computational cost must be limited for integration on embedded computing solutions.

The waveform in Figure 4 represents a typical acceleration signal of a forward fall on three axes and all phases are indicated according to the taxonomy proposed in [16]. For the features extraction process, both critical and postfall phases are of interest. In the former, the shock is measured due to the impact toward the ground, and a dynamic acceleration changing is registered. In the latter (the body is already lying on the floor) the static acceleration value records a great change due to the new position of the individual with respect to the calibration phase.

The 5 sec sliding window considered for features extraction is split in three parts as follow:

(1) from 0 sec to 0.5 sec: the maximum and minimum values of the acceleration are found, and the amplitude dynamics ΔA is calculated. It is proportional to the shock strength;

(2) from 0.5 sec to 2 sec: the system is in a transitory regime (samples are discarded);

(3) from 2 sec to 5 sec: the static-averaged acceleration value $\overline{A}$ is calculated to evaluate the new user position. Acceleration data is averaged to filter out tremors and the little movements of the end user.

During the system calibration phase it is verified that the device is worn correctly (see previous section), and the position changing is calculated as the difference between the value of the 3D-static acceleration after the fall (in the third part of the sliding window) and the one stored in the calibration phase. This difference, called Changing Position Offset (CPO), has a value in the range of 0 to 2, and it is proportional to the user displacement. In this way, for feature extraction, a study of posture was not made: only the relative varying posture analysis was considered causing a computational cost reduction and improving the robustness of the setup. The feature vector is made up of three parameters (one for each axis), coming from the modulation of the CPO and the dynamic acceleration peak, due to the impact of the fall.

If the routine to compensate the device misplacement is enabled, for the CPO calculation, the axes angles of the sensor in the initial condition need to be taken into account (αA_{x0}, αA_{y0}, and αA_{z0}, stored during the calibration phase). Using (1), the CPO value for x-axis is obtained as

$$\mathrm{CPO}_x = \left|\sin\left(\alpha\overline{A}_x - \alpha A_{x0}\right)\right|, \tag{5}$$

where $\alpha\overline{A}_x$ is the x-axis angle of the sensor in the third part of the sliding window; it is calculated considering the static averaged acceleration $\overline{A}_x$ and the same trigonometric equation used in (4) as follows:

$$\alpha\overline{A}_x = \arctan\left(\frac{\overline{A}_x}{\sqrt{\left(\overline{A}_y\right)^2 + \left(\overline{A}_z\right)^2}}\right). \tag{6}$$

Let A_0 initial acceleration value measured after the calibration phase
Let αA_0 initial axis angle of the sensor measured after the calibration phase

ΔA → shock dynamics due to the impact calculated in the first part of the sliding window
$\overline{A}$ → Static averaged acceleration in the third part of the sliding window

If *compensation device misplacement procedure is enabled*
$\alpha\overline{A}$ is calculated → axis angle of the sensor in the third part of the sliding window
$\text{CPO} = \left|\sin\left(\alpha\overline{A} - \alpha A_0\right)\right|$ → Changing Position Offset
else
$\text{CPO} = \left|\overline{A} - A_0\right|$ → Changing Position Offset
End if

Let $\Delta A_x, \Delta A_y, \Delta A_z, \text{CPO}_x, \text{CPO}_y, \text{CPO}_z$, ΔA and CPO calculated for each axis
$[X, Y, Z] = [\Delta A_x \cdot \text{CPO}_x, \Delta A_y \cdot \text{CPO}_y, \Delta A_z \cdot \text{CPO}_z]$ → feature vector input for SVM

Pseudocode 1: Feature extraction pseudocode.

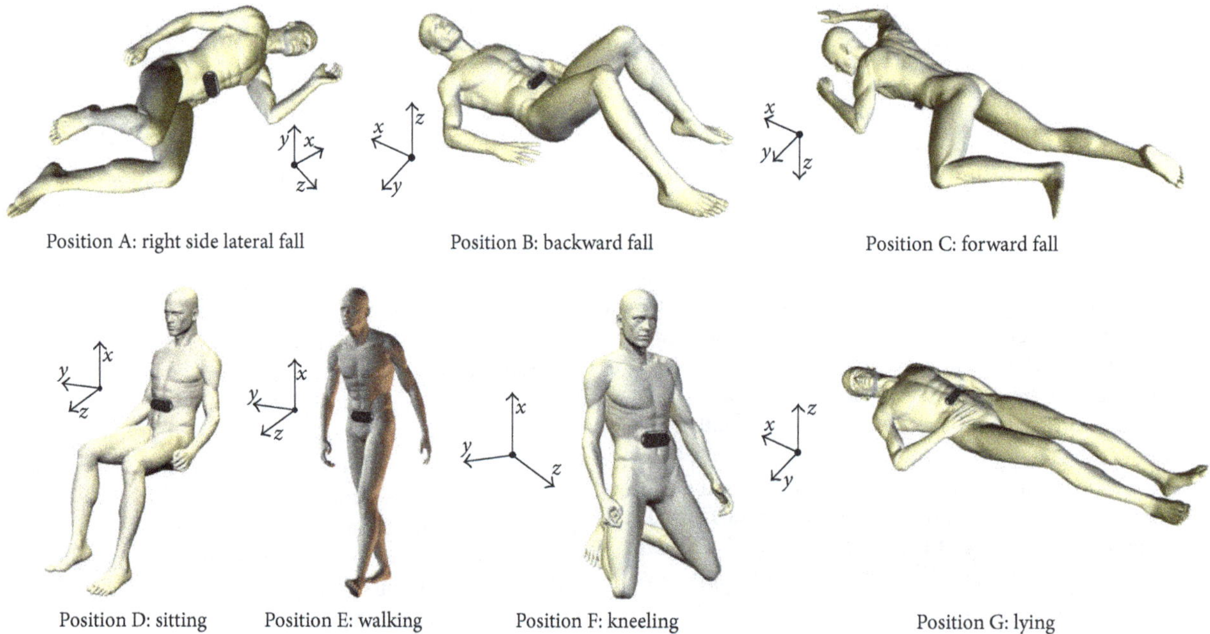

Figure 5: Example of different spatial configurations of the device during falls and ADLs.

The same procedure must be done for the two other axes.

Thanks to this routine, the change in the angle position of the user along the three axes is measured. In this way the information coming from the modulation of the CPO remains highly selective for fall detection recognition even if the device is placed incorrectly and the features remain nearly the same. In this way the system could work efficiently reducing the problems of device positioning at the expense of an increase in the computational costs.

The process of feature extraction, described previous, is summarized in the pseudocode shown in Pseudocode 1.

For the feature extraction, It makes sense to consider the acceleration signal on each axis singularly, because a fall event leads to a change in the value of the static acceleration in at least two of the three acceleration axes (due to the orientation change of sensing axes). This is evident comparing the positions A, B, and C in Figure 5 (postfall phase of the lateral, backward, and forward falls, resp., are depicted) with respect to the initial position (see Figure 3). On the other hand, when a sitting event (D in Figure 5) occurs, the change in static acceleration can be neglected with respect to that of a fall; it is possible to do the same considerations for kneeling and walking (E and F). Instead, the axes orientation changes for a lying event (position G), but the acceleration peak produced is slower and lower than a fall, and through the product between the value of the acceleration peak and CPO it is also possible to discriminate this ADL.

Table 2 shows the nominal values of acceleration along the three axes A_x, A_y, and A_z and the value of CPO corresponding to the positions A–G of Figure 5.

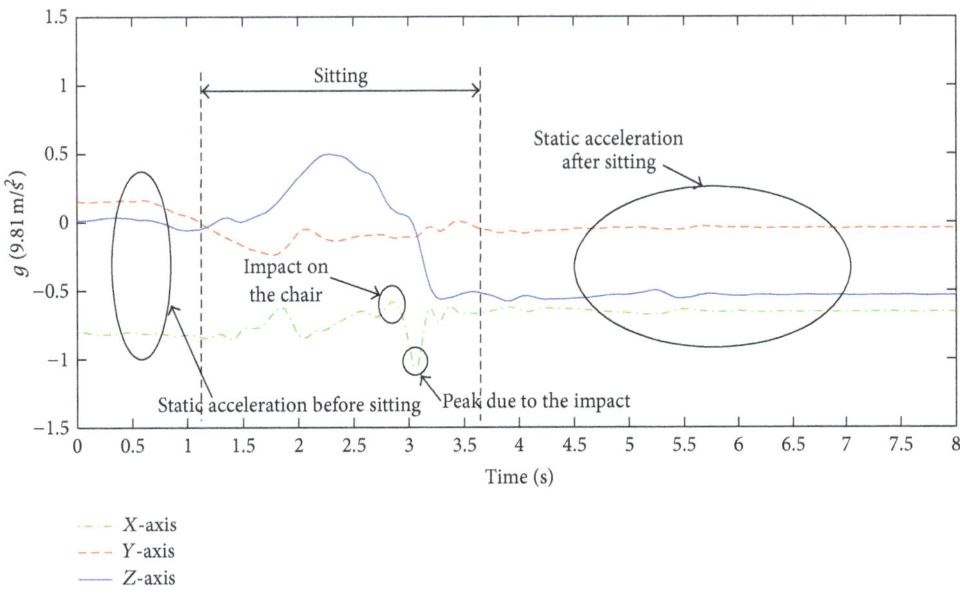

FIGURE 6: Example acceleration waveforms for a sitting event.

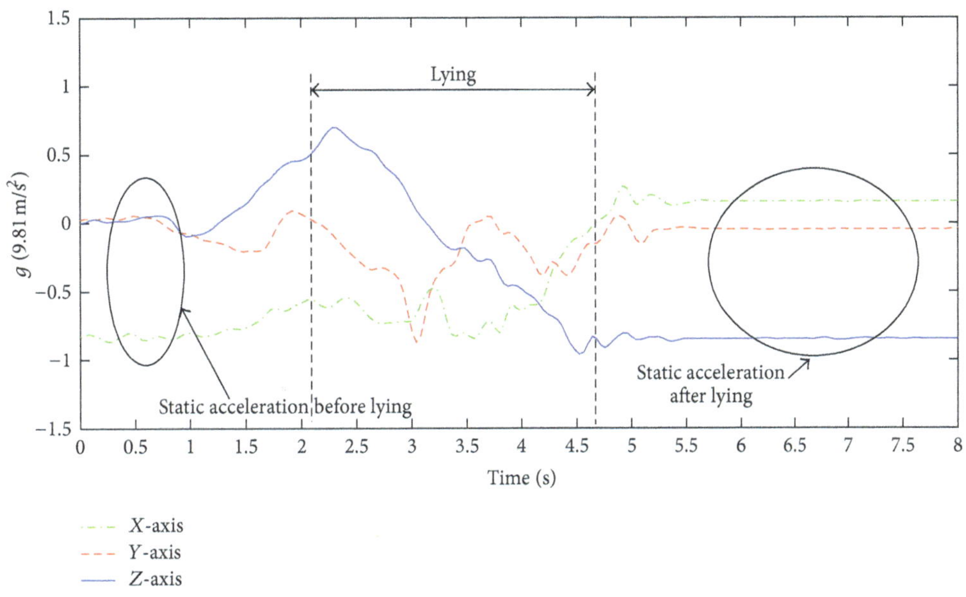

FIGURE 7: Example acceleration waveforms for a lying event.

TABLE 2: CPO coefficients by varying spatial position.

	A_x	A_y	A_z	CPO
Initial condition	−1	0	0	
Position A	0	−1	0	(1, 1, 0)
Position B	0	0	−1	(1, 0, 1)
Position C	0	0	1	(1, 0, 1)
Position D	−1	0	0	(0, 0, 0)
Position E	−1	0	0	(0, 0, 0)
Position F	−1	0	0	(0, 0, 0)
Position G	0	0	−1	(1, 0, 1)

When the orientation of acceleration does not change, the values of CPO are (close to) zero, and the elements of the features also become (close to) zero. In Figures 6 and 7 the waveforms of sitting and lying events are reported to make a comparison with that of the fall (shown in Figure 4), when the device is worn on the abdomen.

The difference among the features of the fall of Figure 4 and the sitting and lying events of Figures 6 and 7 is apparent in Figures 8, 9, and 10.

It was actually verified that all the consideration made thus so far remains valid whether the device is placed on the

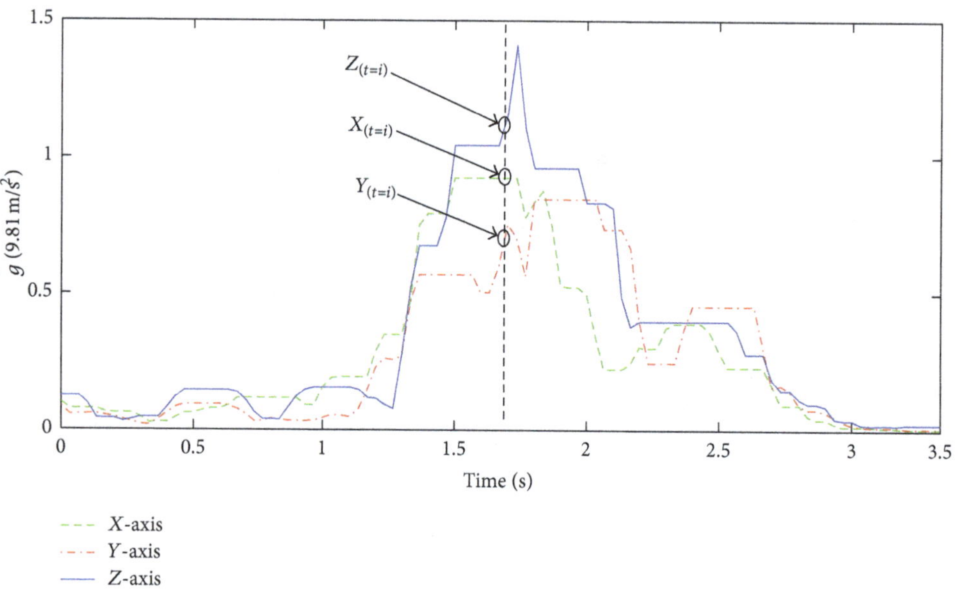

FIGURE 8: Features extracted of the forward fall in Figure 4.

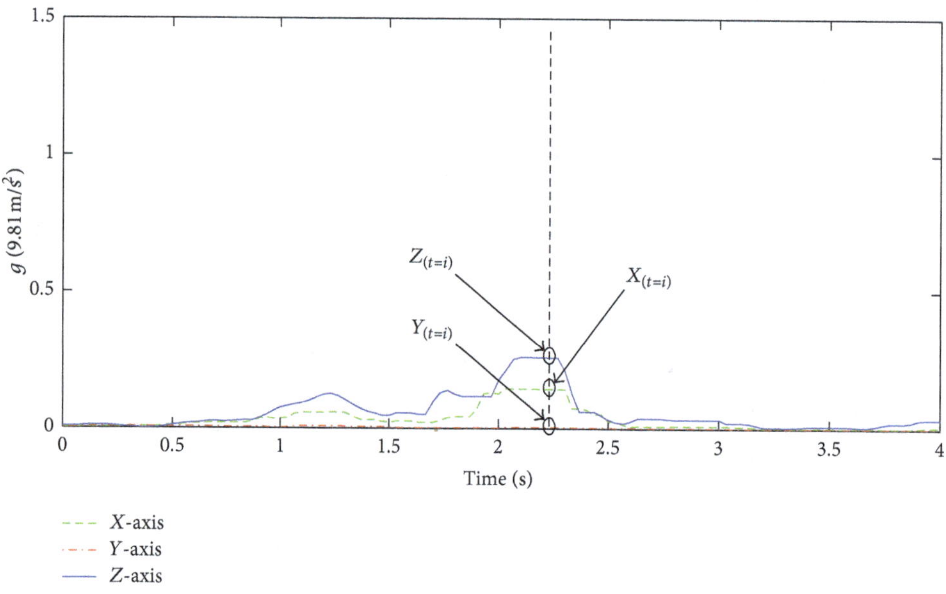

FIGURE 9: Features extracted of the sitting event in Figure 6.

waist or on the chest. In Figures 11 and 12 the features of a sequence of falls and daily events, when the device is worn on the waist and on the chest, are shown. So it is also evident that the features obtained discriminate the falls from ADLs when the device is placed in other area of the torso.

It is important to highlight that the measurements of the actions as shown in Figures 11 and 12 have been carried out by simulating critical behaviours, that is, sitting with a strong impact on the chair; bending down, lying down, and standing up quickly; moderate backward fall.

2.5. Fall Detection Algorithm. Once features are extracted, the fall events are detected by a one-class support vector machine (OC-SVM). SVM is a robust classification tool (in the presence of outliers too) with a good generalization ability. Furthermore it is less computationally intensive than other algorithms like neural networks [21].

One class SVM divides all samples into an objective field and a nonobjective field and then nonlinearly maps those sample into a higher dimensional features space with some efficient operators, called kernel function. The target is to find

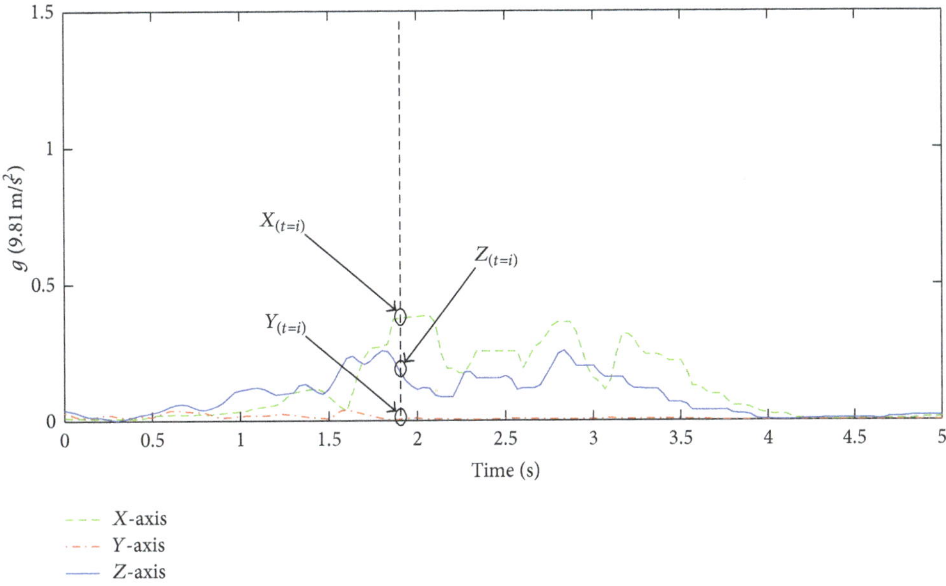

FIGURE 10: Features extracted of the lying event in Figure 7.

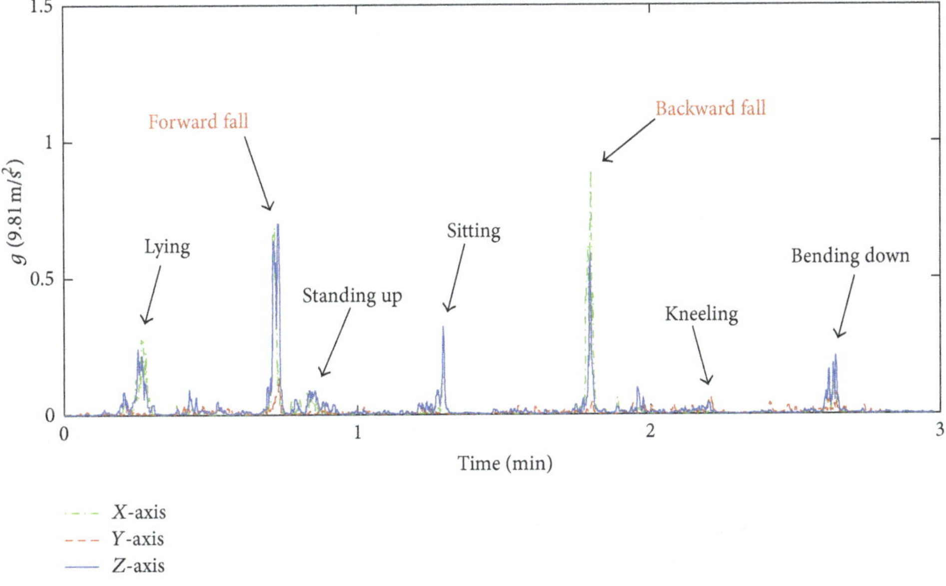

FIGURE 11: Features extracted of falls and ADLs simulated wearing the device on the waist.

a sphere that contains most of the normal data such that the corresponding radius R can be minimized as follows:

$$\begin{aligned} \min \quad & R^2 + C\sum_{i=1}^{n} \xi_i, \\ \text{s.t.} \quad & \|a - v_i\|^2 \leq R^2 + \xi_i, \quad \xi_i \geq 0, \end{aligned} \tag{7}$$

where a is the centre of the sphere and v_i a positive sample set. The slack variables ξ_i allow some data points to lie outside the sphere and the parameter C controls the trade-off between the volume of the sphere and the number of errors. The objective function is

$$\begin{aligned} \max \quad & \sum_{i=1}^{n} \alpha_i \langle v_i, v_i \rangle - \sum_{i,j=1}^{n} \alpha_i \alpha_j \langle v_i, v_j \rangle, \\ \text{s.t.} \quad & 0 \leq \alpha_i \leq C, \quad \sum_{i=1}^{n} \alpha_i = 1. \end{aligned} \tag{8}$$

The original data points are first mapped into a feature space, because some data are not spherically distributed in the input

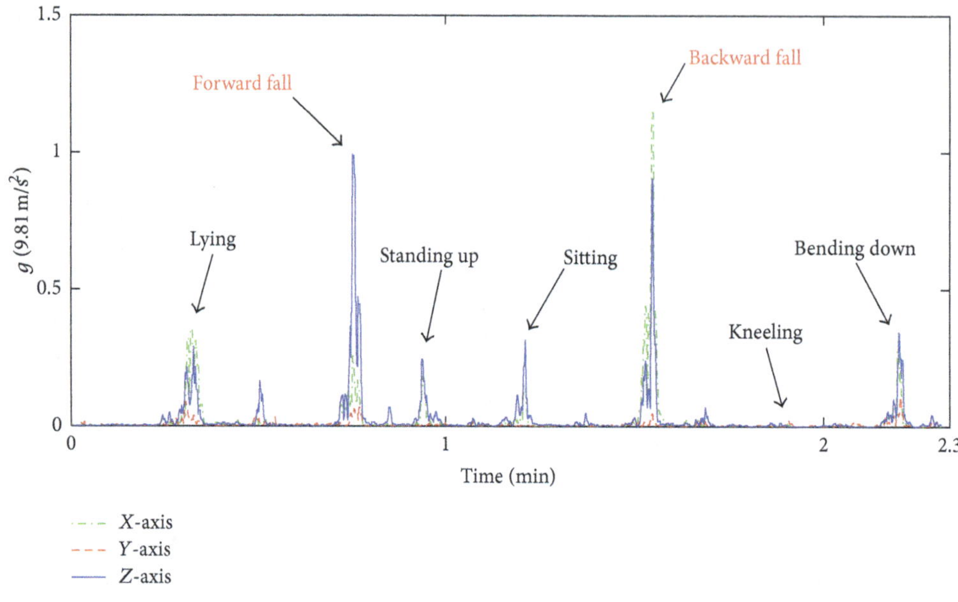

FIGURE 12: Features extracted of falls and ADLs simulated wearing the device on the chest.

TABLE 3: Types of tested kernels.

Kernel	Function
Linear	$K(v_i, v_j) = 1 + v_i^T v_j$
Polynomial	$K(v_i, v_j) = (1 + v_i^T v_j)^p$
Gaussian Radial Basis Function	$K(v_i, v_j) = \exp\left(-\gamma\left\|v_i - v_j\right\|^2\right)$
Sigmoid	$K(v_i, v_j) = \tanh\left(kv_i^T v_j - \delta\right)$

space. For this reason the kernel function $k(\cdot,\cdot)$ is used as follows:

$$\max \quad \sum_{i=1}^{n} \alpha_i k\left(v_i, v_i\right) - \sum_{i,j=1}^{n} \alpha_i \alpha_j k\left(v_i, v_j\right). \tag{9}$$

The common types of kernel found in the literature are tested (in Table 3 the functions are shown). They were analyzed using the MATLAB tool STPRtool [22], changing the values of their parameters.

In the literature the Gaussian Radial Basis Function (GRBF) is often used in fall detectors [23–25]. For the features extracted, GRBF and polynomial kernel functions give the best results in terms of performance, even the polynomial kernel shows a lower computational cost (relative number of support vectors and relative execution time are considered). The comparison with other kernel functions will be shown in the following section.

The parameters of kernel were chosen to detect as many falls as possible, and a simple post-processing step has been implemented in order to reduce false negative events.

2.6. Postprocessing. In choosing the parameters of the tested kernel functions, the value of sensitivity was given priority. In this way some nonfall events are detected from SVM. To reduce this problem a simple filtering by voting was added: it detects an event such as a fall if the alarm coming from the SVM classifier remains high a given time t. With this approach the spikes or temporary anomalies are filtered too. The optimum time t for postprocessing is obtained from a parametric analysis. The results are shown in the next paragraph.

3. Results

This section will show a discussion about the measurements of the experimental data processed in MATLAB. To validate the implemented algorithm, the dataset previously described has been considered.

Since the output of the fall detection step is binary, there are four possible results.

(i) *True Positive (TP)*: a fall happens, and the algorithm detects it.

(ii) *False Positive (FP)*: a fall does not occur, and the algorithm reveals a fall.

(iii) *True Negative (TN)*: a daily event is performed, and the algorithm does not detect it.

(iv) *False Negative (FN)*: a fall occurs, but the algorithm does not detect it.

The performance of the system, with respect to both different kernel functions of SVM and their parameters, can be evaluated considering the following metrics:

$$\text{sensitivity} = \frac{\text{TP}}{\text{TP} + \text{FN}}, \qquad \text{specificity} = \frac{\text{TN}}{\text{TN} + \text{FP}}. \tag{10}$$

TABLE 4: Performance by varying the kernel functions.

	Sensitivity (%)	Specificity (%)	Relative support vectors	Relative execution time
Linear ($C = 10$)	98.6	52.3	0.51x	0.2x
Linear ($C = 100$)	97.2	46.4	0.44x	0.2x
Polynomial ($C = 20$, $P = 2$)	97.5	81.25	0.94x	0.8x
Polynomial ($C = 2.8$, $P = 3$)	97.7	95.8	1x	1x
GRBF ($C = 2.6$, $\gamma = 2$)	98.1	83.1	1.1x	1.3x
GRBF ($C = 2$, $\gamma = 3$)	97.4	95.2	1.3x	1.5x
Sigmoid ($C = 1$, $k = 1$, $\delta = 2$)	98.8	36	0.3x	4x
Sigmoid ($C = 100$, $k = 5$, $\delta = 2$)	99.1	42	0.4x	4.1x

For completeness the results obtained with postprocessing elaboration have been added. The algorithm is tested when the device is placed on the waist, abdomen, or chest.

The one class SVM classifier has been trained by using about 40 falls and 50 ADLs belonging to a large dataset in which more than 250 falls and 200 daily events were performed. The remaining 210 falls and 150 ADLs have been used for testing. Table 4 shows the specificity and the sensitivity for the kernel functions already described. The complexity of the various kernel functions can be studied by considering the relative number of vectors required and also the relative execution time [26, 27]. The kernel with the most significant values for the parameters is considered.

From Table 4 the polynomial kernel and GRBF kernel give better results. Their capacity to detect a fall is higher (more than 95% for sensitivity and specificity) than others. GRBF (with $\sigma = 3$ and $C = 2$) and the polynomial (with $P = 3$ and $C = 2.8$) provide similar values of specificity and sensitivity, but the last one works faster, and its number of vector is slightly lower, thereby it was chosen for this work. Misclassifications are for falls presenting slow dynamics or falls with partial recovery. It is important to see the result of the linear function: its workload is very low (less support vectors number and very low execution time), but it gives many false positive, and its value of the specificity is low. The function of GRBF is the most common used in the literature for the detection of falls, due to its high ability to separate the classes, even in the presence of many outliers (at the expense of the convergence time). In this work it has been also possible to obtain a high accuracy with the polynomial function thanks to features extracted: they have a high degree of intrinsic separability.

The implemented OC-SVM shows improvements in the specificity and sensitivity with respect to a threshold-based approach, and this can be verified in comparison with both the frameworks presented in [17, 28], choosing the algorithms where the same parameters of fall are used (the impact detection and posture monitoring). For the comparison of the fall detection algorithms were used the same hardware

TABLE 5: Comparison of the proposed OC-SVM method and threshold-based algorithm.

	Sensitivity (%)	Specificity (%)
OC-SVM	97.7	94.8
Threshold based	89.2	85.7

TABLE 6: Performance after filtering by voting.

Polynomial ($C = 2.8$, $P = 3$)	Sensitivity (%)	Specificity (%)
$t = 0.4$ s	97.7	94.8
$t = 0.8$ s	96.2	97.2
$t = 1.2$ s	86.4	98.1

(shown in Figure 1), benchmark dataset and training/test sets described above. The values of thresholds were tuned according to the falls presented in the dataset used to train the OC-SVM scheme. The best results for threshold based algorithm were obtained with Bourke's algorithm and are compared with the proposed OC-SVM in Table 5: the latter seems more efficient, demonstrating its higher capacity to generalize, detecting a bigger number of falls with low impact magnitude than threshold-based algorithm. However, the computational cost of thresholdbased is lower.

Due to (a) the reduction in the number of the features, (b) the low computational cost of extracted features, and (c) the used kernel, the overall system workload implemented is compatible with an integration in embedded low-power solutions (DSP, FPGA, and microcontroller).

A reduction of false positive (and so an improving of specificity value), for the polynomial chosen, has been measured with filtering by voting with a temporal window of prefixed dimension (0.8 sec); see Table 6. Finally it was verified that the values of sensitivity and specificity remain almost the same even if the axes of the sensor are placed differently with respect to that shown in Figure 3 (the values of testing dataset have been changed simulating same misplacements of the device).

4. Conclusions

The proposed supervised scheme overcomes the limitation of well-known threshold-based approaches in which a heuristic choice of the parameters is accomplished. High performance in controlled conditions (events simulated) in terms of sensitivity and specificity was obtained using only the 20% of dataset for training. Performance metrics of different kernels in one class SVM are compared: best results are obtained with polynomial function and Gaussian Radial Basis Function. The polynomial kernel is used in order to limit the computational workload. A study of posture was not made, but only posture changing analysis was used, limiting the computational cost. Through the calibration step, the approach allows to generalize the detection of fall events leading invariance to physical characteristic of the end users. Future work will be devoted to validate the solution in real conditions, test the methodology with a large set of different MEMS accelerometers, and port the framework on embedded mobile solutions as FPGA or DSP.

References

[1] N. Noury, P. Rumeau, A. K. Bourke, G. ÓLaighin, and J. E. Lundy, "A proposal for the classification and evaluation of fall detectors," *IRBM*, vol. 29, no. 6, pp. 340–349, 2008.

[2] M. C. Chung, K. J. McKee, C. Austin et al., "Posttraumatic stress disorder in older people after a fall," *International Journal of Geriatric Psychiatry*, vol. 24, no. 9, pp. 955–964, 2009.

[3] S. Sadigh, A. Reimers, R. Andersson, and L. Laflamme, "Falls and fall-related injuries among the elderly: a survey of residential-care facilities in a Swedish municipality," *Journal of Community Health*, vol. 29, no. 2, pp. 129–140, 2004.

[4] A. Shumway-Cook, A. M. Ciol, J. Hoffman, J. B. Dudgeon, K. Yorkston, and L. Chan, "Falls in the medicare population: incidence, associated factors, and on health Care," *Physical Therapy*, vol. 89, no. 4, pp. 324–332, 2009.

[5] S. Elliott, J. Painter, and S. Hudson, "Living alone and fall risk factors in community-dwelling middle age and older adults," *Journal of Community Health*, vol. 34, no. 4, pp. 301–310, 2009.

[6] X. Yu, "Approaches and principles of fall detection for elderly and patient," in *Proceedings of the 10th IEEE International Conference on E-Health Networking, Applications and Service (HEALTHCOM '08)*, pp. 42–47, Singapore, July 2008.

[7] D. M. Karantonis, M. R. Narayanan, M. Mathie, N. H. Lovell, and B. G. Celler, "Implementation of a real-time human movement classifier using a triaxial accelerometer for ambulatory monitoring," *IEEE Transactions on Information Technology in Biomedicine*, vol. 10, no. 1, pp. 156–167, 2006.

[8] J. Y. Hwang, J. M. Kang, Y. W. Jang, and H. C. Kim, "Development of novel algorithm and real-time monitoring ambulatory system using Bluetooth module for fall detection in the elderly," in *Proceedings of the 26th Annual International Conference of the IEEE Engineering in Medicine and Biology Society (EMBC '04)*, pp. 2204–2207, San Francisco, Calif, USA, September 2004.

[9] H. J. Luinge and P. H. Veltink, "Inclination measurement of human movement using a 3-D accelerometer with autocalibration," *IEEE Transactions on Neural Systems and Rehabilitation Engineering*, vol. 12, no. 1, pp. 112–121, 2004.

[10] G. Anania, A. Tognetti, N. Carbonaro et al., "Development of a novel algorithm for human fall detection using wearable sensors," in *Proceedings of the IEEE Sensors (SENSORS '08)*, pp. 1336–1339, Lecce, Italy, October 2008.

[11] A. Leone, G. Diraco, and P. Siciliano, "Detecting falls with 3D range camera in ambient assisted living applications: a preliminary study," *Medical Engineering and Physics*, vol. 33, no. 6, pp. 770–781, 2011.

[12] J. Chen, K. Kwong, D. Chang, J. Luk, and R. Bajcsy, "Wearable sensors for reliable fall detection," in *Proceedings of the 27th Annual International Conference of the Engineering in Medicine and Biology Society (IEEE-EMBS '05)*, pp. 3551–3554, Shanghai, China, September 2005.

[13] M. Kangas, A. Konttila, P. Lindgren, I. Winblad, and T. Jämsä, "Comparison of low-complexity fall detection algorithms for body attached accelerometers," *Gait & Posture*, vol. 28, no. 2, pp. 285–291, 2008.

[14] A. K. Bourke, J. V. O'Brien, and G. M. Lyons, "Evaluation of a threshold-based tri-axial accelerometer fall detection algorithm," *Gait & Posture*, vol. 26, no. 2, pp. 194–199, 2007.

[15] A. K. Bourke, P. Van de Ven, A. E. Chaya, G. M. Olaighin, and J. Nelson, "Testing of a long-term fall detection system incorporated into a custom vest for the elderly," in *Proceedings of the 30th Annual International Conference of the IEEE Engineering in Medicine and Biology Society (EMBC '08)*, pp. 2844–2847, Vancouver, Canada, August 2008.

[16] A. K. Bourke, P. van de Ven, M. Gamble et al., "Evaluation of waist-mounted tri-axial accelerometer based fall-detection algorithms during scripted and continuous unscripted activities," *Journal of Biomechanics*, vol. 43, no. 15, pp. 3051–3057, 2010.

[17] F. Bagalà, C. Becker, A. Cappello et al., "Evaluation of accelerometer-based fall detection algorithms on real-world falls," *PLoS One*, vol. 7, Article ID e37062, 2012.

[18] S. -H. Liu and W. -C. Cheng, "Fall detection with the support vector machine during scripted and continuous unscripted activities," *Sensor*, vol. 12, no. 9, pp. 12301–12316, 2012.

[19] A. Leone, G. Diraco, C. Distante et al., "A multi-sensor approach for people fall detection in home environment," in *Proceedings of the Workshop on Multi-Camera and Multi-Modal Sensor Fusion Algorithms and Applications (M2SFA2 '08)*, pp. 1–12, 2008.

[20] N. Noury, A. Fleury, P. Rumeau et al., "Fall detection—principles and methods," in *Proceedings of the 29th Annual International Conference of IEEE-EMBS, Engineering in Medicine and Biology Society (EMBC '07)*, pp. 1663–1666, Lyon, France, August 2007.

[21] L. M. Manevitz and M. Yousef, "One-class SVMs for document classification," *Journal of Machine Learning Research*, vol. 2, no. 1, pp. 139–154, 2001.

[22] V. Franc and V. Hlavac, "Statistical Pattern Recognition Toolbox (STPRtool)".

[23] T. Zhang, J. Wang, L. Xu, and P. Liu, "Fall detection by wearable sensor and one-class SVM algorithm," *Lecture Notes in Control and Information Sciences*, vol. 345, pp. 858–863, 2006.

[24] J. Yin, Q. Yang, and J. J. Pan, "Sensor-based abnormal human-activity detection," *IEEE Transactions on Knowledge and Data Engineering*, vol. 20, no. 8, pp. 1082–1090, 2008.

[25] H. Cheng, H. Luo, and F. Zhao, "A fall detection algorithm based on pattern recognition and human posture analysis," in *Proceedings of the IET International Conference on Communication Technology and Application (ICCTA '11)*, vol. 2011, pp. 853–857, Beijing, China.

[26] R. Sangeetha and B. Kalpana, "A comparative study and choice of an appropriate kernel for support vector machines," in *Information and Communication Technologies*, V. V. Das and R. Vijaykumar, Eds., vol. 101 of *Communications in Computer and Information Science*, pp. 549–553, Springer, Heidelberg, Germany, 2010.

[27] R. Sangeetha and B. Kalpana, "Performance evaluation of kernels in multiclass support vector machines," *International Journal of Soft Computing and Engineering*, vol. 1, no. 5, pp. 138–145, 2011.

[28] M. Grassi, A. Lombardi, G. Rescio et al., "A hardware-software framework for high-reliability people fall detection," in *Proceedings of the IEEE Sensors (SENSORS '08)*, pp. 1328–1331, Lecce, Italy, October 2009.

Predictions of the Compressible Fluid Model and its Comparison to Experimental Measurements of Q Factors and Flexural Resonance Frequencies for Microcantilevers

Jason Jensen and Martin Hegner

Centre for Research on Adaptive Nanostructures and Nanodevices (CRANN) and School of Physics, Trinity College Dublin, Dublin 2, Ireland

Correspondence should be addressed to Jason Jensen, jensenja@tcd.ie and Martin Hegner, hegnerm@tcd.ie

Academic Editor: Sangmin Jeon

The qualitative agreement between experimental measurements of the Q factors and flexural resonance frequencies in air of microcantilevers and calculations based on the compressible fluid model of Van Eysden and Sader (2009) is presented. The Q factors and resonance frequencies observed on two sets of cantilever arrays were slightly lower than those predicted by the model. This is attributed to the individual design and geometry of the microfabricated hinged end of the cantilever beams in the array.

1. Introduction

The introduction of the atomic force microscope [1] and the improvement of silicon fabrication technologies resulted in the ready availability of high-quality, reproducible, and inexpensive silicon cantilevers. Applications for micron-scale cantilevers as a sensing tool have been found in the fields of genomics [2–6], proteomics [7–9], microbiology [10–14], and many others. Many of these applications make use of the microcantilever as a sensitive mass detector. It has been shown that operating the cantilever at higher resonance modes increases the mass sensitivity of the device [15]. This increase in the sensitivity is linked to the increased Q factor observed for the higher flexural resonance modes of the cantilever [15]. Along with increased interest in possible applications came the need for improved understanding of the dynamics of cantilevers on this scale and models which can predict their behaviour in a range of situations. In general the higher the Q factor of the resonance peak the smaller the minimum observable frequency shift is. Thus it is desirable to obtain the highest Q factor possible during experiments to maximise the sensitivity of the experiment. Models indicating the dynamics of the cantilever are useful when planning such experiments and determining the expected minimum response required for successful detection of the target.

Many models detailing the behaviour of microcantilevers have been proposed, including the Elmer-Dreier model [16] and Sader's viscous [17] and extended viscous models [18]. Sader's extended model includes the 3D flow field of the fluid around the cantilever beam and can be applied for arbitrary mode number.

The models mentioned above assume that the fluid in which the cantilever is vibrating is incompressible, and in general have good agreement with experimental results [19]. However, recent papers by Van Eysden and Sader [20, 21] which detail a model for a cantilever beam oscillating in a compressible fluid indicate that this unbounded increase of the quality factor is not always valid. They predict that as the mode number increases and passes a "coincidence point" (which is determined by the thickness to length ratio of the cantilever and the fluid in which the cantilever is vibrating) the Q factor will begin to decrease.

This coincidence point occurs when the length scale of spatial oscillations of the cantilever beam reduces to a point where it is comparable with the acoustic wavelength of the

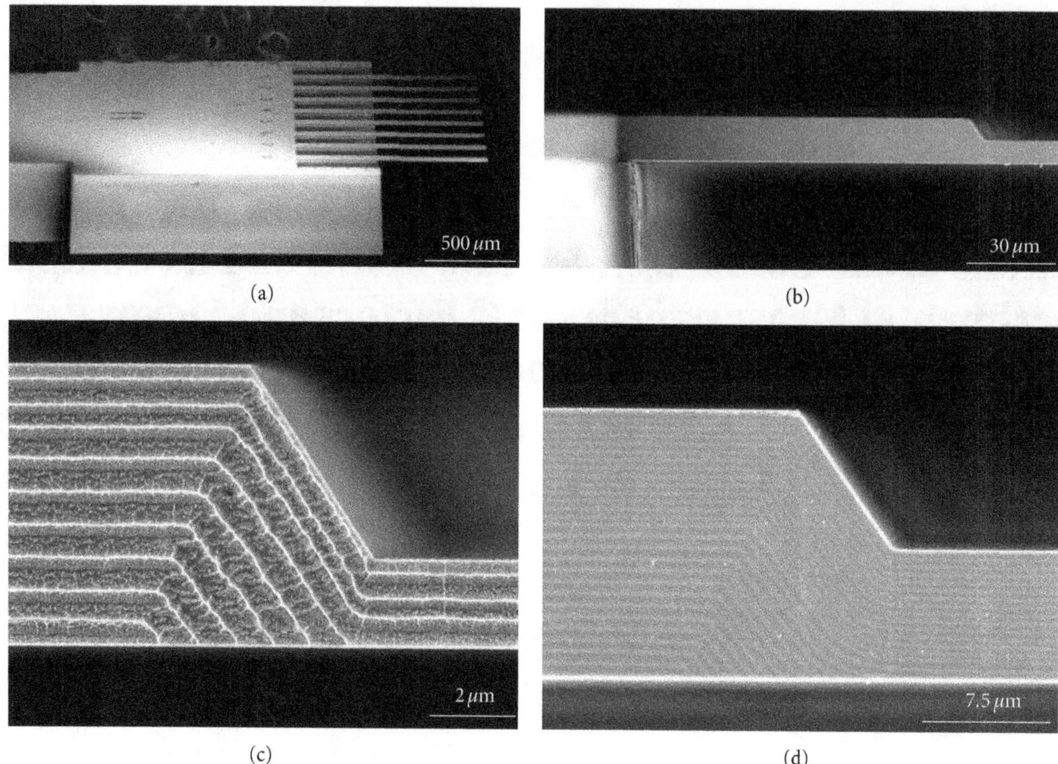

FIGURE 1: (a) SEM image of typical cantilever array used for these measurements. (b) Closer view of the hinged end of one of the 7 μm thick cantilever showing the 120 μm long hinged design that connects the cantilever with the main body of the array. (c), (d) Two closer views of the hinged portion of the 2 and 7 μm thick cantilevers, respectively. The hinge is approximately twice the thickness of the cantilever for the 7 μm thick cantilevers and approximately three times the thickness on the 2 μm thick cantilevers.

media in which the cantilever is vibrating. At this point it is possible that energy can be dissipated by the generation of acoustic waves.

For practical applications of microcantilevers (such as mass sensing) this is not an issue when operating the cantilever in liquid. However, if the cantilever is vibrated in air then it can be possible to observe this effect at higher modes. For a cantilever which is 100 μm wide, 500 μm long, and 7 μm thick the scaling analysis from the compressible fluid model [20] predicts that there should be a turning point at the $n = 3$ mode which occurs below 1 MHz. For a 2 μm thick cantilever of the same size the predicted mode is much higher ($n = 12$) and occurs around 3.6 MHz.

2. Materials and Methods

2.1. Cantilevers. The cantilevers used in these experiments are Si cantilever arrays (orientation: 110) with eight cantilevers per array (IBM Research Laboratory, Rüschlikon, Switzerland). The cantilevers had a pitch of 250 μm and were 500 μm long and 100 μm wide. The thickness of the cantilevers was measured in a scanning electron microscope (SEM, Zeiss Ultra, Cambridge, UK) and were found to be 7.2 ± 0.5 μm and 1.972 ± 0.005 μm thick (Figure 1). The variation in the thicknesses of the cantilevers was shown to depend on their position on the production wafer. As shown in Figure 1 the cantilevers are connected to the main body of the chip via a ~120 μm long segment which is approximately twice as thick as the cantilever itself for the 7 μm thick cantilevers and three times the thickness for the 2 μm thick cantilevers. This design was implemented to facilitate better definition between the hinge (clamping point) of the cantilever and the main body of the array.

2.2. Optical Beam Deflection Device. Thermal actuation of the cantilevers does not provide sufficient vibration of the cantilever beam to allow measurement of higher resonance modes in the current device. The cantilevers are clamped on top of a piezo electric actuator (EBL Products Inc., East Hartford, Conn, USA). The energy from the piezo is efficiently transferred to the cantilevers and provides sufficient vibration amplitudes to allow readout of the vibration modes using optical beam deflection. The cantilevers are excited at various vibrating modes by a linear frequency sweep of a sinusoidal signal which is provided by a frequency generator (NI PCI 5406, National Instruments, Tex, USA) which is controlled via a LabVIEW interface. The drive amplitude of the piezo actuator was kept low to avoid nonlinearities in the response of the cantilevers.

Optical beam deflection was used to detect the resonance frequency of the cantilever vibrations. A schematic of the device is shown in Figure 2. A single wavelength fibre coupled laser (632.99 nm, Free space power >2.4 mW, SWL 7504-P;

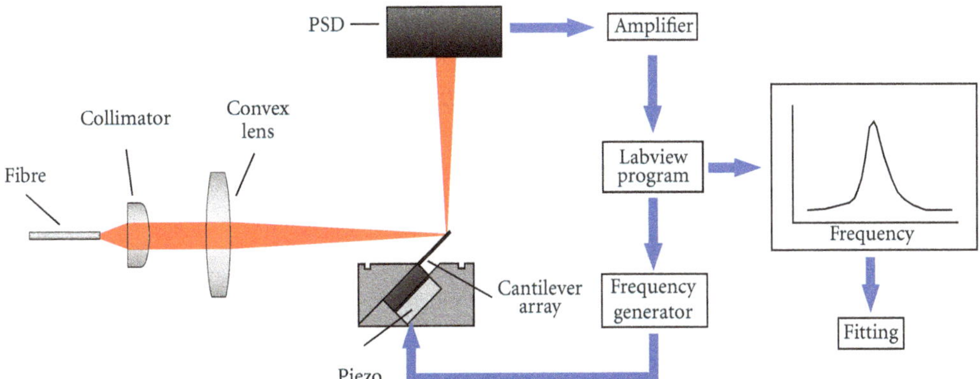

FIGURE 2: Schematic of the optical beam deflection readout procedure. The cantilever array is actuated by a piezo electric ceramic which is excited by a signal from the frequency generator. A laser spot focused onto the tip of the cantilever is deflected onto the surface of a PSD. The output from the PSD is then analysed with the signal from the frequency generator in a LabVIEW program which results in a frequency spectrum, the peaks of which correspond to the flexural resonance modes of the cantilever.

Newport, Calif, USA) was collimated into a 3.5 mm beam diameter (F280 APC-B; Thorlabs, Cambridgeshire, UK) and then focussed onto a 12 μm diameter spot on the surface of the cantilever using a 50 mm focal length convex lens (AC254-050-A1-ML; Thorlabs, Cambridgeshire, UK). The output of the laser was attenuated to avoid saturating the PSD using a neutral density filter (OD 1.3 NE513B; Thorlabs Cambridgeshire, UK).

The optic cage system which maintains the optic axis of the collimator, lens and neutral density filter was mounted on a system of two automated translation stages (M110.1DG & M122.2DD; Physik Instrumente, Bedford, UK) which allowed precise readout from each of the cantilevers in the array in a sequential manner. The motion of the stages is controlled by a LabVIEW interface. An additional microtranslation stage (Gothic Arch 9061-XYZ; Newport, Calif, USA) allows initial positioning of the laser spot at the tip of the cantilever prior to the start of an experiment.

The laser beam is deflected from the tip of the cantilever onto a linear position sensitive detector (PSD, Sitek, Partille, Sweden). The current output from the PSD is converted to a voltage with a cutoff frequency of 2 MHz (due to the response time of the optical detector). The output from the PSD is amplified (SR560 Low-Noise Preamplifier; Stanford Research Systems, Calif, USA) then digitised (NI PCI 5112; National Instruments, Tex, USA) before being analysed with the output from the frequency generator in a LabVIEW program where the time domain signal is converted into a frequency spectrum. The peaks of the spectrum correspond to the flexural resonant modes of the cantilever.

The entire device is housed inside a box which is kept at a constant temperature of 23.0 ± 0.1°C to avoid any drifts in the measurement due to temperature changes. The temperature is kept constant by a fuzzy logic controller which is implemented in LabVIEW.

Cantilever arrays were taken at random from the production wafers and multiple measurements of the first four flexural resonance modes were taken for the 7 μm thick cantilevers and of the first seven modes of the 2 μm thick cantilevers.

The resonance peaks obtained can be described by a simple harmonic oscillator model [22]

$$A(f) = A_{bl} + \frac{A_0 f_{R,n}^2}{\sqrt{\left(f^2 - f_{R,n}^2\right)^2 + f^2 f_{R,n}^2/Q^2}}, \qquad (1)$$

where A_{bl} is the amplitude of the baseline, A_0 is the zero frequency amplitude, f is the frequency, $f_{R,n}$ is the resonance frequency of mode n, and Q is the quality factor. The Q factor and resonance frequencies were extracted from the best fit of the resonance peaks with the above model using a Levenberg-Marquardt algorithm [23]. The mean and standard deviation of the resonance frequencies and Q factors of each of the modes was then calculated from the fitted data.

3. Results and Discussion

3.1. Numerical Calculations. Van Eysden and Sader's extended viscous [18] and compressible fluid models [20] were used to predict the resonance frequency and Q factor of modes of the 7 μm thick cantilevers which were below 2 MHz and the modes of the 2 μm thick cantilevers below 1 MHz. The compressible fluid model is very sensitive to the thickness of the cantilever for a given length. As shown above the thickness of the cantilevers in the array can vary significantly across the production wafer. As a result of the variation of thicknesses observed the models were used to predict the Q factors and resonance frequencies predicted for the middle and the limits of the range of thicknesses (7.2 ± 0.5 μm for the 7 μm thick cantilevers and 1.972 ± 0.005 μm for the 2 μm thick cantilevers).

The material and fluid properties were chosen to match the experimental conditions. Young's Modulus of Si: 169 GPa; density of Si (ρ_{Si}): 2330 kg/m^3; density of air (ρ_{air}) (at RT): 1.1839 kg/m^3; viscosity of air (at RT): 1.78 × 10^{-5} kg/(m s); speed of sound in air (at RT): 346.18 m/s.

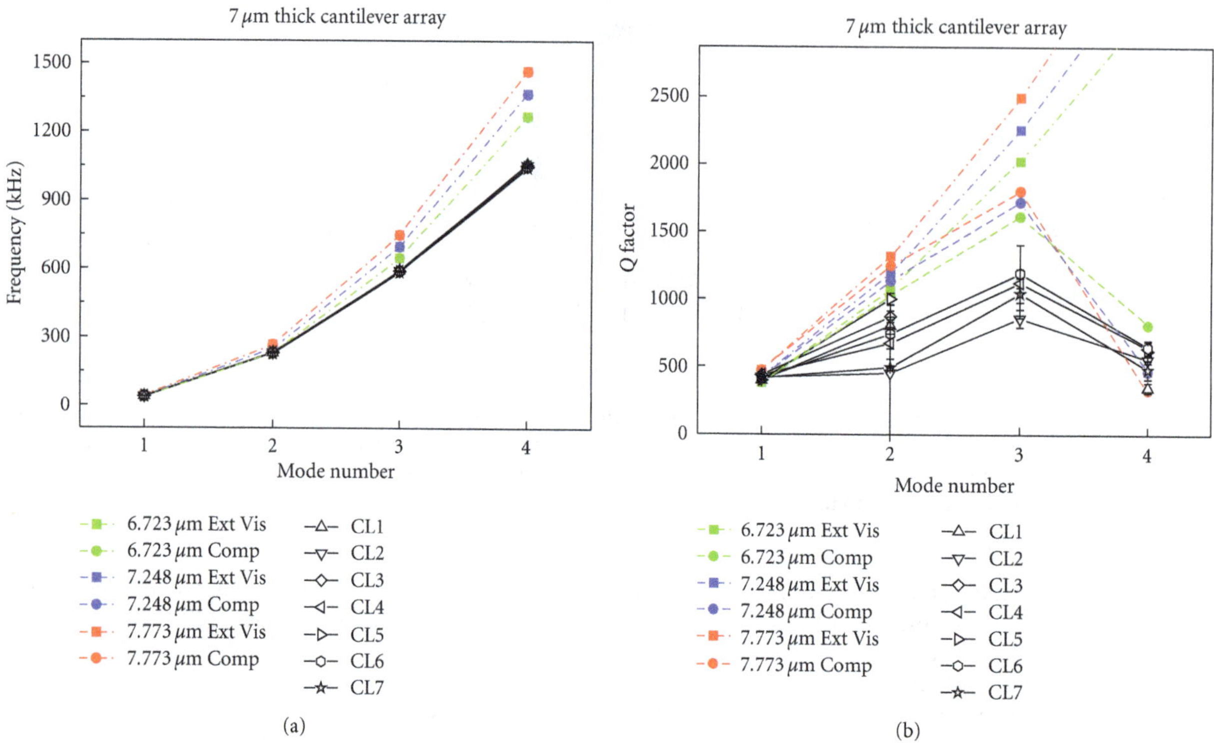

FIGURE 3: Resonance frequency and Q factor versus mode number comparison between theory and experiment for the 7 μm thick cantilevers. The solid square symbols with dotted lines correspond to the extended viscous model, while the solid circles with dashed lines correspond to the compressible fluid model. The open symbols with solid lines correspond to the experimental data. In the frequency plot the experimental data are overlapping.

The general equations for the resonance frequency and Q factor of a given mode are [18, 20]

$$f_{R,n} = \frac{f_{\text{vac},n}}{\sqrt{1 + (\pi \rho_{\text{air}} b / 4 \rho_{\text{Si}} t) \Gamma_r (f_{R,n}, n)}},$$
$$Q = \frac{(4 \rho_{\text{Si}} t / \pi \rho_{\text{air}} b) + \Gamma_r (f_{R,n}, n)}{\Gamma_i (f_{R,n}, n)}, \quad (2)$$

where t is the thickness of the cantilever, b is the width of the cantilever, $f_{\text{vac},n}$ is the vacuum resonance frequency of mode n of the cantilever, and $\Gamma(f_{R,n}, n)$ is the dimensionless hydrodynamic function and the subscripts r and i refer to the real and imaginary components, respectively. The calculations of the Q factor and resonance frequencies required finding the hydrodynamic function for each of the models (it is this term that the compressibility of the fluid affects). This involved solving the systems of linear equations given in [18, equation (11)] and in [20, equation (7)]. The integer M described in the models was chosen to be 36 and was shown to provide sufficient convergence of the solution for the higher modes of vibration (data not shown). For further information on the characteristics of these functions and their convergence see references [18, 20, 24]. Mathematica 8.0 was used to perform the calculations.

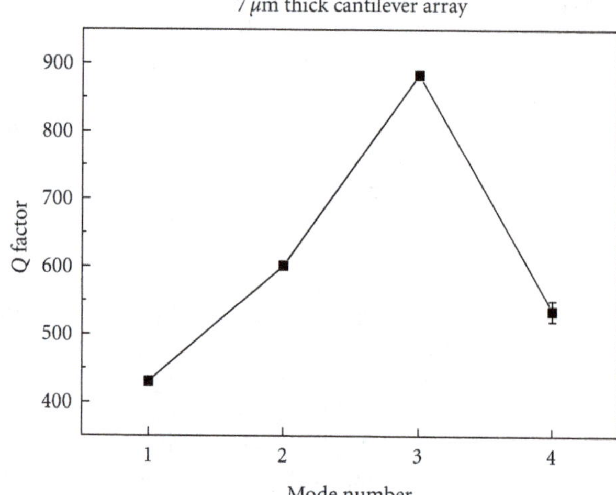

FIGURE 4: Repeated measurement of the Q factor versus mode number for a single cantilever without removing the array between measurements. The standard deviation of the Q factors for modes 1–3 is 0.003% and the standard deviation for mode 4 is 0.02%. This indicated that the previously observed larger standard deviations were due to difference in the coupling between the cantilever and the piezo between experiments.

3.2. Comparison between Theory and Experiment

3.2.1. 7 μm Thick Cantilevers. It was found that there was a decrease in the Q factor of the seven cantilever beams

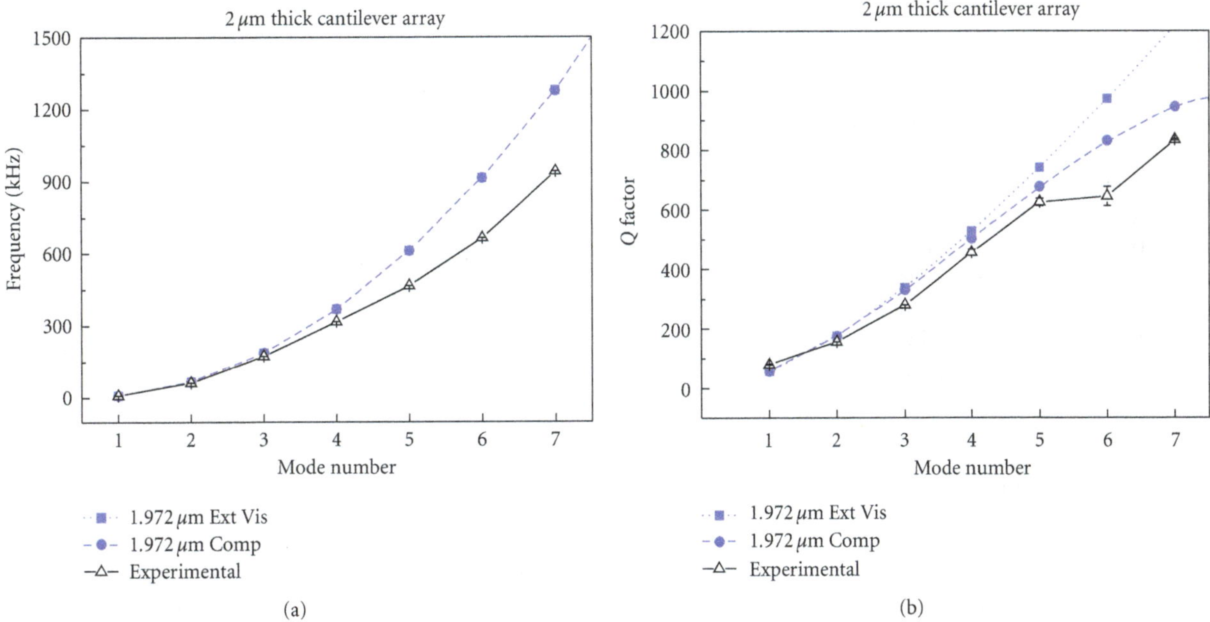

FIGURE 5: Resonance frequency and Q factor versus mode number comparison between theory and experiment for the 2 μm thick cantilever. The solid square symbols with dotted lines correspond to the extended viscous model, while the solid circles with dashed lines correspond to the compressible fluid model. The open symbols with solid lines correspond to the experimental data.

between the third and fourth modes which agrees with the prediction of the scaling analysis mentioned above that the "coincidence point" being the $n = 3$ mode. This decrease in Q factor can be clearly seen in Figure 3.

The large error bars of the experimental data shown in Figure 3 correspond to the standard deviation of the Q factor as measured from five experiments where the cantilever was unclamped and reclamped between experiments and is an indication of the coupling between the piezo and the cantilever. The standard deviation of the resonance frequencies measured for each cantilever were on the order of 0.001%. There was a difference of ~400 Hz in the fundamental frequency between cantilever 1 and cantilever 7. A similar decrease was also noted in the higher modes of the cantilevers and is an indication that there is a noticeable difference in thickness of the cantilevers within the array.

The resonance frequencies measured experimentally at the first mode agreed well with the models, however as the mode number increased the measured frequencies were increasingly lower than those predicted by the models. The lower frequencies observed are consistent with a cantilever which is longer than the cantilevers used here, but shorter than the cantilever and hinge section added together (~620 μm total length, data not shown). The strong dependence of the predictions of the compressible fluid model on the thickness of the cantilever can be observed in Figure 3.

A repeat of the experiment where the cantilever was not removed from the holder between measurements is shown in Figure 4. Here ten measurements were taken and it should be noted that the standard deviation is considerably smaller.

3.2.2. 2 μm Thick Cantilevers. The "coincidence point" predicted for a 2 μm thick cantilever was mode 12 with a resonance frequency of ~3.6 MHz. Using the current device it is not possible to observe the flexural resonance modes at such a high frequency, and therefore only the modes up to 1 MHz were observed. The hinge portion of the array is relatively thicker for these arrays than for the 7 μm thick cantilever arrays and as such should have less of an effect on the dynamics of the cantilever.

Figure 5 shows the comparison between the experimental data and the predictions of the extended viscous and compressible fluid models. It is clear that the resonance frequencies are still below those predicted by the models; however they match better than for the 7 μm thick cantilever array (16% compared to 29% difference at mode 4). This is an indication that the comparatively thinner hinge portion of the array is making a significant contribution to the dynamics of the 7 μm thick cantilever arrays.

It is clear from Figure 5 that there are differences in the predicted Q factors of the two models well below the "coincidence point". The experimental values match well with those predicted by the compressible fluid model (20% lower for the compressible fluid model compared to 75% lower for the extended viscous at mode 7).

3.3. Discussion. It is clear that the experimental data agrees qualitatively with the predictions of the compressible fluid model of Van Eysden and Sader, but that absolute quantitative agreement is not demonstrated here. Deviations of the resonance frequency and Q factors of the cantilevers between the predictions from the compressible fluid model and the experimental findings could be due to the hinge of the cantilever being only approximately twice or three times the thickness of the cantilever (Figure 1). This may not provide a sufficiently stiff support and there may be some degree of mechanical coupling between the hinge and the cantilever beam. This is significantly more noticeable for the 7 μm thick cantilevers where the hinge is comparatively thinner and as such is an indication that the hinge is the cause of the observed deviations from the compressible fluid model predictions. The models are based on an ideal cantilever extending from a fixed support and as such should not be expected to predict exactly the behaviour of cantilevers with a hinge design such as the one used here, however theoretical geometric assumptions are not always translatable into physical microfabricated devices.

Another possible reason for a qualitative and not a quantitative result could be that the model is based on thermal actuation of the cantilever beam and here a piezo actuator is used to amplify the motion of the cantilever, and while efforts are made to keep the cantilever operating within the linear regime of the vibrations this may not be 100% successful. It should also be noted that the model is valid for cantilevers with a large aspect ratio and here the cantilevers used to conduct the experiment have a ratio of 5 which places them very near the boundary for which the theory is valid.

In conclusion it was observed that there is at least qualitative agreement with the compressible fluid model for practical microcantilevers with a thickness to length ratio of $\sim$7 : 500 and an aspect ratio of 5. The prediction from the scaling analysis of Van Eysden and Sader of a "coincidence point" at mode 3 for the 7 μm thick cantilever is accurate and is clearly observed in the experimental data. The lower than predicted Q factors and resonant frequencies are likely attributed to the geometry and design of the hinge portion of the cantilever. The compressible fluid model should be considered when planning experiments involving the use of higher resonant modes of relatively thick microcantilevers in air.

Acknowledgments

The authors would like to thank John Sader and Anthony Van Eysden for discussions regarding their compressible fluid model. The SEM images were taken in the Advanced Microscopy Laboratory, Trinity Technology and Enterprise Campus, Dublin 2, Ireland under the framework of the INSPIRE program, funded by the Irish Government's Programme for Research in Third Level Institutions, Cycle 4, National Development Plan 2007–2013. This work was supported by Science Foundation Ireland under the CSET scheme SFI08/CE/I1432 and PI scheme SFI/09IN/1B2623.

References

[1] G. Binnig, C. F. Quate, and C. Gerber, "Atomic force microscope," *Physical Review Letters*, vol. 56, no. 9, pp. 930–933, 1986.

[2] R. McKendry, J. Zhang, Y. Arntz et al., "Multiple label-free biodetection and quantitative DNA-binding assays on a nanomechanical cantilever array," *Proceedings of the National Academy of Sciences of the United States of America*, vol. 99, no. 15, pp. 9783–9788, 2002.

[3] F. Huber, M. Hegner, C. Gerber, H. J. Guntherodt, and H. P. Lang, "Label free analysis of transcription factors using microcantilever arrays," *Biosensors & Bioelectronics*, vol. 21, no. 8, pp. 1599–1605, 2006.

[4] L. M. Lechuga, J. Tamayo, M. Alvarez et al., "A highly sensitive microsystem based on nanomechanical biosensors for genomics applications," *Sensors and Actuators B*, vol. 118, no. 1-2, pp. 2–10, 2006.

[5] J. Mertens, C. Rogero, M. Calleja et al., "Label-free detection of DNA hybridization based on hydration-induced tension in nucleic acid films," *Nature Nanotechnology*, vol. 3, no. 5, pp. 301–307, 2008.

[6] D. Ramos, M. Arroyo-Hernandez, E. Gil-Santos et al., "Arrays of dual nanomechanical resonators for selective biological detection," *Analytical Chemistry*, vol. 81, no. 6, pp. 2274–2279, 2009.

[7] P. S. Waggoner, M. Varshney, and H. G. Craighead, "Detection of prostate specific antigen with nanomechanical resonators," *Lab on a Chip*, vol. 9, no. 21, pp. 3095–3099, 2009.

[8] N. Backmann, C. Zahnd, F. Huber et al., "A label-free immunosensor array using single-chain antibody fragments," *Proceedings of the National Academy of Sciences of the United States of America*, vol. 102, no. 41, pp. 14587–14592, 2005.

[9] T. Braun, M. K. Ghatkesar, N. Backmann et al., "Quantitative time-resolved measurement of membrane protein-ligand interactions using microcantilever array sensors," *Nature Nanotechnology*, vol. 4, no. 3, pp. 179–185, 2009.

[10] B. Ilic, Y. Yang, and H. G. Craighead, "Virus detection using nanoelectromechanical devices," *Applied Physics Letters*, vol. 85, no. 13, pp. 2604–2606, 2004.

[11] N. Nugaeva, K. Y. Gfeller, N. Backmann, H. P. Lang, M. Duggelin, and M. Hegner, "Micromechanical cantilever array sensors for selective fungal immobilization and fast growth detection," *Biosensors & Bioelectronics*, vol. 21, no. 6, pp. 849–856, 2005.

[12] K. Y. Gfeller, N. Nugaeva, and M. Hegner, "Rapid biosensor for detection of antibiotic-selective growth of Escherichia coli," *Applied and Environmental Microbiology*, vol. 71, no. 5, pp. 2626–2631, 2005.

[13] K. Y. Gfeller, N. Nugaeva, and M. Hegner, "Micromechanical oscillators as rapid biosensor for the detection of active growth of Escherichia coli," *Biosensors & Bioelectronics*, vol. 21, no. 3, pp. 528–533, 2005.

[14] D. Ramos, J. Tamayo, J. Mertens, M. Calleja, L. G. Villanueva, and A. Zaballos, "Detection of bacteria based on the thermomechanical noise of a nanomechanical resonator: origin of the response and detection limits," *Nanotechnology*, vol. 19, no. 3, Article ID 035503, 2008.

[15] M. K. Ghatkesar, V. Barwich, T. Braun et al., "Higher modes of vibration increase mass sensitivity in nanomechanical microcantilevers," *Nanotechnology*, vol. 18, no. 44, Article ID 445502, 2007.

[16] F.-J. Elmer and M. Dreier, "Eigenfrequencies of a rectangular atomic force microscope cantilever in a medium," *Journal of Applied Physics*, vol. 81, no. 12, pp. 7709–7714, 1997.

[17] J. E. Sader, "Frequency response of cantilever beams immersed in viscous fluids with applications to the atomic force microscope," *Journal of Applied Physics*, vol. 84, no. 1, pp. 64–76, 1998.

[18] C. A. Van Eysden and J. E. Sader, "Frequency response of cantilever beams immersed in viscous fluids with applications to the atomic force microscope: arbitrary mode order," *Journal of Applied Physics*, vol. 101, no. 4, Article ID 044908, 2007.

[19] M. K. Ghatkesar, T. Braun, V. Barwich et al., "Resonating modes of vibrating microcantilevers in liquid," *Applied Physics Letters*, vol. 92, no. 4, Article ID 043106, 2008.

[20] C. A. Van Eysden and J. E. Sader, "Frequency response of cantilever beams immersed in compressible fluids with applications to the atomic force microscope," *Journal of Applied Physics*, vol. 106, no. 9, Article ID 094904, 2009.

[21] C. A. V. Van Eysden and J. E. Sader, "Compressible viscous flows generated by oscillating flexible cylinders," *Physics of Fluids*, vol. 21, no. 1, Article ID 013104, 2009.

[22] J. W. M. Chon, P. Mulvaney, and J. E. Sader, "Experimental validation of theoretical models for the frequency response of atomic force microscope cantilever beams immersed in fluids," *Journal of Applied Physics*, vol. 87, no. 8, pp. 3978–3988, 2000.

[23] K. Levenberg, "A method for the solution of certain problems in least squares," *Quarterly of Applied Mathematics*, vol. 2, pp. 164–168, 1944.

[24] C. A. Van Eysden and J. E. Sader, "Resonant frequencies of a rectangular cantilever beam immersed in a fluid," *Journal of Applied Physics*, vol. 100, no. 11, Article ID 114916, 2006.

18

Force-Sensor-Based Estimation of Needle Tip Deflection in Brachytherapy

Thomas Lehmann,[1] Mahdi Tavakoli,[1] Nawaid Usmani,[2] and Ronald Sloboda[2]

[1] *Department of Electrical and Computer Engineering, University of Alberta, 9107-116 Street, Edmonton, AB, Canada T6G 2V4*
[2] *Cross Cancer Institute, University of Alberta, 11560 University Avenue, Edmonton, AB, Canada T6G 1Z2*

Correspondence should be addressed to Thomas Lehmann; lehmann@ualberta.ca

Academic Editor: Aiguo Song

A virtual sensor is developed for the online estimation of needle tip deflection during permanent interstitial brachytherapy needle insertion. Permanent interstitial brachytherapy is an effective, minimally invasive, and patient friendly cancer treatment procedure. The deflection of the needles used in the procedure, however, undermines the treatment efficiency and, therefore, needs to be minimized. Any feedback control technique to minimize the needle deflection will require feedback of this quantity, which is not easy to provide. The proposed virtual sensor for needle deflection incorporates a force/torque sensor, mounted at the base of the needle that always remains outside the patient. The measured forces/torques are used by a mathematical model, developed based on mechanical needle properties. The resulting estimation of tip deflection in real time during needle insertion is the main contribution of this paper. The proposed approach solely relies on the measured forces and torques without a need for any other invasive/noninvasive sensing devices. A few mechanical models have been introduced previously regarding the way the forces are composed along the needle during insertion; we will compare our model to those approaches in terms of accuracy. In order to conduct experiments to verify the deflection model, a custom-built, 2-DOF robotic system for needle insertion is developed and discussed. This system is a prototype of an intelligent, hand-held surgical assistant tool that incorporates the virtual sensor proposed in this paper.

1. Introduction

Permanent interstitial brachytherapy is a cancer treatment procedure, in which radioactive seeds are implanted in tissue (e.g., the prostate, see Figure 1) in order to eliminate the cancer from inside. This procedure has emerged as an effective, minimally invasive, patient friendly, and cost-effective treatment option. For maximum treatment efficiency, the seeds have to be placed at exact locations inside and around the tumor that are determined in the preoperative planning stage. The radioactive seeds are initially loaded inside special needles. Intraoperatively, the seed-carrying needles are manually advanced toward planned locations where the seeds are deposited.

Two critical assumptions in the above procedure are that the needles will remain parallel across the entire length of their insertion in tissue and that the tissue will not deform as the needles penetrate it. However, in practice, neither assumption holds well, causing the actual needle trajectories to not pass through the planned locations. In fact, current manual needle insertion techniques for prostate treatment can place seeds with an accuracy of only 5 mm, which is a substantial error given the average prostate size. As a result, due to delivery of a different radiation dose, the radiation might not have the desired effect on the cancerous tissue and could instead undermine healthy tissue.

The predominant causes of inaccuracy in seed placement are needle deflection and tissue deformation during needle insertion/retraction [1, 2]. In general, needle bending is a function of the needle geometry [3, 4]. For instance, needles experience more bending with smaller diameters and with beveled tips. Beveled tips (Figure 2) are needed for easily cutting into tissue [5]. With a beveled tip, needle bending is larger for smaller bevel angles [6].

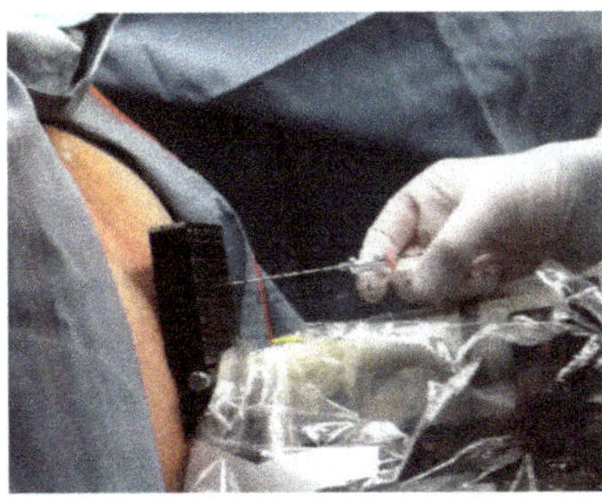

FIGURE 1: View of a prostate brachytherapy procedure as it is currently performed.

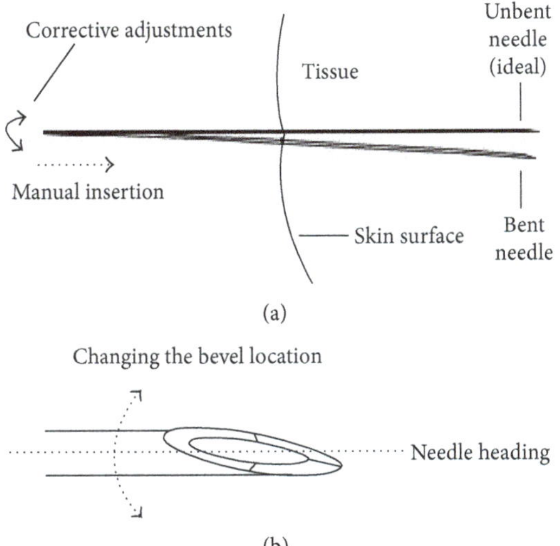

FIGURE 2: (a) A needle's bevel tip causes it to have a curved trajectory in tissue. (b) The needle's bevel location, which is the angular position of the bevel relative to the needle axis, can be adjusted to steer the needle.

Needle deflection and tissue deformation are coupled effects and influence each other. For simplicity, past studies have tried to decouple these effects by considering one of two extreme cases: a rigid needle in deformable tissue or a flexible needle in rigid tissue. For the second case, a bicycle-like kinematic model predicting the needle tip position was proposed [7]. A result pertinent to this paper is that when a beveled-tip flexible needle is pushed through tissue, the asymmetry of the tip causes the needle to bend (see Figure 2(a)). This is intuitively understood because the asymmetric geometry of the beveled tip leads to a force on the tip in the same direction as the bevel and a displacement of tissue to that side [5].

To improve needle targeting accuracy despite the bevel effect, there is active research on high-resolution imaging, tissue modeling, and robotic insertion preplanning. However, the resulting systems are still too complex and costly for the operating room. There is currently a need for a simple, inexpensive, and effective instrument in the surgeons' hands that can enable them to place the seeds more accurately.

We are working toward a low-cost and effective hand-held instrument for use in prostate needle insertion that automatically corrects the needle tip's bevel location during manual insertion in order to improve needle targeting and seed placement accuracy; the bevel location is defined as the angular position of the bevel tip relative to the needle's longitudinal axis (Figure 2(b)). A main idea related to needle steering is to leverage the very tendency of the bevel-tip needle for traveling on a curve to bring the needle tip back on the intended path once the needle has been bent, regardless of what caused the needle deflection [8]. To this end, the needle's bevel location can be changed by a feedback control algorithm based on the *real time feedback of needle deflection*. The feedback control algorithm will steer the needle into the opposite direction when needle deflection exceeds a set threshold by simply turning the needle about its longitudinal axis by 180°. This causes the forces acting at the needle tip to point in the opposite direction and steer the needle back toward the unbent ideal path. Since the surgeon is fully in charge of inserting the needle and the computerized system only modifies the bevel location, the hand-held instrument will be intrinsically safe to use.

The contribution of this paper lies in real time measurement of needle deflection using our proposed novel *virtual sensor*; incorporating a *physical sensor* for measuring this quantity is next to impossible in a clinical setting at least with the current technology. To measure the needle deflection during insertion into soft tissue, one may consider using 3D ultrasound imaging. However, the detection of the typical needle tip deflections in the order of several millimeters will be difficult due to the low resolution of ultrasound images and the associated high computational demands, notwithstanding the high costs of integrating a 3D imager and the required image processors in a computerized needle insertion system. Using 2D ultrasound imaging is not suitable because deflection measurement requires that the needle be always visible in the image plane while in practice the surgeon needs to intermittently switch between various sagittal and axial planes to monitor the insertion progress. Another possibility to measure the needle deflection is to use an electromagnetic tracking system, such as the Aurora Electromagnetic Tracking System by NDI, which involves a sensor embedded in the needle. However, this is only appropriate for in vitro testing where invasiveness is not an issue; sterilization requirements and the preloading of seeds inside the needle make in vivo utilization of a tracker inside the needle prohibitive. Therefore, in this paper, we pursue a noninvasive virtual sensing approach to needle deflection measurement in soft-tissue needle insertion. The foundation of our proposed virtual sensor is the hypothesis that the transverse force and the bending moment at the needle base predict the needle's longitudinal deflection inside the tissue. Therefore, in this paper, as the needle is being inserted into tissue, the forces and moments reflected at the needle base are measured by a physical sensor. Then, a needle tip deflection model and virtual sensor are developed solely based on this force/moment at the needle base. This virtual needle deflection sensor can be applied to a wide range of

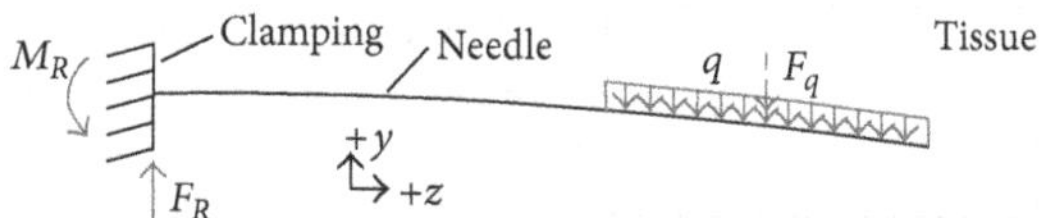

FIGURE 3: The distributed load q acting along the inserted portion of the needle as a reaction force to the needle, as proposed by Kataoka et al. [3]. F_R is the resulting force at the needle base. The 2D diagram plane is the deflection plane of the needle. The dashed arrow represents the resulting force acting at the geometric center (centroid) of the area of the distributed load q.

percutaneous medical procedures that involve needles prone to deflection.

2. Methods and Materials

2.1. The Needle Deflection Model. The developed needle model is fundamentally based on a static model for deflection of a beam. The needle can be regarded as a cantilever beam since its only fixation is at the base. The equations are based on the Euler-Bernoulli beam theory, which relates the loads or forces applied to the beam to its deflection. As the needle deflects increasingly during insertion in soft tissue, it acts as a loaded spring that tries to return to its initial unbent state but is kept in place by the tissue. The needle exerts a distributed load perpendicular to the needle axis onto the tissue. To keep the needle in its bent state, that is, to maintain the equilibrium condition, the tissue in return reacts with a distributed load along the needle (Figure 3). In general, distributed loads can be replaced by a resultant point force (concentrated load) that acts at a specific point along the needle. This reduction allows for a simplification of the equations used to calculate the deflection. The location of the resultant force is the *geometric center* or *centroid* of the area of the distributed load [9]. The magnitude of the force acting on the needle base (F_R) is the integral of the load and can be expressed as

$$F_R = F_q = \int_{L-l_0}^{L} q(z)\,dz \tag{1}$$

or

$$F_R = F_q = q \cdot l \tag{2}$$

for an equally distributed load, where q is the load per unit length, l_0 represents length portion outside of tissue, and l is the length portion of the needle, which is inserted into tissue. It should be noted that the indicated force at the needle clamping is the resistance force exerted by the clamp onto the needle. While the measured force (F_S) is equal in magnitude to F_R, the directions are opposed. The relation between F_R and F_S can be expressed as

$$F_R = -F_S. \tag{3}$$

Since the only support of the needle is at its base, the full magnitude of the resulting force and moment as well as the direction are reflected at the needle base and can be

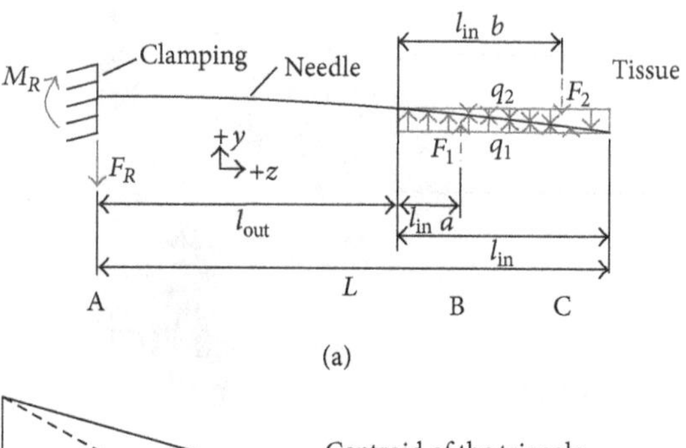

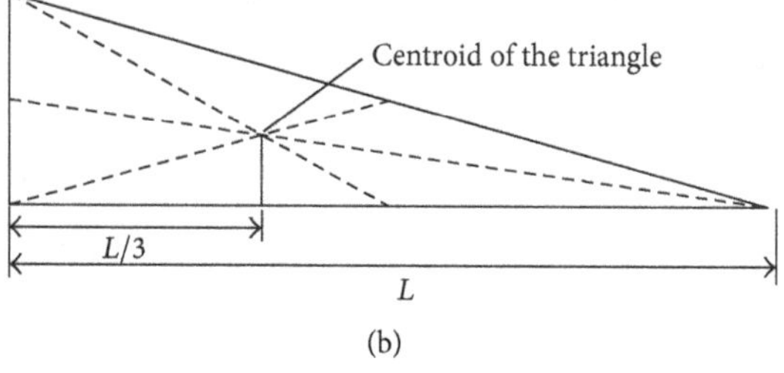

FIGURE 4: (a) The triangularly distributed loads q_1 and q_2 acting along the needle while inserting into tissue. F_R is the resulting force at the needle base. F_1 and F_2— shown with dashed lines—are the resulting forces from the distributed loads q_1 and q_2. (b) The location of the centroid of a triangle.

measured. This means that the expected force measurement results should have the same direction as the deflection. Insertion experiments, however, showed a force at the needle base in opposite direction. Figure 8(a) shows forces and torques for a deflection in negative y-direction. This means that the forces exerted by the tissue cannot only be one distributed load acting in negative y-direction (cf. Figure 3) but must be comprised of multiple loads acting in positive and negative y-direction. Tissue below the needle can be considered as acting as a support, and tissue above the needle still exerts a distributed reaction force on the needle.

The triangular distribution of q_1 in Figure 4(a) can be explained by assuming that tissue is pushed down more towards the side where the needle enters the tissue. As the needle penetrates deeper into tissue, the deflection increases, which leads to pressure being applied on the tissue in negative y-direction. Tissue, which is horizontally closer to the point of insertion (here: left side, see Figure 4(a)), is exposed to this pressure for an increased period. Therefore, the distribution of the load applied on the needle by tissue is at its maximum at the insertion point and decreases in z-direction. This assumption was also made by Abolhassani et al. [10]. They however did not include it in their model, as it had very little impact on their case. For the load above the needle, a reverse effect applies. As the tissue closer to the needle entry point is pushed towards the direction of deflection increasingly, the tissue above the needle in this area cannot apply any resistance force from above.

As before mentioned, the loads q_1 and q_2 can be reduced to the concentrated loads F_1 and F_2, which act at the centroid of the triangularly distributed load. The centroid of a triangle and its location is illustrated in Figure 4(b). The figure shows that the centroid of a triangularly distributed load is at $L/3$, where L is the length of the triangle's leg. Thus, the factors a and b should have the values 1/3 and 2/3, respectively (see Figure 4(a)). It should be noted that the triangular distribution of loads q_1 and q_2 is an assumption.

A common method to finding deflections in beam structures is twice integrating the bending moment equation:

$$M = EI\frac{d^2v}{dz^2}, \quad (4)$$

where M and v are the moment and deflection at a distance z from the base and E and I are the Young's modulus of stainless steel (200 GPa) and the area moment of inertia of the needle. The equation for the moment M can be found by analysis of the free body diagram. In our case there are multiple forces acting simultaneously (F_1 and F_2) at different locations. Therefore, the method of superposition needs to be applied. This method allows for regarding multiple forces acting separately on the beam and superimposing the deflections resulting from each force [11]. The deflections for each force are

$$\delta_1 = \frac{F_1(l_{\text{out}} + l_{\text{in}}a)^2(3L - l_{\text{out}} - l_{\text{in}}a)}{6EI},$$
$$\delta_2 = -\frac{F_2(l_{\text{out}} + l_{\text{in}}b)^2(3L - l_{\text{out}} - l_{\text{in}}b)}{6EI}, \quad (5)$$

([11, p. 1084]) where δ_1 and δ_2 are the deflections at the needle tip for F_1 and F_2, respectively, l_{in} and l_{out} are the length proportions inside and outside of tissue, L represents the total needle length, and a and b are factors to adjust the points at which forces F_1 and F_2 act. Superimposing the two deflections results in

$$\delta_{1,2} = \frac{F_1(l_{\text{out}} + l_{\text{in}}a)^2(3L - l_{\text{out}} - l_{\text{in}}a)}{6EI} - \frac{F_2(l_{\text{out}} + l_{\text{in}}b)^2(3L - l_{\text{out}} - l_{\text{in}}b)}{6EI}, \quad (6)$$

where $\delta_{1,2}$ is the deflection at the needle tip for forces F_1 and F_2 combined. The up to now unknown forces F_1 and F_2, with which a relation to F_R and M_R can be established, can be obtained from equilibrium conditions. These equilibrium conditions apply in the force and moment system of the free body diagrams in Figure 5. The cut in the free body diagram in Figure 5(a) was chosen because, in this case, the force F_1 will cancel out at both sides, as F_1 acts in the same direction at both sides of the cut. Force F_1 at point B can also be regarded as support, which means that no shear forces are acting at the cut. Since in every point during the insertion equilibrium conditions are assumed, the moments acting at both the left and right sides of B must sum up to zero:

$$\text{left side:} \quad \sum M = -M_R - M + F_R(l_{\text{out}} + l_{\text{in}}a) = 0$$
$$\Longleftrightarrow M = -M_R + F_R(l_{\text{out}} + l_{\text{in}}a),$$
$$\text{right side:} \quad \sum M = M - F_2 l_{\text{in}}(b - a) = 0 \quad (7)$$
$$\Longleftrightarrow M = F_2 l_{\text{in}}(b - a).$$

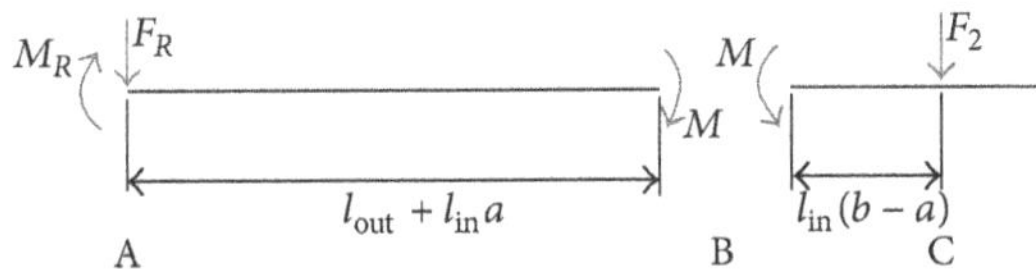

(a) The free body diagram of the bending moments acting at a cut at point B. F_R and M_R are the resistance force and moment of the clamping and are both measured by the attached sensor

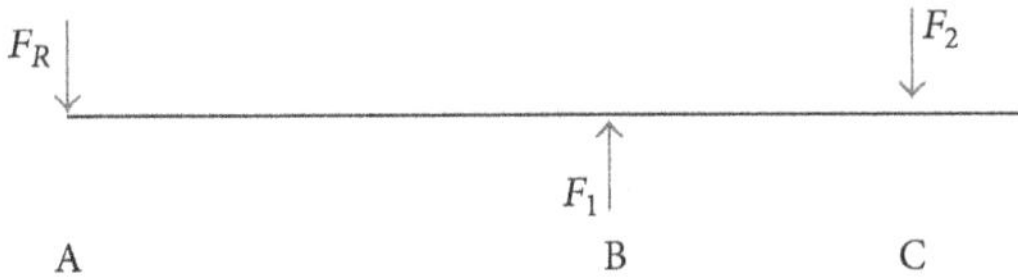

(b) The free body diagram of the forces acting perpendicularly along the needle

FIGURE 5: Free body diagrams of the needle while inserting into tissue.

As the bending moments at each side of the cut must be equal, (7) can be equalized to

$$-M_R + F_R(l_{\text{out}} + l_{\text{in}}a) = F_2 l_{\text{in}}(b - a)$$
$$\Longleftrightarrow F_2 = \frac{-M_R + F_R(l_{\text{out}} + l_{\text{in}}a)}{l_{\text{in}}(b - a)}, \quad (8)$$

and F_2 can be obtained. The force F_1 can be obtained from the free body diagram in Figure 5(b). The sum of the forces acting perpendicularly along the needle must also be zero:

$$\sum F = F_1 - F_R - F_2 = 0$$
$$\Longleftrightarrow F_1 = F_R + F_2. \quad (9)$$

Although here only deflection in the 2D plane is regarded, the model can be easily transferred to 3D, provided that a 4 DOF force/torque sensor is used. Also the deflections can be measured in both directions. If the needle deflects into positive y-direction, all forces and moments, including forces/torques measured at the needle base, will simply point into the opposite direction.

2.2. Experimental Setup. The setup for conducting insertion experiments into soft tissue consists of a robotic system with two degrees of freedom (DOF), which is capable of translational and rotational motion (see Figure 6). The translational motion is along the needle axis and rotational about the needle axis. The restriction to translational rotation along the needle axis is granted by a linear stage, which consists of a ball bearing mounted rail-carriage system. The system provides high precision while maintaining very low friction between rail and carriage (see Figure 6(a)). The carriage, which holds the rotational part, can be connected to a motor, which ensures linear insertion velocities (see Figure 6(b)). To carry out manual insertion experiments, the carriage needs to be entirely decoupled from the motor such that motor inertia does not affect the manual insertion procedure. A belt is used to connect motor and carriage as this provides the

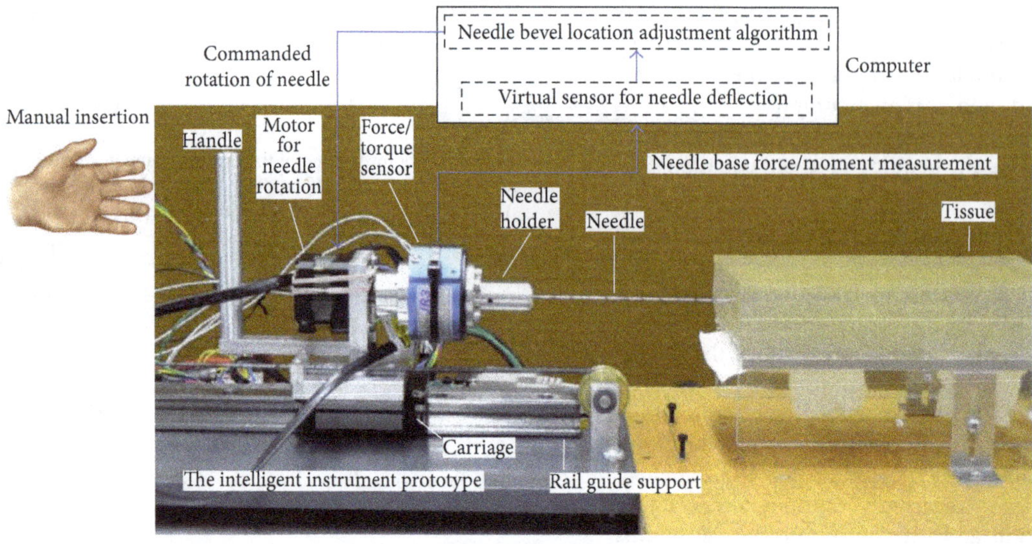

(a) Side view

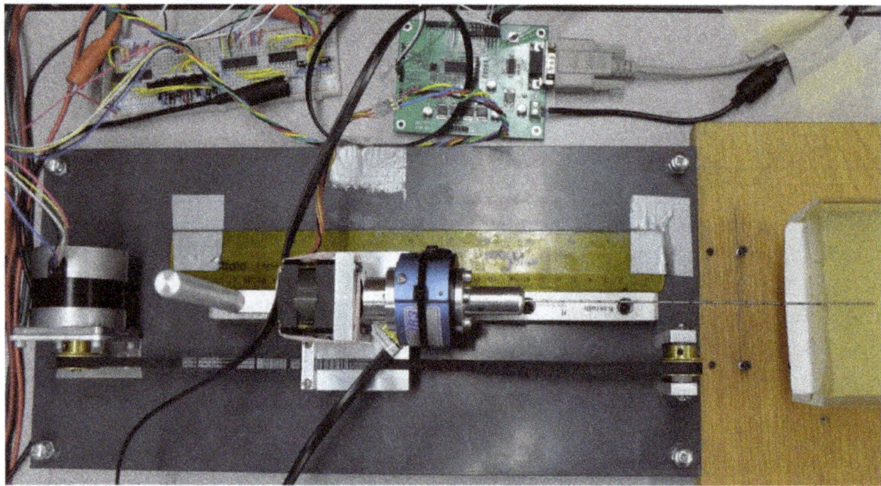

(b) Top view

FIGURE 6: The setup of the intelligent instrument prototype.

most convenient way to couple or completely decouple motor and carriage by simply clamping the belt to the carriage when necessary. Mounted on the carriage is the motor for needle rotation, which holds a force/torque sensor, and the needle holder, which fixates the hand piece of the needle via two opposing set screws. Also attached to the carriage is a handle to perform manual insertions.

The resulting needle bevel location correction system is a marked improvement over current practice, which typically involves nonoverlapping insertion-only and rotation-only phases; simply making rotations at predetermined insertion depths is an example of the strategies used by implanters. The same is the case in robotic needle insertion research. For instance, in [12], the robot maintains a constant insertion velocity for the needle except when the insertion is paused to rotate the needle by a fixed amount. Or, in [13], high-speed rotation reversal at half the insertion depth was tested. In contrast to the above, we adjust the needle bevel location during the insertion and believe that such controlled, smooth, and incremental corrections can maximize needle tip targeting accuracy. Our research should not be confused with continuous, drilling-like rotation of a needle during insertion. Given that a bevel-tip needle's trajectory in soft tissue is curved, constantly spinning the needle at a fast rate helps to keep the needle straight but also increases tissue trauma. As opposed to this, we will only adjust the needle's rotational position on a calculated and intelligent basis.

For sensing forces, a JR3 6 DOF sensor of type *50M31A* is used. The maximally admissible forces are 100, 100, and 200 N in x, y, and z direction, and maximum torques are 5 Nm for all three axes. The sensor's ADC has a resolution of 14 bits. The moments are measured about the center of the sensor. Since the relevant moments for deflection measurement occur at the tip of the needle holder (see Figure 6), the measured moments have to be recalculated in our setup. The moment

(M_H) acting at the tip of the needle holder can be expressed as

$$M_H = M_S - F_S l, \tag{10}$$

where M_S and F_S are the moment and force measured by the sensor and l represents the distance from the center of the force sensor to the tip of the needle holder (52.75 mm).

Both motors for rotational and translational motion are unipolar stepper motors. Stepper motors are used due to their ease of control and availability. The motors are controlled via L297 stepper controller ICs. The clock, enable, and direction signal are provided by a HILINK data acquisition (DAQ) board (see Figure 6(b)). The HILINK board interfaces a PC via RS232 and is controlled in MATLAB/Simulink.

The force sensor is interfaced via a C API. To read the data provided by the force sensor in Simulink, a C S-Function was written. Simulink S-Functions provide a way to run C/C++-code in Simulink models. The S-Function block outputs two 3D vectors for forces and torques. By including the reading of sensor data in Simulink, the setup can entirely be controlled in Simulink. Real time control of the assembly is however only possible in soft real time, as the force sensor's S-Function cannot be used together with MATLAB's Real Time Windows kernel. This limits the maximally admissible sampling rate to roughly 20 Hz, which is a relatively low but sufficient rate for our purposes.

The used phantom tissue for the insertion experiments is liquid plastic, which is made of plastisol and produced by M-F Manufacturing Co. The stiffness of the plastic can be adjusted by the amount of added plastic softener. The thickness of the homogeneous tissue is roughly four centimeters, which provides enough weight to prevent too much shifting along the insertion axis during insertion. A key factor of the phantom tissue is its transparency since the needle needs to be tracked while being inserted in order to measure its deflection. To determine the deflection, images of the needle inside tissue are recorded from above by a Logitech C270 webcam during insertion. Since the needle deflection can only be monitored in the horizontal two-dimensional plane, the deflection also needs to be kept to this plane. This can be achieved by aligning the bevel vertically. This way it can be safely assumed that the needle will deflect in the horizontal plane only. Potentially occurring gravity effects, which could lead to deflection in the vertical plane, can be neglected as the magnitude is below the range of the sensor. Furthermore, the supporting effect of the tissue prevents the needle from being pulled down by gravity.

To track the needle tip during insertion for comparison to estimated deflection, template matching is used. The template is illustrated in Figure 7(a). A MATLAB implementation by *Dirk-Jan Kroon* is used. The used matching method is normalized cross-correlation (NCC). This approach provides a high robustness against brightness variations and specifically bubbles in the tissue (see Figure 7(b), bright spots on the needle). The images have a resolution of 800 $\times$ 448 pixels and the camera observes a width of 80 mm at the height of the needle. This results in a resolution of 0.179 mm/pixel. The deflection is measured in respect to the unbent needle in air.

The used needle is a standard 18 G brachytherapy needle with an outer diameter of 1.27 mm and inner diameter of 1 mm. The bevel angle is roughly 20°. The length of the needle is 20 cm, but since a small part of the needle is clamped to the holder, the effective needle length, which can bend, is 19.1 cm.

3. Results

In this section we present experimental results for the validation of the proposed deflection model, and for the illustration of the impact that needle turning has on the deflection.

3.1. Model Verification. To verify the developed deflection model, insertion experiments were conducted consisting of six trials. Manual insertions with varying velocities and automated insertion with 10 mm/s and 15 mm/s were performed with two homogeneous tissue samples, which varied in stiffness. For this set of experiments, the needle was not turned at any point during the insertion. Forces and moments at the needle base and images were recorded during the insertion. The insertion depth was 120 mm throughout all the trials. Each trial consisted of 6 runs, in each of which a new point of insertion was used and the needle bevel was adjusted to deflect into the right direction as seen from the side of the robot. The recorded forces and moments were then used to estimate the deflection. The measured tip deflection was finally compared to the estimated tip deflection. For the manual insertion it was tried to keep the velocity at a constant level since the insertion speed was not measurable in real time. Before the start of insertion, the needle was inserted approximately 10 mm into the tissue.

Figure 8 shows an unfiltered sample deflection curve of the needle tip (see Figure 8(b)) and forces and moments at the needle base during insertion (see Figure 8(a)). The sensor data was filtered offline by a zero-phase lowpass filter, using MATLAB's *filtfilt*(·) function. Figure 8(b) also shows the deflections estimated by a model proposed by Abolhassani et al. [10] and the estimation of our proposed model. As the plot shows, the estimations of both models are very similar until later in the insertion process. At this point, our model maintains a relatively high precision whereas Abolhassani's model increasingly overestimates the deflection.

In Figures from 9(a) to 9(f), the mean estimations (n = 6) at insertion depths of 60 mm and 120 mm are illustrated for all 6 trials. The data shows that the aforementioned overestimation later on during the insertion can be observed throughout all the trials. It furthermore shows that our proposed model also slightly overestimates the deflection at a depth of 120 mm. At a depth of 60 mm, the estimations of both models are very close to the measured deflection throughout most of the trials.

To show where the estimations are in fact significantly different to the measured deflection, a paired t-test was carried out using MATLAB's function *ttest*(·). The null hypothesis of a paired t-test is that the mean of the difference of two data sets is *not* significantly different at a significance level of 5%. The test was carried out between measured and estimated samples (n = 6) for each model, for each

(a) The template used for tracking the needle tip

(b) Tracked needle tip, indicated by the red cross at the tip

FIGURE 7: Tracking of the needle tip.

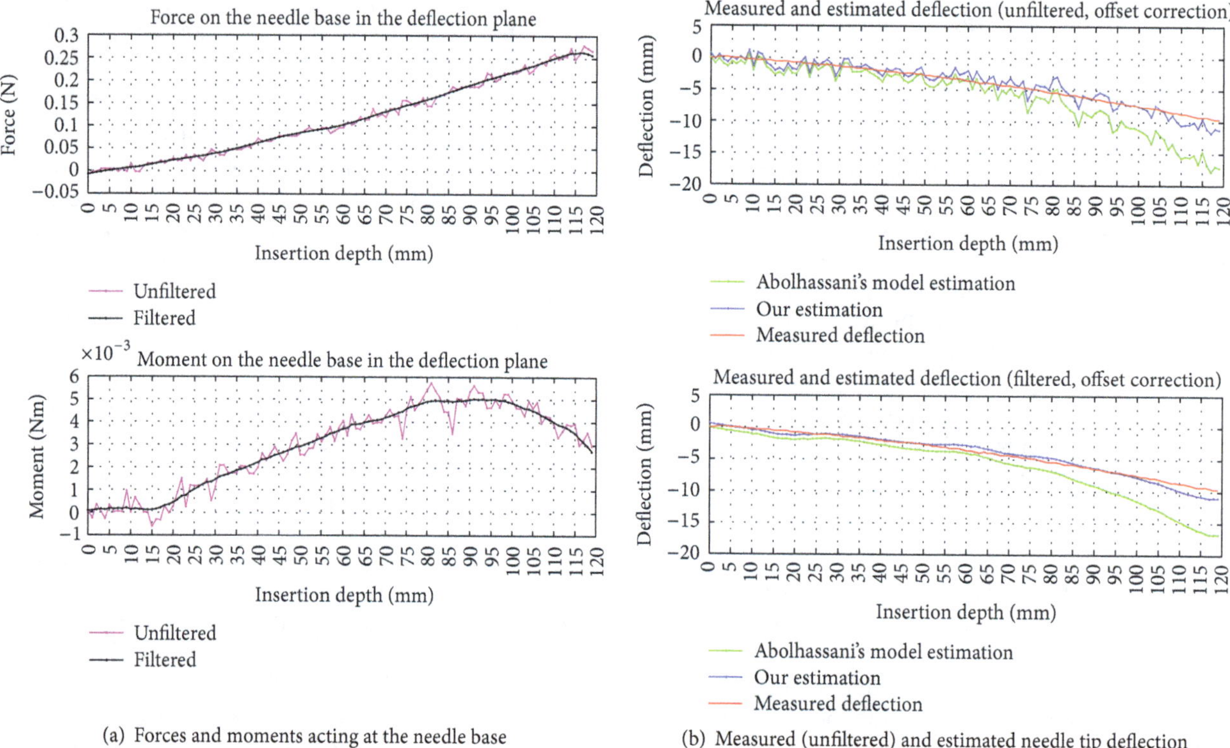

(a) Forces and moments acting at the needle base

(b) Measured (unfiltered) and estimated needle tip deflection

FIGURE 8: Sample plots for forces and moments at the needle base and tip deflection during insertion into tissue 1 with a velocity of 10 mm/s.

trial, and for the two observed insertion depths separately. The results of the t-test can be found in Table 1. They show that, at the depth of 60 mm, all trials do not reject the null hypothesis, which means that both model estimations are not significantly different at this point. At the depth of 120 mm however, the test shows that both model estimations are significantly different from the measured deflection for most cases. Figures 9(a) to 9(f) and the t-test confirm the previously made assertion based on Figure 8(b).

TABLE 1: Results of a paired t-test performed on estimated and measured data.

Model	Tissue 1			Tissue 2		
	10[a]	15	Man	10	15	Man
Abl[b] 60[c]	$\overline{r}$[d]	$\overline{r}$	$\overline{r}$	$\overline{r}$	$\overline{r}$	$\overline{r}$
Our 60	$\overline{r}$	$\overline{r}$	$\overline{r}$	$\overline{r}$	$\overline{r}$	$\overline{r}$
Abl 120	r	r	r	r	r	r
Our 120	r	$\overline{r}$	r	r	r	r

[a]Insertion speed in mm/s or "man" for manual.
[b]Model identifier, "Abl" for Abolhassani.
[c]Insertion depth in mm.
[d]"$\overline{r}$" if null hypothesis is *not* rejected.

3.2. Correcting the Deflection. A second type of experiment was performed to study the corrective effect of turning the needle about 180° when a certain estimated deflection threshold is reached. The needle is turned with an angular velocity of 180 deg/s. As the needle is being turned, the insertion is not paused but continues at the same speed. To estimate the threshold, the virtual sensor, which utilizes the developed model, was used. Two trials were executed, also consisting of six runs each. In each trial the threshold was set to a different level, the first being 1 mm and the second 5 mm. The needle also deflected in each run to the right side, or negative y-direction, until it was turned. One of the tissue samples, which was used for the first experiment (tissue 1), was used for this set of experiments. The runs involving turning the needle were conducted with a velocity of 5 mm/s, and, for the comparison with deflection without turning, data from the model verification

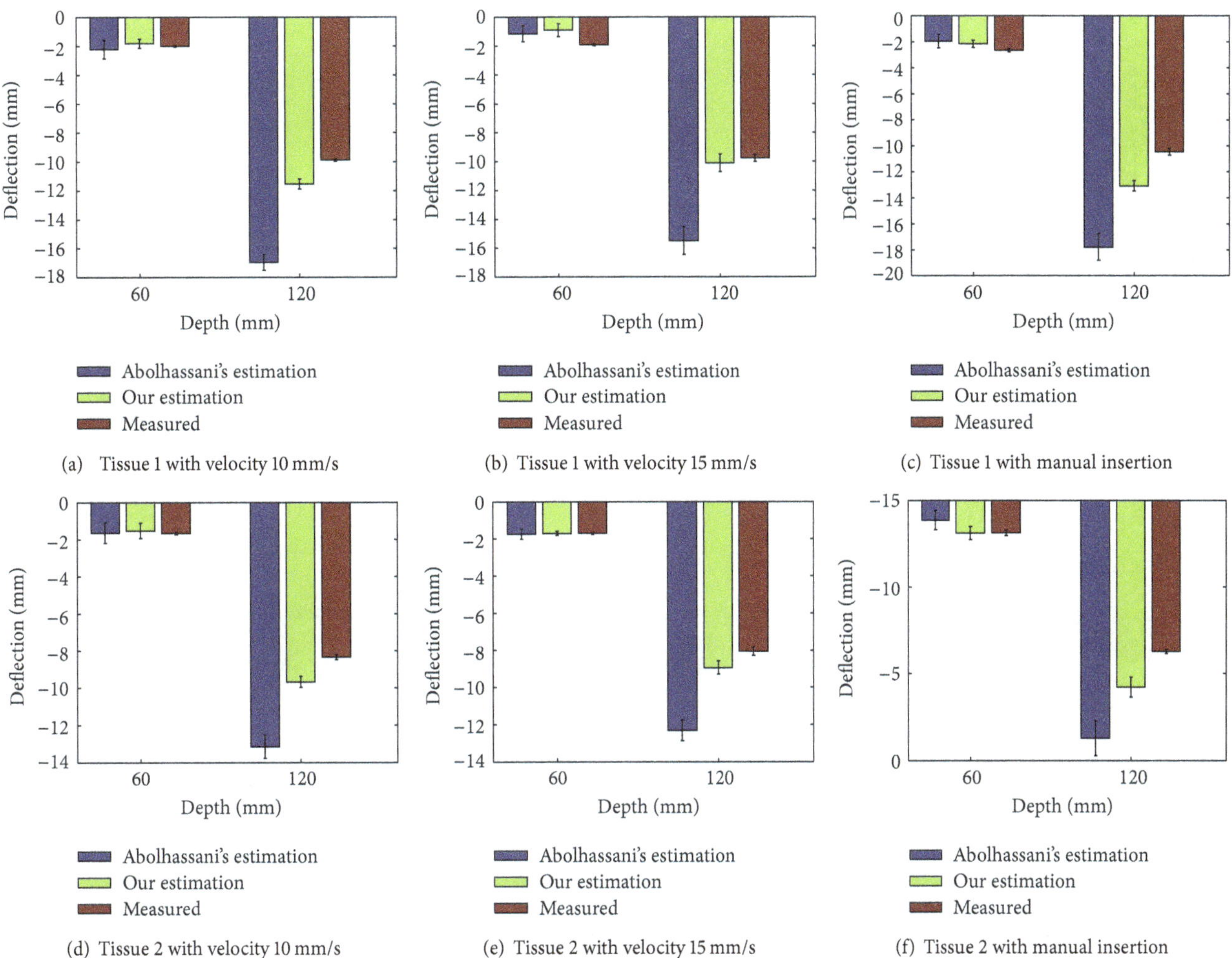

(a) Tissue 1 with velocity 10 mm/s

(b) Tissue 1 with velocity 15 mm/s

(c) Tissue 1 with manual insertion

(d) Tissue 2 with velocity 10 mm/s

(e) Tissue 2 with velocity 15 mm/s

(f) Tissue 2 with manual insertion

FIGURE 9: Measured and estimated tip deflection with tissue 1 in (a), (b), and (c) and tissue 2 in (d), (e), and (f). The insertion velocities were for (a) and (d) 10 mm/s automatic, for (b) and (e) 15 mm/s automatic, and for (c) and (f) manual with alternating velocities. The error bars show the standard deviation ($n = 6$).

experiment was used. Although the speed differs in between the experiments from 5 to 10 mm/s, judging by Figure 9, the insertion velocity does not influence the amount of deflection. This was also concluded by Webster et al. in 2005 [6].

Real time force readings are used to calculate the deflection with the proposed model in real time. Figure 10(a) shows three insertions: one without rotation, rotated at 5 mm deflection threshold and one rotated at 1 mm deflection threshold. The plot shows that turning the needle at a specific point has a very noticeable impact on the deflection. According to Figure 10(a), the best point for turning the needle is relatively early. The best correction results are achieved when the needle is turned when 1 mm deflection is reached. Figure 10(b) reaffirms this claim. Here the mean ($n = 6$) of each trial (turning at 1 mm, 5 mm and not turning) is shown for an insertion depth of 60 and 120 mm. The error bars, which show the standard deviation, indicate that the results for each turning point are significantly different. None of the error bars overlap.

4. Discussion

Several studies have been published, which show the effects of different parameter modifications or relate the force and moment at the needle base to the deflection in soft tissue. Kataoka et al. [3] proposed a first model, which uses a distributed load along the inserted section of the needle to estimate the deflection at the needle tip. The model is based on the Euler-Bernoulli beam theory, which provides a method for calculating the deflection of beams under load. Their results however showed an offset in the estimated tip deflection. Their conclusion was that their model underestimates the deflection outside of tissue, which leads to the offset. Abolhassani et al. [10] proposed a different model, which also relates the forces along the needle to the base forces and moments. Their model is also based on the Euler-Bernoulli beam theory but with different force assumptions along the needle.

Our deflection model, which is based on mechanical properties of the needle introduces a new approach as to

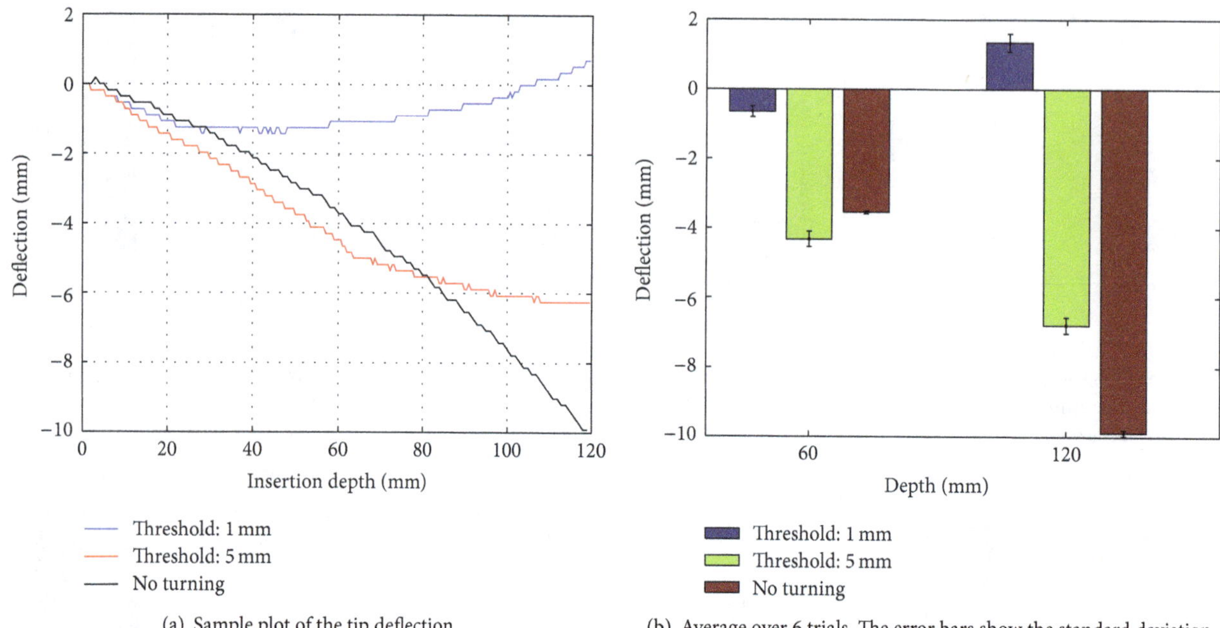

(a) Sample plot of the tip deflection

(b) Average over 6 trials. The error bars show the standard deviation

FIGURE 10: The impact of needle turning about 180° at multiple deflection thresholds.

how the forces are composed along the axis of the inserted needle into tissue. This new approach was proven to be more accurate than previously proposed mechanical deflection models.

The distribution of the loads described in Section 2.1, which are applied by the tissue onto the needle, are an assumption, as mentioned previously. Further investigation would be necessary to examine whether the load distribution is triangular or has a rather different shape. If the shape is in fact different, the factors a and b in (6) (see also Figure 4(a)) can be adjusted to fit the load distribution. The parameters a and b could also be fit to measured data in order to obtain a higher precision for the effective position of F_1 and F_2. This could further improve the model's performance, especially as the insertion depth increases.

It should be noted that the model can only estimate the deflection while the needle has not been turned. After the needle is turned, the distributed loads along the needle change.

In our experiments, homogeneous phantom tissue samples were used. In our future work, the model's precision will also be tested with tissue which has inhomogeneities.

The results show that a force sensor with a very low range is necessary, as the maximum forces are in the sub-Newton range. The sensor used for these experiments is very close to its lower limit.

5. Conclusion

A virtual sensor was introduced in this paper, which is able to precisely sense the needle tip deflection during needle insertion procedures without relying on any other invasive or noninvasive devices than a 4 DOF force sensor, which measures the forces and moments at the needle base. Higher precisions can be maintained over higher insertion depths with the proposed model than other models before.

It was also shown that adjusting the bevel location during the insertion can drastically increase the tip targeting accuracy with only one rotation during insertion.

This virtual deflection sensor can be applied to a wide range of medical procedures, which involve needles which are prone to deflection.

Conflict of Interests

The authors of this paper do not have any affiliations to or relationships with the companies NDI, M-F Manufacturing Co., JR3 Inc., Zeltom, Logitech, and Mathworks.

References

[1] R. S. Sloboda, N. Usmani, J. Pedersen, A. Murtha, N. Pervez, and D. Yee, "Time course of prostatic edema post permanent seed implant determined by magnetic resonance imaging," *Brachytherapy*, vol. 9, no. 4, pp. 354–361, 2010.

[2] V. Lagerburg, M. A. Moerland, J. J. W. Lagendijk, and J. J. Battermann, "Measurement of prostate rotation during insertion of needles for brachytherapy," *Radiotherapy and Oncology*, vol. 77, no. 3, pp. 318–323, 2005.

[3] H. Kataoka, T. Washio, M. Audette, and K. Mizuhara, "A model for relations between needle deflection, force, and thickness on needle penetration," in *Proceedings of the International Conference on Medical Image Computing and Computer Assisted Intervention (MICCAI '01)*, pp. 966–974, Springer, Heidelberg, Germany, 2001.

[4] R. Alterovitz, A. Lim, K. Goldberg, G. S. Chirikjian, and A. M. Okamura, "Steering flexible needles under Markov motion uncertainty," in *Proceedings of the IEEE IRS/RSJ International*

Conference on Intelligent Robots and Systems (IROS '05), pp. 120–125, August 2005.

[5] A. M. Okamura, C. Simone, and M. D. O'Leary, "Force modeling for needle insertion into soft tissue," *IEEE Transactions on Bio-Medical Engineering*, vol. 51, no. 10, pp. 1707–1716, 2004.

[6] R. J. Webster, J. Memisevic, and A. M. Okamura, "Design considerations for robotic needle steering," in *Proceedings of the IEEE International Conference on Robotics and Automation (ICRA '05)*, pp. 3588–3594, April 2005.

[7] R. J. Webster III, J. S. Kim, N. J. Cowan, G. S. Chirikjian, and A. M. Okamura, "Nonholonomic modeling of needle steering," *The International Journal of Robotics Research*, vol. 25, no. 5-6, pp. 509–525, 2006.

[8] N. Abolhassani, R. Patel, and M. Moallem, "Trajectory generation for robotic needle insertion in soft tissue," in *Proceedings of the International Conference of the IEEE Engineering in Medicine and Biology Society*, vol. 4, 2004.

[9] R. Hibbeler, *Engineering Mechanics: Statics*, Engineering Mechanics, Prentice Hall, 6th edition, 2010.

[10] N. Abolhassani, R. Patel, and F. Ayazi, "Needle control along desired tracks in robotic prostate brachytherapy," in *Proceedings of the IEEE International Conference on Systems, Man, and Cybernetics (ISIC '07)*, pp. 3361–3366, October 2007.

[11] J. M. Gere and B. J. Goodno, *Mechanics of Materials*, CL Engineering, 2012.

[12] N. Abolhassani, R. V. Patel, and F. Ayazi, "Minimization of needle deflection in robot-assisted percutaneous therapy," *International Journal of Medical Robotics and Computer Assisted Surgery*, vol. 3, no. 2, pp. 140–148, 2007.

[13] G. Wan, Z. Wei, L. Gardi, D. B. Downey, and A. Fenster, "Brachytherapy needle deflection evaluation and correction," *Medical Physics*, vol. 32, no. 4, pp. 902–909, 2005.

19

Welding Diagnostics by Means of Particle Swarm Optimization and Feature Selection

J. Mirapeix,[1] P. B. García-Allende,[2] O. M. Conde,[1] J. M. Lopez-Higuera,[1] and A. Cobo[1]

[1] *Photonics Engineering Group, University of Cantabria, 39005 Santander, Spain*
[2] *Hamlyn Centre for Robotic Surgery, Institute of Global Health Innovation, and Department of Surgery and Cancer, Imperial College London, London SW7 2AZ, UK*

Correspondence should be addressed to J. Mirapeix, mirapeixjm@unican.es

Academic Editor: Francesco Baldini

In a previous contribution, a welding diagnostics approach based on plasma optical spectroscopy was presented. It consisted of the employment of optimization algorithms and synthetic spectra to obtain the participation profiles of the species participating in the plasma. A modification of the model is discussed here: on the one hand the controlled random search algorithm has been substituted by a particle swarm optimization implementation. On the other hand a feature selection stage has been included to determine those spectral windows where the optimization process will take place. Both experimental and field tests will be shown to illustrate the performance of the solution that improves the results of the previous work.

1. Introduction

Welding processes play an important role in today's industry as they are employed in a wide range of industrial scenarios. Some typical examples to be mentioned are the fabrication of heavy components for nuclear power stations (e.g., steam generators), automobiles, engines for aeronautics, and tubes for different energy applications or civil engineering. In some of these applications the demands in terms of welding quality are very restrictive: a porosity produced during the tube-to-tubesheet welding process of a steam generator is a good example in this regard.

One of the main problems to be faced by engineers in the early stages of the definition of a specific welding procedure is the complexity of the physics involved in the process [1, 2]. Although both theoretical and experimental works have been attempted, experience indicates that the determination of the optimal parameters for a given scenario requires to perform previous studies in the laboratory and, afterwards, to carry out welding trials on coupons to verify the predicted behavior. In spite of all these efforts, defects will appear during the process even if all the variables are carefully controlled. This implies the use of both destructive and nondestructive evaluation techniques to examine the resulting seams and to verify that they comply with the required standards.

Different monitoring approaches have been proposed for both laser and arc-welding processes based on the use of electric [3–5], acoustic [6, 7], and optical sensors [8, 9]. Industrial cameras within the visible range have been also employed, typically with the aid of filters and illumination sources [10, 11] and infrared thermography has been also used for both online and offline inspection [12, 13]. Among all these alternatives, only the first has been seriously commercialized as it allows to establish a reliable process window. However, some defects, like the identification of spurious materials in the joint, are impossible to be detected with this approach. Apart from the sensor technology chosen, a great effort has been also developed in processing strategies designed for defect detection [14, 15] and classification [16–18].

Plasma optical spectroscopy has been also studied for its application in welding diagnostics [19, 20], and it is currently one of the most promising solutions in this area. The immunity of the optical fiber to the strong electromagnetic interference generated during the process, the robustness of the spectroscopic analyses of the different species to be found in the plasma, and the possibility to identify spurious materials in the weld pool are some of its most relevant advantages. The typical approach when using plasma spectroscopy in this context has been the determination of the plasma electronic temperature by means of two or more emission lines of the same species [19–22]. However, it has its limitations, as the uncertainty in the identification of the plasma emission lines, what has led to the exploration of other monitoring parameters [23–25].

Recently new analysis alternatives have been proposed: for example Sibillano et al. [26] introduced the so-called *Covariance Mapping Technique* to the analysis of the plasma dynamics in laser welding and Groslier et al. [27] studied the application of the pitch analysis to the voltage and current signals of a lap-welding (MIG-MAG) process. Another method is based on the generation of synthetic spectra to be matched to real experimental data by means of optimization algorithms [28]. In this way the resulting participating profiles of the chosen species showed a clear correlation to the quality events. However, some issues have given rise to a revision and improvement of the previous model. On the one hand the optimization algorithm previously selected, the CRS6 (Controlled Random Search-6), has been substituted by a simple implementation of the PSO (Particle Swarm Optimization) [29]. On the other hand, it was remarkable that the Ar II profiles exhibited a lack of sensitivity to some defects in some experimental results discussed in [28], what was supposed to be related to the use of the relative intensities from the NIST [30] local database to generate the synthetic spectra. To solve this problem a feature selection algorithm has been considered within the model to provide a selection of a narrower spectral range where the optimization will take place.

2. Spectroscopic Monitoring Parameters for Online Welding Inspection

The plasma electronic temperature T_e is the spectroscopic parameter commonly employed as monitoring parameter in this framework. Although a more precise estimation of this temperature can be obtained by means of the Boltzmann-plot [31], this solution, which implies the consideration of several emission lines and an additional regression process, is typically substituted by a simplified expression [19]:

$$T_e = \frac{E_m(2) - E_m(1)}{k \ln[(I(1)A(2)g_m(2)\lambda(1))/(I(2)A(1)g_m(1)\lambda(2))]}, \quad (1)$$

where E_m is the upper level energy, k the Boltzmann constant, I the emission line intensity, A the transition probability, g the statistical weight, and λ the wavelength associated with its corresponding emission line. For the particular case of arc-welding, (1) varies, including in the logarithm of the denominator the quotient between the emission line upper level energies [32].

The appearance of defects is related to the occurrence of perturbations on the T_e profile, but, although the correlation between this spectroscopic parameter and the quality of the seams has been proved [19–22], there are some issues, like the selection of the emission lines to participate in the T_e estimation, that have led to the investigation of alternative approaches.

The analysis of the wavelength associated with the maximum intensity of the plasma continuum radiation [23], the plasma RMS signal [25], and the line-to-continuum method used with a feature selection algorithm [24] are some solutions that have been recently investigated. A completely different approach was suggested in [28], where a model based on the determination of the so-called participation profiles of the plasma species was built by generating synthetic spectra and, afterwards, using optimization algorithms to try to match the real welding spectra. The synthetic spectra are created after the identification of the most significant species participating in the process and employing a local copy of a database with spectroscopic information of the required elements. Both central wavelengths and relative intensities are used in this process, but the latter give rise to convergence problems of the optimization stage if a wide spectral range is considered. This problem was identified in [28] as the Ar II, the predominant species in our scenario in the wavelength range under analysis (195–535 nm), did not show the expected response to some defects, while other profiles (Fe I, Mn I, Ar I) allowed a correct flaw detection.

A possible solution to this issue lies in the definition of narrower spectral windows where the optimization process and, consequently, the generation of the participation profiles will be performed. Obviously, this gives rise to the uncertainty in the selection of the most suitable spectral ranges in terms of defect detection. A similar problem was studied in [24], where a feature selection algorithm (SFFS) [33] was used to determine those emission lines most discriminating in terms of defect detection. Results showed a high dependency between the selected spectral band and the associated output monitoring profile. Apart from this modification, a simple implementation of the PSO (Particle Swarm Optimization) algorithm will be used instead of the original CRS6, as it will be demonstrated that the former exhibits an improved computational performance.

3. Modifications to the Original Model

3.1. Optimization Algorithm: PSO. In the original implementation (see Figure 1) a controlled random search algorithm, the CRS6, was employed to perform the optimization stage. A natural evolution of the model lies in the inclusion of a better algorithm in terms of the computational performance of the whole solution. In this regard, it is worth mentioning that this model is not originally intended to be used in a real-time analysis scenario, but to better understand the dynamics of the different species within the plasma and their behaviour when different defects appear in the welding process. However, it could be used as a support for other spectroscopic approaches for online monitoring (e.g., in feasibility studies), what justifies the search for more efficient implementations.

After some initial studies the PSO was chosen as a good candidate, given its simplicity and widespread use in several scenarios. In the field of welding some authors have chosen

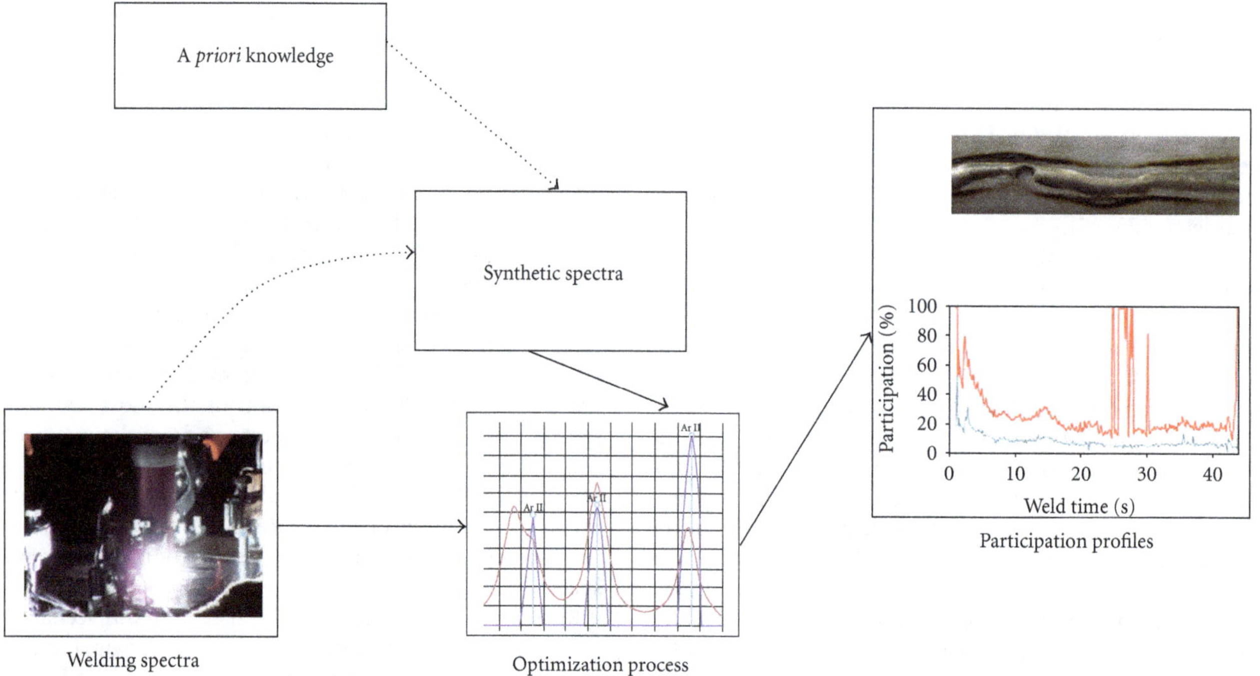

Figure 1: Schematic representation of the generation of the participation profiles from the creation of the synthetic spectra and the optimization process.

PSO algorithms to solve the optimization of key welding parameters [34] or for the training stage of neural networks [35]. PSO was originally proposed in 1995 [29], and it is inspired in the social behaviour of bird flocking and fish schooling, having suffered many changes since its original formulation, with new versions and applications. Apart from those already mentioned, typical fields of application for PSO have been image and video analysis, antenna design or power generation and systems, just to mention some examples.

The original PSO algorithm can be summarized as follows.

(1) Initialize a population array of particles with random position and velocities on D dimensions in the search space.

(2) Evaluate the predefined optimization fitness function for each particle.

(3) Compare the latest fitness evaluation of the current particle with its "previous best" p_{best}. If the current value is better, then p_{best} will be updated and p_i (previous best position) will be updated to the current location x_i.

(4) Determine the particle within the swarm with the best success so far (g_{best}) and assign its location to p_g.

(5) Proceed to change velocity and position of each particle within the swarm according to the following expression:

$$\begin{aligned} v_{\text{id}}(t+1) &= w v_{\text{id}}(t) + c_1 r_1 \left(p_{\text{id}}(t) - x_{\text{id}}(t)\right) \\ &\quad + c_2 r_2 \left(p_{\text{gd}}(t) - x_{\text{id}}(t)\right), \\ x_{\text{id}}(t+1) &= x_{\text{id}}(t) + v_{\text{id}}(t+1). \end{aligned} \tag{2}$$

(6) If the stopping condition is met, then exit with the best result so far; otherwise repeat from point 2.

Each particle within the swarm is defined by its position X_i and velocity V_i within the D-dimensional search space, where

$$X_i = (x_{i1}, x_{i2}, \ldots, x_{iD}), \tag{3}$$

$$V_i = (v_{i1}, v_{i2}, \ldots, v_{iD}). \tag{4}$$

In (2) w is the inertia weight, c_1, and c_2 are positive constants, typically defined as learning rates, and r_1 and r_2 are random functions in the range $[0, 1]$. Equation (2) describes a basic PSO algorithm, where the values of parameters w, c_1 and c_2 may significantly affect the behaviour of the algorithm [15], even making it unstable. The inertia weight can be interpreted as the fluidity of the medium where the swarm particles move, and typical values can be found between 0.4 and 0.9. Parameters c_1 and c_2 are typically assigned to 2, although they may have a significant influence on the

Table 1: Performance of CRS6 algorithm.

Condition (%)	Ar II (% participation)		Ar I (% participation)		Global iterations	Local iterations	Processing time (s)
	Mean	Std	Mean	Std	Mean	Mean	Mean
50	16.42	7.33	19.21	9.33	519	17	0.11
10	19.91	2.11	13.33	3.43	1156	113	0.23
5	20.75	1.78	13.18	1.93	1326	157	0.27
1	22.12	0.66	13.56	0.83	1600	243	0.34
0.5	22.27	0.51	13.59	0.87	1740	268	0.37
0.1	22.51	0.50	13.89	0.35	2049	363	0.44
0.05	22.69	0.46	13.97	0.16	2211	410	0.48
0.01	22.95	0.22	13.99	0.08	2547	534	0.57
0.001	23	0	13.99	0.08	3008	710	0.69
0.0001	23	0	14	0	3341	891	0.79

Table 2: Performance of PSO algorithm.

Particles	Iterations	Ar II (% participation)		Ar I (% participation)		Processing Time (s)
		Mean	Std	Mean	Std	Mean
20	20	24.02	5.86	14.32	4.91	0.0352
20	40	23.47	2.18	14.72	2.13	0.0657
20	60	23.80	0.73	14.79	0.68	0.0964
20	80	23.92	0.44	14.86	0.52	0.1261
20	100	23.96	0.21	14.96	0.21	0.1550
20	120	23.96	0.26	14.97	0.20	0.1883
20	140	23.99	0.08	15	0	0.2197
20	160	24	0	15	0	0.2480
10	100	23.97	0.18	14.95	0.21	0.08
10	120	23.99	0.08	14.99	0.11	0.09
10	140	24	0	15	0	0.11
10	160	23.99	0.08	15	0	0.12

search results. In addition, it is recommended to keep particle velocities within the range $[-V_{\max}, +V_{\max}]$, but the optimal value of $V_{\max}$ depends on the specific problem under analysis. An alternative to (2) is the use of the so-called constriction method [36]:

$$\begin{aligned} v_{\mathrm{id}}&(t+1) \\ &= \chi\Big(v_{\mathrm{id}}(t) + c_1 r_1 (p_{\mathrm{id}}(t) - x_{\mathrm{id}}(t)) + c_2 r_2 \big(p_{\mathrm{gd}}(t) - x_{\mathrm{id}}(t)\big)\Big), \\ &\qquad x_{\mathrm{id}}(t+1) = x_{\mathrm{id}}(t) + v_{\mathrm{id}}(t+1), \end{aligned} \tag{5}$$

where

$$\chi = \frac{2}{\phi - 2 + \sqrt{\phi^2 - 4\phi}}, \qquad \phi = \phi_1 + \phi_2 > 4. \tag{6}$$

Typical values for these parameters are $\phi = 4.1$, $\phi_1 = \phi_2$, and $\chi = 0.7298$. Although not necessary, it is recommended to establish $V_{\max} = X_{\max}$.

Once the solution described by (5) was implemented, some tests were performed to compare the performance of PSO with the results offered by CRS6 that are summarized in Table 1, where *Condition* is the stopping condition ε of the algorithm:

$$\left| f^* - f(\hat{x}^*) \right| \le \varepsilon, \tag{7}$$

where $f(x)$ is the function to minimize, f^* the minimum, x^* the value to be found in the optimization process, and $\hat{x}^*$ an approximation to x^*.

Using a welding plasma spectrum captured during the experimental tests in the laboratory, both the convergence and the processing times of the PSO were determined under different conditions described in Table 2. *Particles* is the number of particles considered in the swarm for the optimization process, *Iterations* the number of iterations considered in each search, *Participation* the relative concentration of the species (neutral atoms and ions) participating in the plasma, and *Processing time* the overall estimated computational time of the optimization process. Both mean and standard deviation (std) values of the Ar I and Ar II participation have been calculated, indicating the ability of PSO to converge to the expected solution. It should be mentioned that the optimization process was performed over a set of 150 identical spectra, thus simulating a perfect seam without any defect.

In terms of the computational performance, it can be observed that PSO offers in this case processing times from 0.035 to 0.248 s (using 20 particles), while the results for CRS6 in Table 1 ranged from 0.11 to 0.79. It is also worth mentioning that the convergence values for both Ar I and Ar II are quite similar, but the standard deviation (std) is clearly higher for CRS6 and, although the parameters to be adjusted in both cases are different, it seems clear that the computational performance of PSO exceeds the one presented by CRS6, what justifies the inclusion of the former in the model under analysis.

3.2. Use of the SFFS Algorithm for Spectral Range Selection. In the art of pattern recognition, that is in the automatic recognition, description, classification, and grouping of patterns in disciplines ranging from biology and psychology to computer vision or remote sensing [37], dimensionality reduction techniques are employed prior to recognition/classification. These attempts to find the minimum number of dimensions a data set can be expressed in without significant loss of information reduces the number of variables of the pattern representation (i.e. the number of features) required for the analysis. There are two main reasons to keep the number of features as small as possible: measurement cost and classification accuracy. A small number of features can alleviate the curse of dimensionality [38] if the number of training samples is limited, but what is more classification hit-rate could be greatly enhanced too if class separability or the distance among patterns belonging to different clusters is simultaneously maximized. There exists a wide variety of characterization methods [37] that achieve these objectives essentially by two different ways. Feature selection algorithms select the (hopefully) best subset of the input feature set while methods that create new features based on transformations or combination of the original feature set are called feature extraction algorithms. Although both alternatives are aimed at maximizing class separability, feature selection is preferable when dealing with spectral data since it also provides a physical insight of the problem [39]. Moreover, dimensionality reduction could be performed inversely or in advance to identify the spectral bands that best separate the classes (correct seams and flaws) and use them to construct the monitoring signal. In this way the signal to noise ratio of the latter, and as a consequence defect sensitivity, would be clearly increased. The feasibility of this approach was demonstrated in a previous work [24], where the line-to-continuum method (i.e., the ratio between intensity lines and their adjacent background radiation) was used to generate the output monitoring profiles and Bhattacharyya distance [40] was employed as the criterion to measure class separability for wavelength selection. This probabilistic distance is very convenient to evaluate class separability for normal distributions, but even for nonnormal cases it seems to be a reasonable equation [41]. The Mahalanobis distance, given by (8), is a particular case of the Bhattacharyya distance that assumes equal covariances of the classes:

$$J_M = (\mu_2 - \mu_1)^T \Sigma^{-1} (\mu_2 - \mu_1), \tag{8}$$

where μ_i is the mean of the i class and Σ the covariance matrix. It is widely used as dissimilarity measure too because it requires about p^2 flops for a multivariate feature characterized by its mean vector $\mu \in R^p$ and covariance matrix $\Sigma \in R^{p\times p}$, while the computation of the Bhattacharyya distance involves $p^3/3 + 2p^2$ flops [42].

Given the necessity mentioned above of constricting the optimization process to spectral ranges narrower than the one provided by the spectrometer, the use of the SFFS algorithm to identify suitable spectral regions seems interesting in this scenario, as it constitutes an automatic procedure instead of having to perform specific studies for new processes or spectrometers. In this case, the performance of the employment of the Mahalanobis distance for retrieving the most appropriate wavelengths that will make up the output monitoring signal will be evaluated here. Let X be the number of spectral regions where the optimization process and, consequently, the generation of the participation profiles will be performed, that is, X is the number of bands to be selected. At a certain point of the selection process, S is the current set of previously selected bands and R is the set of remaining or unselected bands. The selection process starts being $S = \emptyset$. The pseudocode that describes the selection procedure is as follows:

```
while |S| < X do
    select band S_inc = arg max[J_M(SUS_inc)]
    S = SUS_inc
    R = R \ S_inc
    while |S| > 2
        select band S_exc = arg max[J_M(S \ S_exc)]
        if J_M(S \ S_exc) > J_M(S)
        S = S \ S_exc
        R = RUS_exc
        else
        break
        end
    end
end,
```

where "SUS_{inc}" denotes that the band S_{inc} is included into the set S, "$S \setminus S_{exc}$" denotes that the band S_{exc} is excluded from set S, and ϕ is the empty set.

4. Experimental and Field Test Validation

The first studies were aimed at improving the results obtained for the Ar II species in [28], given the already commented lack of response for some defects. After an initial analysis via SFFS, some spectral bands were chosen by the algorithm as the most suitable in terms of discrimination among spectra associated with correct seams and with defects, respectively. The details of the experimental tests are described in the previous work, but it should be mentioned that a GTAW (Gas Tungsten Arc Welding) process was employed to weld AISI-304 stainless steel plates. Defects

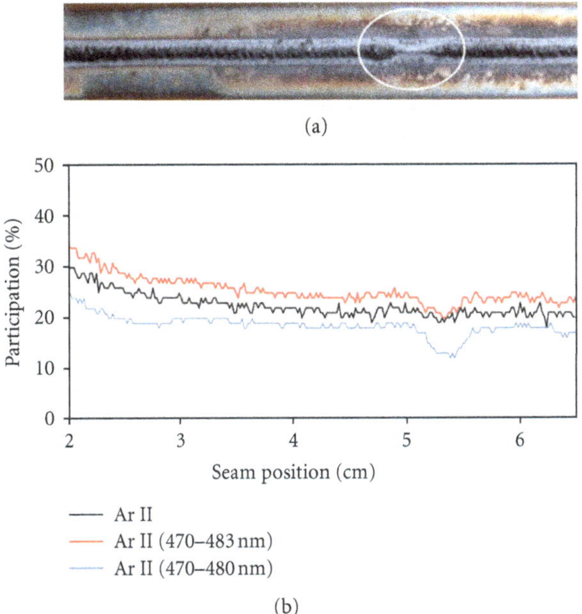

FIGURE 2: (a) Weld seam with defect. (b) Participation profiles of Ar II (whole spectral range), Ar II (470–483 nm), and Ar II (470–480 nm).

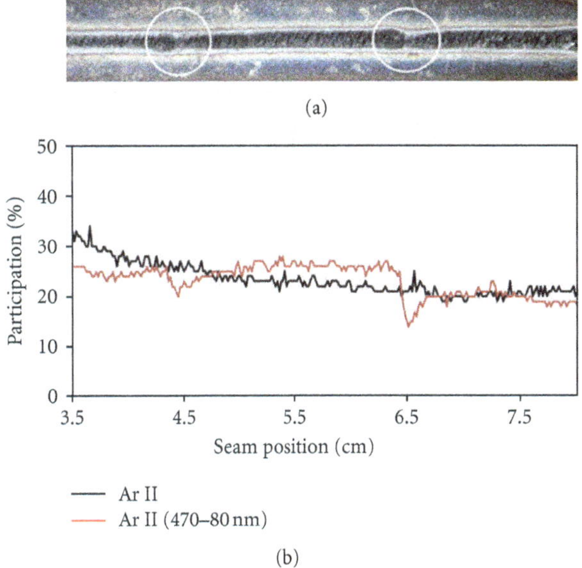

FIGURE 3: (a) Weld seam with defects. (b) Participation profiles of Ar II (whole spectral range), and Ar II (470–480 nm).

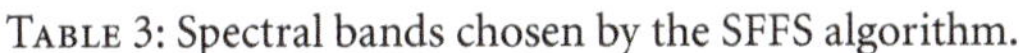

TABLE 3: Spectral bands chosen by the SFFS algorithm.

Order	AISI-304 Band (nm)	Inconel-718 Band (nm)	Ti 6Al-4V Band (nm)
1	393.20	520.33	393.35
2	356.75	360.49	396.14
3	396.81	390.55	334.87
4	356.58	344.15	416.27
5	442.15	402.67	375.95
6	355.06	517.93	453.32
7	476.54	484.10	422.68
8	487.69	340.72	397.61
9	396.97	357.28	430.00
10	480.63	403.81	497.80

were provoked intentionally by introducing perturbations on the shielding gas (argon) flow rate. The first spectral bands chosen by the SFFS algorithm using the Mahalanobis distance are presented in Table 3.

From these wavelengths, those in the range from 460 to 490 nm are related to Ar II emission lines, what suggests the suitability of selecting that spectral window for the optimization process. Figure 2 depicts the result of using the windows between 470 and 483 nm and 470 and 480 nm, respectively, in comparison to the original Ar II participation profile derived from the use of the whole spectral range of the spectrometer (*Ocean Optics* USB2000: 195 to 535 nm). It can be observed that the correlation between the defect in the seam (provoked by a perturbation on the shielding gas flow rate) and the resulting Ar II participation profile is significantly enhanced if the spectral range of the optimization process is reduced. The result is better for the narrower spectral range, what can be explained by the poor match obtained during the optimization process between the synthetic and the real spectra for the Ar II emission line located at 480.5 nm.

A similar comparison is established in Figure 3, where another seam with two discontinuities can be observed. Again, the defects at $x \approx 4.5$ and 6.5 cm are not clearly detected using the whole spectral range, but the employment of the 470 to 480 nm window gives rise to a more sensitive monitoring signal.

To extened the analysis to other processes and materials several studies have been performed on data from field tests [25]. In this case the materials to be welded were Inconel-718 and Titanium 6Al-4V, with 2 and 1.6 mm of thickness, respectively. Filler wire was used for the former, and Ar was used as shielding gas (10 L/min), being also guided to the bottom side of the plates (30 L/min). The optical setup was basically constituted of a 600 μm core diameter optical fiber connecting the spectrometer (again the USB2000) and the fiber end acting as input optics located at approximately 10 cm from the electrode tip. Apart from correct seams, different defects were provoked during the analyses to obtain the desired spectroscopic data.

Figure 4 shows an Inconel-718 seam cataloged as correct after visual and X-ray inspection. The Ar II participation profiles depicted does not show any clear perturbation, although both signals exhibit a significant noise level. It is worth mentioning that other spectroscopic parameters, like the plasma RMS profile [25], also show that behavior. A possible explanation to this can be found in perturbations affecting the process that do not give rise to defects. The associated heat input profile (acquired by Tecnalia [43] with an electric sensor system) is also constant (Figure 4(c)), as expected for a seam free of defects.

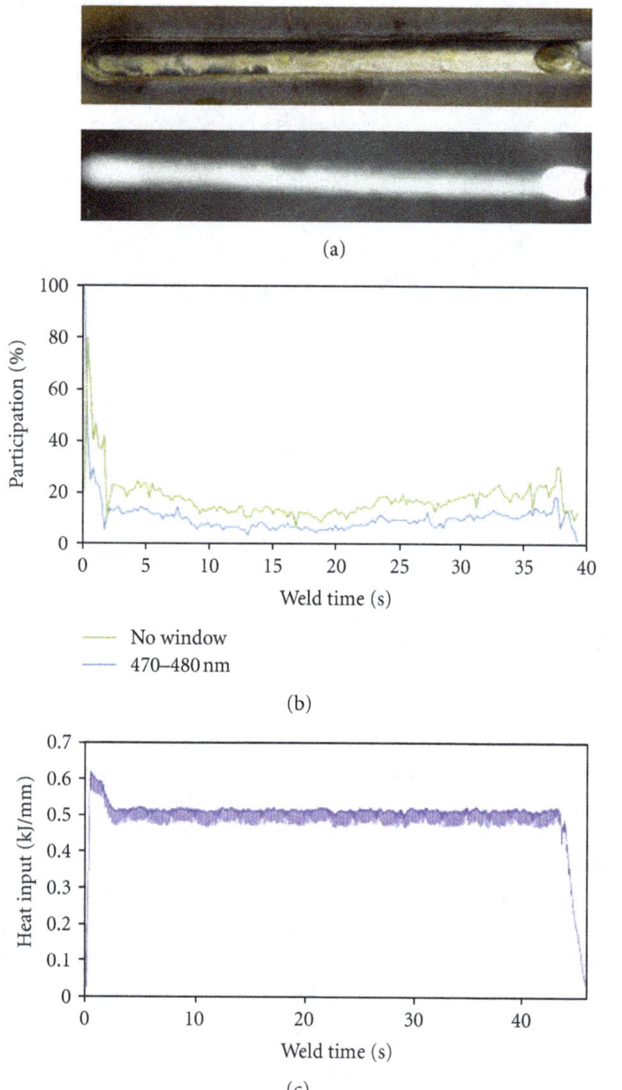

Figure 4: (a) Correct weld seam (including X-ray image) (Inconel-718). (b) Participation profiles of Ar II (whole spectral range) and Ar II (470–480 nm). (c) Heat input profile.

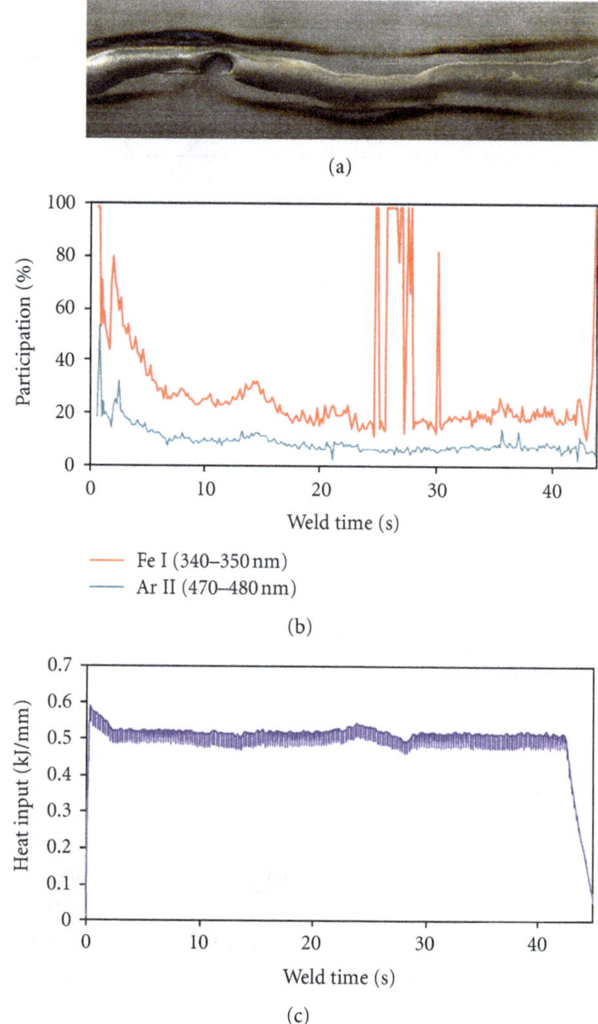

Figure 5: (a) Defective seam (trajectory deviation) (Inconel-718). (b) Participation profiles of Fe I (340–350 nm) and Ar II (470–480 nm). (c) Heat input profile.

A defective seam is analyzed in Figure 5, where the trajectory of the welding torch over the joint was deviated (Figure 5(a)). It can be appreciated that the heat input signal gives an indication of defect at $x \approx 25$ to 30 s, while the rest of the profile is almost constant. Two different participation profiles have been depicted in Figure 5(b), corresponding to two spectral ranges: 340 to 350 nm and 470 to 480 nm. The first band was chosen taking into account that the feature selection algorithm indicates the 344.15 band, being in this case the Fe I species the one selected for the process. This window generates a monitoring signal with a strong perturbation correlated to the one observed in the heat input signal, although other regions also indicate the occurrence of defects. In comparison, the Ar I profile do not exhibit in this case so clear perturbations.

The seam showed in Figure 6 was performed to join two Inconel-718 plates with a misalignment of approximately 1 mm, being the maximum allowed in this case 0.3 mm (15% of the plate thickness). The heat input signal depicted in Figure 6(c) does not exhibit any perturbation, being constant during the whole process. Almost the same situation can be observed in the T_e profile presented in Figure 6(d), calculated using the Ar II emission lines located at 460.96 and 487.99 nm, respectively. However, the participation profile in this case is somewhat noisier, suggesting that a defective situation has taken place.

Defects provoked by lack of cleanliness have also been studied on Ti6Al-4V plates, simulating this situation by applying oil on the joint before the welding process. In the test described in Figure 7 the presence of oil gives rise to a clear defect at $x \approx 30$ s, which is signaled by the heat input profile. The reduction of the spectral range to generate the Ar II participation profile (Figure 7(b)) produces a similar response to the one derived from the whole spectrometer range, although the sensitivity of the latter appears to be somewhat worse, specially around $x = 40$ s. It must be noted

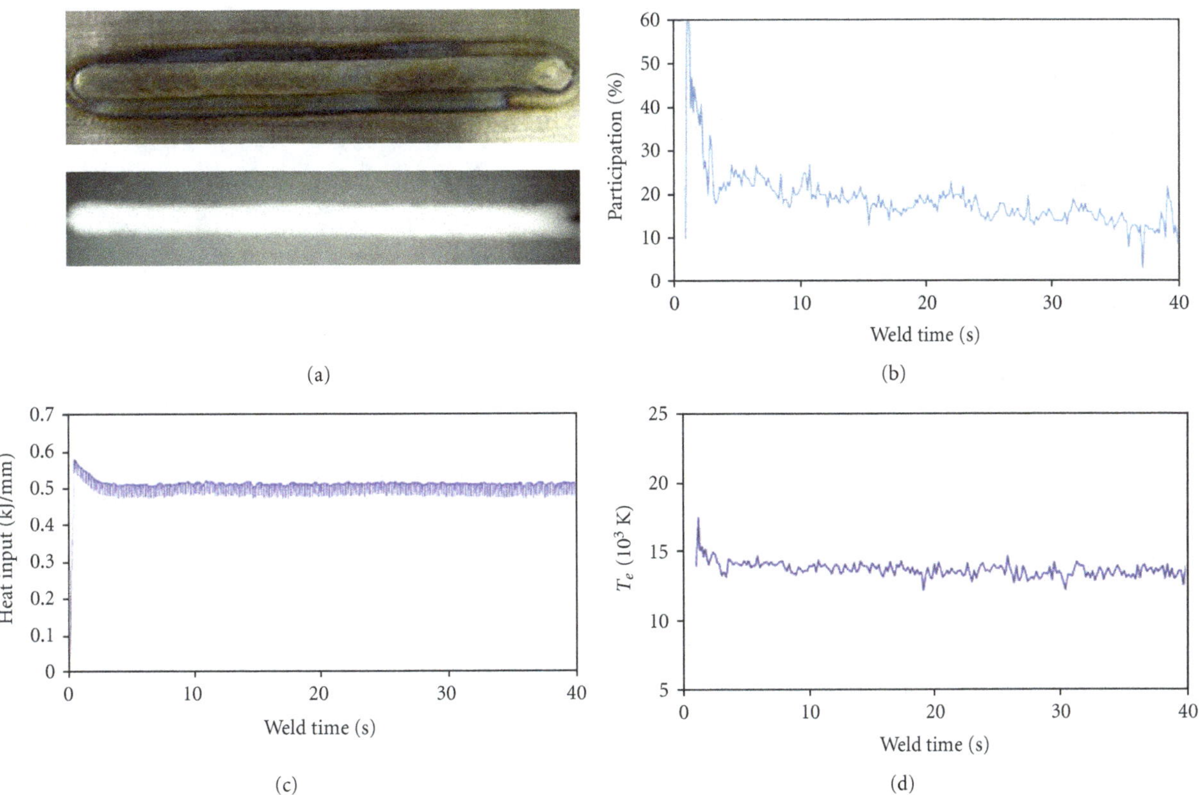

FIGURE 6: (a) Defective seam (misaligment) (Inconel-718)n. (b) Participation profiles of Ar II (470–480 nm). (c) Heat input profile. (d) Plasma temperature profile.

the change on the Ar II signal at $x \approx 18$ s, what can be associated with the application of oil at the middle of the welding path. The interpretation to the appearance of the defect later in the seam can be explained by the dragging of the oil by the welding arc up to the defect location.

The same defect was tried to be repeated for the seam depicted in Figure 8, but no defects were observed in this case after the visual inspection. Again, the heat input signal remains constant through the process, while the T_e profile exhibits a clear slope and some subtle perturbations. The signal offered by the Ar II species (470–480 nm) clearly indicates the occurrence of defects, what seems to be in good agreement with the scenario under analysis.

5. Conclusion

An evolution on a spectroscopic model proposed in a previous contribution for welding diagnostics has been presented and discussed in this paper. The original proposal was based on the generation of synthetic spectra and the employment of optimization algorithms to generate participation profiles of those species contributing to the welding plasma. It was demonstrated that a direct correlation existed between these profiles and the resulting seam quality, that is, appearance of defects. However, the experimental tests demonstrated that Ar II, the predominant species within the spectral range under analysis, did not exhibit the same response associated with some defects correctly signaled by other species. A revision of the proposed model suggested that the problem could be motivated by the use of the relative intensities from the NIST spectroscopic database for the creation of the synthetic spectra. Particularly, the use of wide spectral ranges with those intensities seemed to give rise to the mentioned lack of sensitivity to be found in the Ar II participation profiles. A possible solution to this issue lies in the reduction of the spectral window where the optimization process takes place, what has been implemented in this paper with the aid of a feature selection algorithm that helps to indicate the suitable spectral bands to be used. It has been demonstrated that this new approach has significantly improved the results obtained in the original work, given that now the Ar II participation signal shows a good correlation with the defects studied in the experimental tests. In addition, to extened the validity of the model, field tests on both Inconel-718 and Ti6Al-4V samples have been included in the analysis, also allowing to detect different weld defects: trajectory deviation, misalignment, and lack of cleanliness.

Apart from the use of the SFFS algorithm with the Mahalonobis distance to perform the spectral range reduction for the optimization process, the CRS6 algorithm used to perform this task in the original contribution has also been substituted by a simple implementation of the PSO, improving in this way the computational performance of the processing scheme.

Some issues remain still unsolved and should be dealt with in the future to improve the proposed model. On

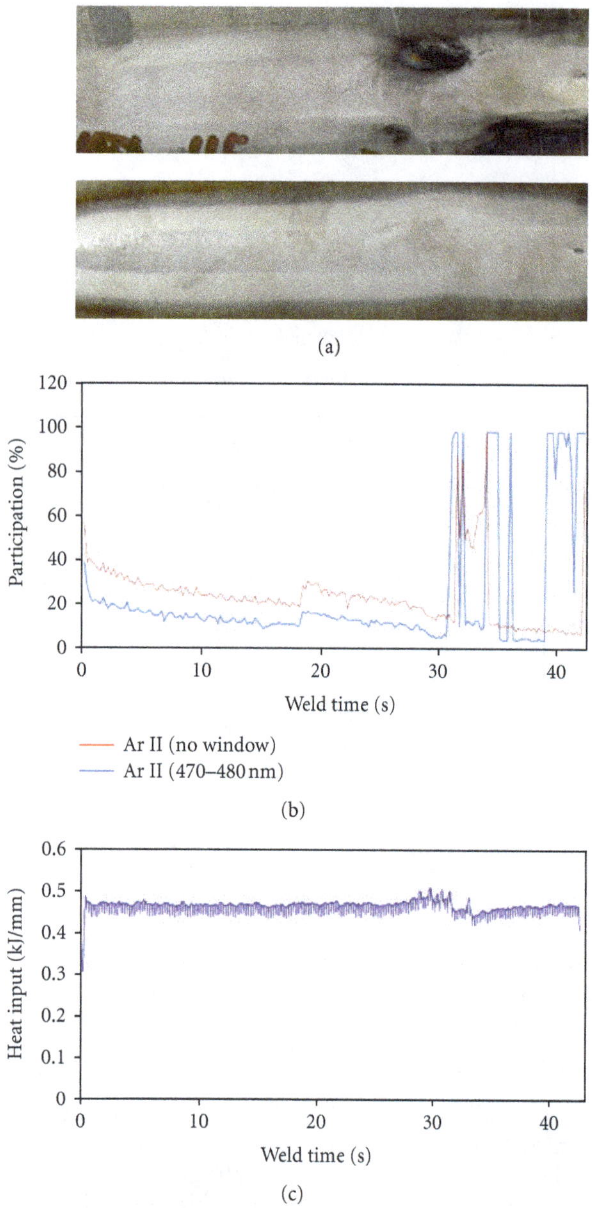

(a)

(b)

(c)

FIGURE 7: (a) Top and bottom views of defective seam (lack of cleanliness) (Ti6Al-4V). (b) Participation profiles of Ar II (470–480 nm) (blue) and Ar II (whole spectral range) (red). (c) Heat input profile.

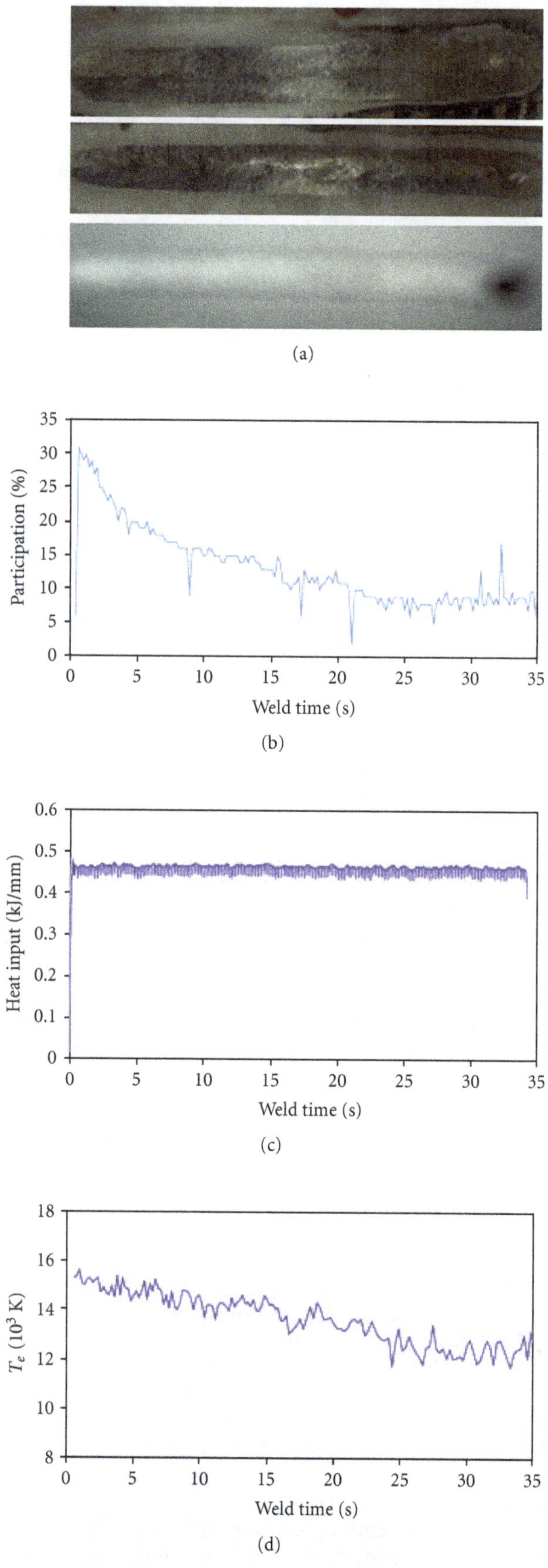

the one hand the employment of the relative intensities in the generation of the synthetic spectra should be avoided: a solution to be explored might be based on a feedback scheme where the intensities of the chosen emission lines could be calculated from the estimation of a spectroscopic parameter, like the plasma temperature T_e, using different species in the process. It could be also interesting to try to relate the relative participation profiles of consecutive ionization stages for a given element via the Saha equation, although it should be studied whether this approach would be excessively costly in terms of the computational performance of the model. An application of this method might lie in the framework

(a)

(b)

(c)

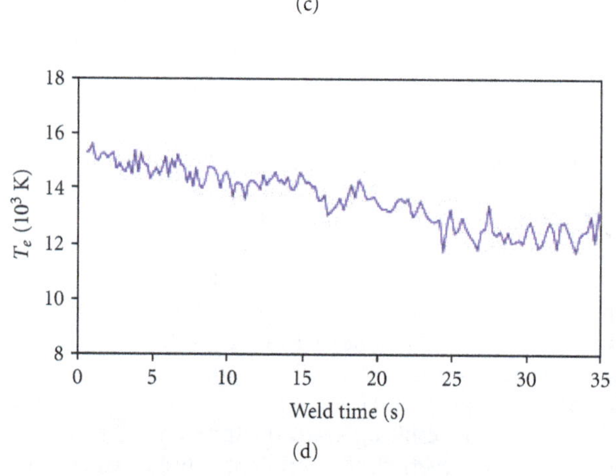

(d)

FIGURE 8: (a) Defective seam (lack of cleanliness) (Ti6Al-4V). (b) Participation profiles of Ar II (470–480 nm). (c) Heat input profile. (d) Plasma temperature profile.

of LIBS (Laser Induced Breakdown Spectroscopy), where it might be used for a quantitative estimation of the composition of samples.

Acknowledgments

This work has been supported by the TEC2010-20224-C02-02 and OPENAER CENIT 2007–2010 projects. The authors want to thank Jose J. Valdiande (Photonics Engineering Group), Juan M. Etayo (Tecnalia) for the heat input data and Marta Davila and Raul Llorente (ITP) for his help and support during the field tests. In addition, the authors also thank ITP for allowing to publish the field test data, which were obtained within the framework of the OPENAER CENIT 2007–2010 project.

References

[1] C. S. Wu, M. Ushio, and M. Tanaka, "Analysis of the TIG welding arc behavior," *Computational Materials Science*, vol. 7, no. 3, pp. 308–314, 1997.

[2] T. W. Eagar, "Physics of arc welding processes," in *Advanced Joining Technologies*, T. H. North, Ed., London, UK, 1990.

[3] L. Li, D. J. Brookfield, and W. M. Steen, "Plasma charge sensor for in-process, non-contact monitoring of the laser welding process," *Measurement Science and Technology*, vol. 7, no. 4, pp. 615–626, 1996.

[4] W. Lu, Y. M. Zhang, and J. Emmerson, "Sensing of weld pool surface using non-transferred plasma charge sensor," *Measurement Science and Technology*, vol. 15, no. 5, pp. 991–999, 2004.

[5] D. Groslier, S. Pellerin, F. Valensi, F. Richard, and F. Briand, "Explorative approach of the spectral analysis tools to the detection of welding defects in lap welding," *Nondestructive Testing and Evaluation*, vol. 26, no. 1, pp. 13–33, 2011.

[6] H. Gu and W. W. Duley, "A statistical approach to acoustic monitoring of laser welding," *Journal of Physics D*, vol. 29, no. 3, pp. 556–560, 1996.

[7] D. F. Farson and K. R. Kim, "Generation of optical and acoustic emissions in laser weld plumes," *Journal of Applied Physics*, vol. 85, no. 3, pp. 1329–1336, 1999.

[8] A. Cobo, F. Bardin, J. Mirapeix, D. P. Hand, J. D. C. Jones, and J. M. López-Higuera, "Optoelectronic device for non-invasive focal point measurement and control of the laser welding process," *Measurement Science and Technology*, vol. 16, no. 3, pp. N1–N6, 2005.

[9] J. O. Connolly, G. J. Beirne, G. M. O'Connor, J. T. Glynn, and A. J. Conneely, "Optical monitoring of laser generated plasma during laser welding," in *Laser Plasma Generation and Diagnostics*, vol. 3935 of *Proceedings of the SPIE*, pp. 132–138, San Jose, Ca, USA, 2000.

[10] G. J. Zhang, Z. H. Yan, and L. Wu, "Visual sensing of weld pool in variable polarity TIG welding of aluminium alloy," *Transactions of Nonferrous Metals Society of China*, vol. 16, no. 3, pp. 522–526, 2006.

[11] R. Kovacevic, Y. M. Zhang, and S. Ruan, "Sensing and control of weld pool geometry for automated GTA welding," *Journal of Engineering for Industry-Transactions of the ASME*, vol. 117, no. 2, pp. 210–222, 1995.

[12] B. A. Chin, N. H. Madsen, and J. S. Goodling, "Infrared thermography for sensing the arc welding process," *Welding Journal*, vol. 62, no. 9, 1983.

[13] H. C. Wikle, S. Kottilingam, R. H. Zee, and B. A. Chin, "Infrared sensing techniques for penetration depth control of the submerged arc welding process," *Journal of Materials Processing Technology*, vol. 113, no. 1-3, pp. 228–233, 2001.

[14] A. Ancona, T. Maggipinto, V. Spagnolo, M. Ferrara, and P.M. Lugara, "Optical sensor for real-time weld defect detection," in *Sensors and Camera Systems for Scientific, Industrial, and Digital Photography Applications III*, M.M. Blouke, J. Canosa, and N. Sampat, Eds., vol. 4669 of *Proceedings of the SPIE*, p. 217, San Jose, CA, USA, 2002.

[15] D. Bebiano and S. C. A. Alfaro, "A weld defects detection system based on a spectrometer," *Sensors*, vol. 9, pp. 2851–2861, 2009.

[16] I. S. Kim, J. S. Son, S. H. Lee, and P. K. D. V. Yarlagadda, "Optimal design of neural networks for control in robotic arc welding," *Robotics and Computer-Integrated Manufacturing*, vol. 20, no. 1, pp. 57–63, 2004.

[17] H. Luo, H. Zeng, L. Hu, X. Hu, and Z. Zhou, "Application of artificial neural network in laser welding defect diagnosis," *Journal of Materials Processing Technology*, vol. 170, no. 1-2, pp. 403–411, 2005.

[18] J. Mirapeix, P. B. García-Allende, A. Cobo, O. M. Conde, and J. M. López-Higuera, "Real-time arc-welding defect detection and classification with principal component analysis and artificial neural networks," *NDT and E International*, vol. 40, no. 4, pp. 315–323, 2007.

[19] P. Sforza and D. De Blasiis, "On-line optical monitoring system for arc welding," *NDT and E International*, vol. 35, no. 1, pp. 37–43, 2002.

[20] A. Ancona, V. Spagnolo, P. M. Lugarà, and M. Ferrara, "Optical sensor for real-time monitoring of CO_2 laser welding process," *Applied Optics*, vol. 40, no. 33, pp. 6019–6025, 2001.

[21] J. Mirapeix, A. Cobo, C. Jaúregui, and J. M. López-Higuera, "Fast algorithm for spectral processing with application to on-line welding quality assurance," *Measurement Science and Technology*, vol. 17, no. 10, article no. 013, pp. 2623–2629, 2006.

[22] A. Ancona, P. M. Lugarà, F. Ottonelli, and I. M. Catalano, "A sensing torch for on-line monitoring of the gas tungsten arc welding process of steel pipes," *Measurement Science and Technology*, vol. 15, no. 12, pp. 2412–2418, 2004.

[23] J. Mirapeix, A. Cobo, S. Fernandez, R. Cardoso, and J. M. Lopez-Higuera, "Spectroscopic analysis of the plasma continuum radiation for on-line arc-welding defect detection," *Journal of Physics D*, vol. 41, no. 13, Article ID 135202, 2008.

[24] P. B. Garcia-Allende, J. Mirapeix, O. M. Conde, A. Cobo, and J. M. Lopez-Higuera, "Defect detection in arc-welding processes by means of the line-to-continuum method and feature selection," *Sensors*, vol. 9, no. 10, pp. 7753–7770, 2009.

[25] J. Mirapeix, A. Cobo, J. Fuentes, M. Davila, J. M. Etayo, and J. M. Lopez-Higuera, "Use of the plasma spectrum RMS signal for arc-welding diagnostics," *Sensors*, vol. 9, no. 7, pp. 5263–5276, 2009.

[26] T. Sibillano, A. Ancona, V. Berardi, E. Schingaro, P. Parente, and P. M. Lugarà, "Correlation spectroscopy as a tool for detecting losses of ligand elements in laser welding of aluminium alloys," *Optics and Lasers in Engineering*, vol. 44, no. 12, pp. 1324–1335, 2006.

[27] D. Groslier, S. Pellerin, F. Valensi, F. Richard, and F. Briand, "Explorative approach of the spectral analysis tools to the detection of welding defects in lap welding," *Nondestructive Testing and Evaluation*, vol. 26, no. 1, pp. 13–33, 2011.

[28] J. Mirapeix, A. Cobo, D. A. González, and J. M. López-Higuera, "Plasma spectroscopy analysis technique based on optimization algorithms and spectral synthesis for arc-welding quality assurance," *Optics Express*, vol. 15, no. 4, pp. 1884–1897, 2007.

[29] J. Kennedy and R. C. Eberhart, "Particle swarm optimization," in *Proceedings of the IEEE International Conference on Neural Networks*, vol. 4, pp. 1942–1948, Perth, Australia, 1995.
[30] National Institute for Standards and Technology (NIST), "atomic spectra database," http://physics.nist.gov/cgi-bin/AtData/main_asd.
[31] H. R. Griem, *Principles of Plasma Spectroscopy*, chapter 5, Cambridge University Press, 1997.
[32] A. Marotta, "Determination of axial thermal plasma temperatures without Abel inversion," *Journal of Physics D*, vol. 27, no. 2, pp. 268–272, 1994.
[33] P. Pudil, J. Novovičová, and J. Kittler, "Floating search methods in feature selection," *Pattern Recognition Letters*, vol. 15, no. 11, pp. 1119–1125, 1994.
[34] P. Sathiya, S. Aravindan, A. N. Haq, and K. Paneerselvam, "Optimization of friction welding parameters using evolutionary computational techniques," *Journal of Materials Processing Technology*, vol. 209, no. 5, pp. 2576–2584, 2009.
[35] P. Zhang, L. Kong, W. Liu, J. Chen, and K. Zhou, "Real-time monitoring of laser welding based on multiple sensors," in *Proceedings of the Chinese Control and Decision Conference (CCDC '08)*, pp. 1746–1748, July 2008.
[36] R. Poli, J. Kennedy, and T. Blackwell, "Particle swarm optimization: an overview," *Swarm Intelligence*, vol. 1, pp. 33–57, 2007.
[37] A. K. Jain, R. P. W. Duin, and J. Mao, "Statistical pattern recognition: a review," *IEEE Transactions on Pattern Analysis and Machine Intelligence*, vol. 22, no. 1, pp. 4–37, 2000.
[38] G. F. Hughes, "On the mean accuracy of statistical pattern recognizers," *IEEE Transactions on Information Theory*, vol. 14, pp. 55–63, 1968.
[39] L. Gomez-Chova, J. Calpe, G. Camps-Valls et al., "Feature Selection of Hyperspectral Data Through Local Correlation and SFFS for Crop Classification," in *IEEE International Symposium on Geoscience and Remote Sensing (IGARSS '03)*, IEEE Cat. No.03CH37477, pp. 555–557, July 2003.
[40] A. Bhattacharyya, "On a measure of divergence between tow multinomial population," *The Indian Journal of Statistics*, vol. 7, pp. 401–406, 1946.
[41] K. Fukunaga, *Introduction to Statistical Pattern Recognition*, M. Kaufmann Academic Press, San Diego, Calif, USA, 1990.
[42] D. Comaniciu, P. Meer, K. Xu, and D. Tyler, "Retrieval performance improvement through low rank corrections," in *Proceedings of the IEEE Workshop on Content-Based Access of Image and Video Libraries*, pp. 50–54, 1999.
[43] http://www.tecnalia.com.

Dynamic Compensation for Two-Axis Robot Wrist Force Sensors

Junqing Ma, Aiguo Song, and Dongcheng Pan

Jiangsu Key Lab of Remote Measurement and Control, School of Instrument Science and Engineering, Southeast University, Nanjing 210096, China

Correspondence should be addressed to Aiguo Song; a.g.song@seu.edu.cn

Academic Editor: Guangming Song

To improve the dynamic characteristic of two-axis force sensors, a dynamic compensation method is proposed. The two-axis force sensor system is assumed to be a first-order system. The operation frequency of the system is expanded by a digital filter with backward difference network. To filter high-frequency noises, a low-pass filter is added after the dynamic compensation network. To avoid overcompensation, parameters of the proposed dynamic compensation method are defined by trial and error. Step response methods are utilized in dynamic calibration experiments. Compared to experiment data without compensation, the response time of the dynamic compensated data is reduced by 30%~40%. Experiments results demonstrate the effectiveness of our method.

1. Introduction

Multiaxis robot wrist force sensors are necessary for robotic systems in which contact force information between robots and environments needs to be obtained. There are various kinds of multiaxis force sensors available in commercial and research area, for example, cross-beam type multiaxis force sensors [1, 2], piezoelectric multiaxis force sensors [3], fiber multiaxis force sensors [4], and so on [5]. Multiaxis robot wrist force sensors are always mounted on the wrists of robots to convert multidimensional contact force signals into multichannel voltage signals. Such kinds of applications can be frequently found in assemble robots, teleoperation robotic systems, rehabilitation robots, and so forth [6–9].

During a robot task, the effectiveness of on-line force perception and feedback highly relies on the performances of the multiaxis robot wrist force sensor. The strong real time and rapidity in robot tasks require multiaxis force sensors to perform high dynamic characteristic. However, multiaxis force sensors (hereafter referred to as "force sensors") always have low natural frequency and small damping ratio owing to the low stiffness of elastic body and using of strain gauges. As a result, the dynamic response of the force sensors is more than 0.2 ms, and the adjusting time is relatively long [10]. The A/D converters for force sensors will prolong the response time as well. The disparity of dynamic requirements from robotic tasks and the current performances of force sensors motivate the need to improve dynamic characteristics of force sensors. Improving dynamic characteristic of force sensors by hardware is limited and costly. In the field of measurement, dynamic performances of sensors are often improved by algorithms. Hence, dynamic compensation algorithms need to be designed to improve dynamic behavior of force sensors. Altintas and Park in [11] designed dynamic compensation algorithms by a Kalman filter for a spindle-integrated force sensor. Xu and Li in [10, 12] designed dynamic compensation algorithms by functional link artificial neural network (FLANN) for six-axis wrist force/torque sensor. Yu et al. in [13] designed dynamic compensation algorithms by genetic neutral network for robot wrist force sensor, but the artificial neural network is slow to converge and may be subjected to local minimum [14].

In this paper, we proposed a dynamic compensation method based on a digital filter with backward difference design. Because there are high-frequency noises in the output voltages of force sensors, a low-pass filter is added in the dynamic compensation system. Dynamic calibration experiments are conducted, in which step-response method is utilized. During the dynamic calibration experiments, a two-axis force sensor which is designed and fabricated in our lab is used. Step signals of force are generated, while corresponding step responses of the two-axis force sensor are recorded by

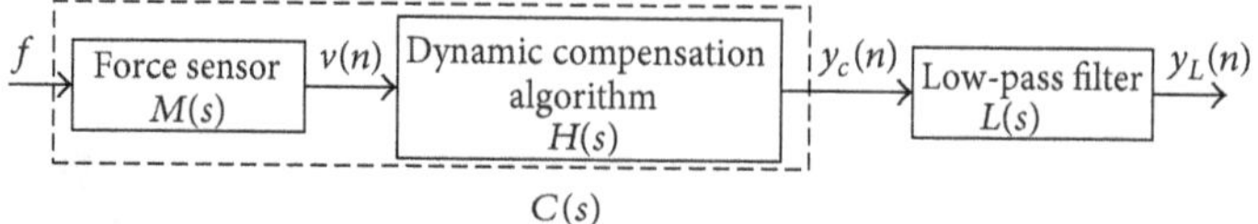

FIGURE 1: Flowchart of dynamic compensation of two-axis force sensors.

high-frequency data acquisition card. The proposed dynamic calibration method is applied to the dynamic calibration experiment data. The response time is greatly reduced with our method. The experiment results demonstrated the correctness and effectiveness of our method.

2. Dynamic Compensation System

2.1. Principle of Dynamic Compensation. The original operation frequency band of force sensors is too narrow to cover all the frequency components of input force signals. This causes the decays of high frequency components. Furthermore, dynamic behavior of force sensors will be decreased. In order to broaden the operation frequency, in this paper, a dynamic compensation part is added in force-sensor systems. With this method, dynamic performance of force sensors can be improved. Dynamic compensation of input forces in each dimension is done separately. Figure 1 shows the flowchart of the dynamic compensation of force sensors.

In Figure 1, f is the input force signal, and $v(n)$ is the output voltage of the two-axis force sensors. $y_c(n)$ is the results of $v(n)$ passing dynamic compensation network, and $y_L(n)$ is the dynamic compensated data after low-pass filter. $M(s)$ represents the transfer function of the force sensor, $H(s)$ represents the transfer function of the dynamic compensation, $L(s)$ represents the transfer function of the low-pass filter, and $C(s)$ represents the transfer function of the whole force sensor system.

We have

$$C(s) = M(s) H(s). \tag{1}$$

2.2. Dynamic Compensation Algorithm. The force sensor system, as shown in Figure 1, is assumed as a first-order system. The transfer function of the force sensor, $M(s)$, can be expressed as

$$M(s) = \frac{1}{1+\tau s} e^{-\lambda s}, \tag{2}$$

where τ is the time constant of the first-order system and λ is the lag time of the force sensor.

The cut-off angular frequency of the force sensor ω can be calculated as

$$\omega = \frac{1}{\tau}. \tag{3}$$

If the frequency band of the force sensor is broadened to k times, the cut-off angular frequency becomes $\omega_c = k\omega$. Then, $C(s)$ can be expressed as

$$C(s) = \frac{1}{1+(\tau/k)s} e^{-\lambda s}. \tag{4}$$

Combinations of (1), (2), and (4) lead to

$$H(s) = \frac{C(s)}{M(s)} = \frac{1+\tau s}{1+(\tau/k)s}. \tag{5}$$

After A/D converter, the output voltages signals of force sensors are digital signals. Equivalent digital filter in z domain $H(z)$ can be obtained from (5) and backward difference method as shown in (6):

$$s = \frac{1}{\tau_{\text{sam}}} \left(1 - z^{-1}\right), \tag{6}$$

where τ_{sam} is the sampling interval.

Combination of (5) and (6) leads to

$$H(z) = k \frac{1 + c\tau_{\text{sam}} - z^{-1}}{1 + b\tau_{\text{sam}} - z^{-1}} = \frac{Y_c(z)}{X(z)}, \tag{7}$$

where $c = 1/\tau$, $b = k/\tau = kc$.

The difference equation of (7) can be calculated as

$$y_c(n) = \frac{1}{1+b\tau_{\text{sam}}} \left[k\left(1 + c\tau_{\text{sam}}\right) x(n) - kx(n-1) + y_c(n-1)\right]. \tag{8}$$

Equation (8) is the dynamic compensation algorithm of force sensors. The proposed dynamic algorithm is simple and fast. It can be easily realized in software.

2.3. Digital Low-Pass Filter. As mentioned in the former section, the operation frequency of the force sensors is broadened in the proposed dynamic compensation algorithm. However, the extension of operation frequency will intensify high-frequency noises in the output voltage signals in each dimension. As a result, a low-pass filter is added after the dynamic compensation part, as shown in Figure 1, to filter out high frequency noise signals.

The common low-pass filter, called "moving average filter," is utilized. Moving average filter is able to reduce random noises while retaining a sharp step response. As a result, moving average filter is the premier filter for time domain encoded signals with random noises. The moving average filter operates by averaging a number of points from input signals to produce every point in output signals [15].

In the dynamic compensation process, a moving average signal average m numbers of points in $y_c(n)$, as shown in (9):

$$y_L(n) = \frac{1}{m} \sum_{i=n-m+1}^{n} y_c(n+i), \quad n = 0, 1, 2 \ldots. \tag{9}$$

FIGURE 2: Prototype of a two-axis force sensor.

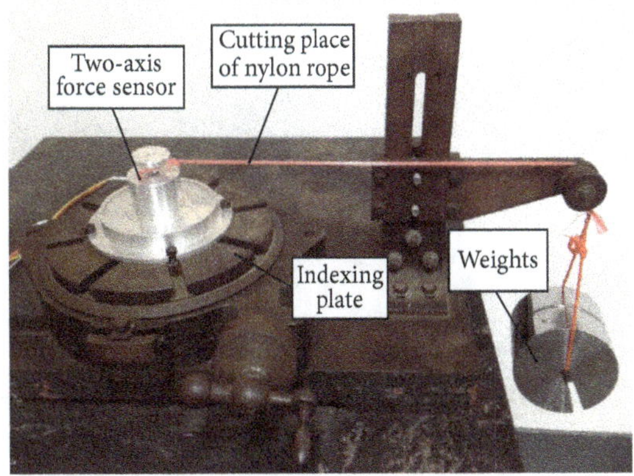

FIGURE 3: Dynamic compensation experiments setup.

3. Dynamic Compensation Results

3.1. Dynamic Calibration Experiments. A prototype two-axis force sensor which is designed and fabricated in our lab is used in the dynamic calibration experiments, as shown in Figure 2.

The key component of the force sensor is the cross-beam elastic body. When external forces are applied, the cross-beam elastic body will be deformed. Eight strain gauges which are pasted on the cross-beam elastic body of the force sensor will detect the deformation and convert the variations of input forces into variations of resistances. Each four strain gauges are connected to establish a Wheatstone bridge circuit. Variations of resistances of strain gauges can be converted into variations of voltages. The voltages are amplified by amplifying circuit [16].

The two-axis force sensor is able to measure horizontal forces in both X direction and Y direction. Because of the symmetrical characteristic of the two-axis force sensor, we only do dynamic calibration experiment in X direction.

There are three kinds of common dynamic calibration methods for sensors, namely, the frequency method, the step-response method, and the impact response method [12]. As for force sensors, a sine wave input force signal is difficult to generate. An impact force can be generated by a hammer with a piezoelectric sensor [17, 18]. However, the knocking position and angle are hard to control. Hence, step-response method [19] is preferred in dynamic calibration experiments for force sensors.

Figure 3 shows the platform of the dynamic calibration experiment setup.

As shown in Figure 3, a two-axis force sensor is mounted on an indexing plate in the center of the calibration table. The indexing plate can be rotated to ensure the directions of the loading forces. External horizontal forces are generated by pulley, nylon ropes and weights.

The cutting place of the nylon rope will be cut by a pair of scissors during the calibration experiment. When the nylon rope which is near the force sensor is cut, weights will fall down to the ground. A negative step force is generated. The cutting process should be quick and decisive.

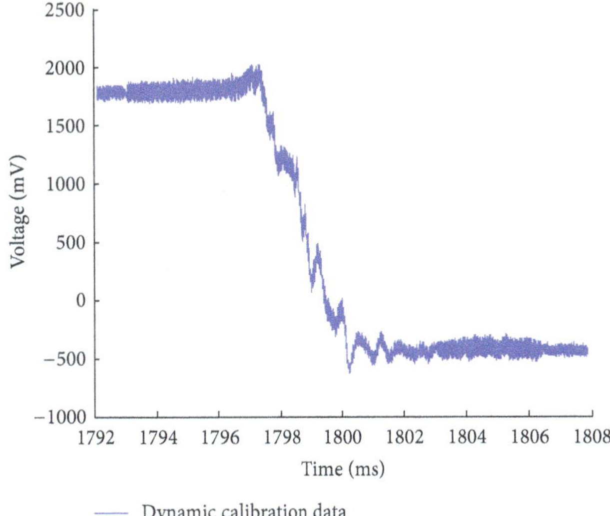

FIGURE 4: Original step response data from dynamic compensation experiments.

The response of the two-axis force sensor is recorded by a data acquisition card with a sampling frequency of 250 KHz.

3.2. Results. The whole dynamic calibration experiment is repeated three times. The best experiment data which show the fewest vibrations are chosen. Figure 4 shows the original negative step response data recorded by the data acquisition card.

The calibration experiment data support the assumption that the two-axis force sensor is a first-order system. The transfer function of the force sensor can be expressed in (2). The time constant $\tau = 1$ ms, and the sampling interval $\tau_{\text{sam}} = 0.004$ ms.

In the dynamic compensation network, parameters in the dynamic compensation algorithm and low-filter pass filter are defined by trial and error. Finally, the factor k is defined as 3; that is, the operation frequency is expanded three times.

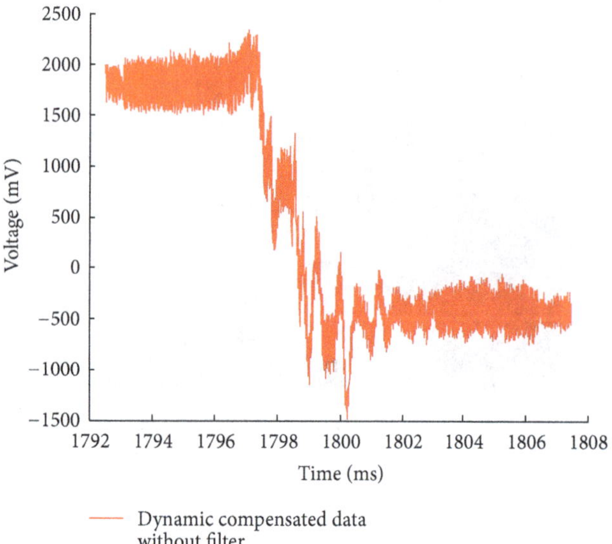

Figure 5: Dynamic compensation results of step response without low-pass filter.

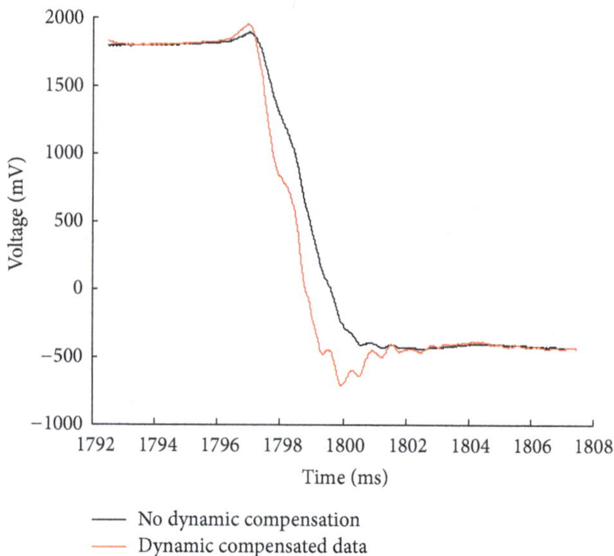

Figure 6: Comparisons between original data and dynamic compensated data after low-pass filter.

The dynamic algorithm of the two-axis force sensor can be obtained as shown in (10):

$$y_c(n) = \frac{1}{1+0.012} \times [3 \times (1+0.004)\, v(n) - 3 \times v(n-1) + y_c(n-1)], \quad n = 1, 2, 3 \ldots . \tag{10}$$

Figure 5 shows dynamic compensated data without the low-pass filter. The red line indicates the step response data after dynamic compensation but without a low-pass filter.

Comparing Figure 5 with Figure 4, the red line indicates that much more high-frequency noises are induced from the dynamic compensation algorithm. As a result, a low-pass filter is necessary after the dynamic compensation network.

The moving average filter is utilized as the low-pass filter. The length of the moving average filter kernel is defined as 100, that is, $m = 100$. The low-pass filter can be expressed as

$$y_L(n) = \frac{1}{100} \sum_{i=n-50+1}^{n+50} y_c(n+i) \quad n = 0, 1, 2, \ldots . \tag{11}$$

In order to make a contrast, both the original data as shown in Figure 4 and the dynamic compensated data as shown in Figure 5 are passed through the moving average filter. Figure 6 shows comparisons between original data and dynamic compensated data after the low-pass filter.

In Figure 6, the black line represents original data filtered by the low-pass filter. The red line represents dynamic compensated data filtered by the same low-pass filter. Figure 6 shows that the response time of the two-axis force sensor is reduced from about 4 ms to about 2.5 ms. The time constant τ is reduced from about 1 ms to about 0.6 ms.

The unit step response function of the dynamic compensated two-axis force sensor can be expressed as

$$f_U(t) = \left(1 - e^{-(t-\lambda)/(6 \times 10^{-4})}\right) U(t-\lambda), \tag{12}$$

where U represents the unit step function.

As a result, the dynamic characteristic of the two-axis force sensor is improved a lot from the proposed dynamic compensation method. The dynamic compensation method is simple and not subjected to local minimum.

4. Conclusions

In this paper, a dynamic compensation method to efficiently improve dynamic characteristic of two-axis force sensors is presented. The dynamic compensation method contains a dynamic compensation network and a low-pass filter. The two-axis force sensor is assumed as a first-order system. The dynamic compensation network, which is based on backward difference method, broadens the operation frequency of the force sensors. High-frequency noises may be intensified from the compensation network. A moving average filter is added to filter the high-frequency noises.

Dynamic calibration experiments are implemented. A negative step force signal is generated by sudden release of applying loads. The step response obtained from the calibration experiment demonstrates the rationality of the assumption of force sensors to be a first-order system. The proposed dynamic compensation method is applied in the experiment data. Experiment results demonstrate the effectiveness of our method. The adjusting time in the step response is reduced from about 4 ms to about 2.5 ms. The time constant τ is reduced from about 1 ms to about 0.6 ms.

Compared to the traditional dynamic compensation algorithm based on artificial neural network, our method is fast and not subjected to local minimum. The proposed dynamic compensation method can also be implemented in other multiaxis force sensors.

Acknowledgment

This work was supported by the National Natural Science Foundation of China (no. 61272379).

References

[1] J. Ma and A. Song, "Fast estimation of strains for cross-beams six-axis force/torque sensors by mechanical modeling," *Sensors*, vol. 13, pp. 6669–6686, 2013.

[2] A. Song, J. Wu, G. Qin, and W. Huang, "A novel self-decoupled four degree-of-freedom wrist force/torque sensor," *Measurement*, vol. 40, no. 9-10, pp. 883–891, 2007.

[3] Y.-J. Li, B.-Y. Sun, J. Zhang, M. Qian, and Z.-Y. Jia, "A novel parallel piezoelectric six-axis heavy force/torque sensor," *Measurement*, vol. 42, no. 5, pp. 730–736, 2009.

[4] M. S. Müller, L. Hoffmann, T. C. Buck, and A. W. Koch, "Fiber bragg grating-based force-torque sensor with six degrees of freedom," *International Journal of Optomechatronics*, vol. 3, no. 3, pp. 201–214, 2009.

[5] R. J. Wood, K.-J. Cho, and K. Hoffman, "A novel multi-axis force sensor for microrobotics applications," *Smart Materials and Structures*, vol. 18, no. 12, Article ID 125002, 2009.

[6] B. Siciliano and O. Khatib, *Springer Handbook of Robotics*, Springer, New York, NY, USA, 2008.

[7] T. Lefebvre, J. Xiao, H. Bruyninckx, and G. De Gersem, "Active compliant motion: a survey," *Advanced Robotics*, vol. 19, no. 5, pp. 479–499, 2005.

[8] G. Xu, A. Song, and H. Li, "Adaptive impedance control for upper-limb rehabilitation robot using evolutionary dynamic recurrent fuzzy neural network," *Journal of Intelligent and Robotic Systems*, vol. 62, no. 3-4, pp. 501–525, 2011.

[9] F. Nagata, Y. Kusumoto, K. Watanabe et al., "Polishing robot for PET bottle molds using a learning-based hybrid position/force controller," in *Proceedings of the 5th Asian Control Conference*, pp. 914–921, July 2004.

[10] K.-J. Xu and C. Li, "Dynamic decoupling and compensating methods of multi-axis force sensors," *IEEE Transactions on Instrumentation and Measurement*, vol. 49, no. 5, pp. 935–941, 2000.

[11] Y. Altintas and S. S. Park, "Dynamic compensation of spindle-integrated force sensors," *CIRP Annals-Manufacturing Technology*, vol. 53, no. 1, pp. 305–308, 2004.

[12] K.-J. Xu, C. Li, and Z.-N. Zhu, "Dynamic modeling and compensation of robot six-axis wrist force/torque sensor," *IEEE Transactions on Instrumentation and Measurement*, vol. 56, no. 5, pp. 2094–2100, 2007.

[13] A. Yu, W. Huang, and G. Qin, "Dynamic modeling and compensation method based on genetic neural network for new type robot wrist force sensor," *Chinese Journal of Mechanical Engineering*, vol. 42, no. 12, pp. 239–244, 2006.

[14] S. Chen and S. A. Billings, "Neural networks for nonlinear dynamic system modelling and identification," *International Journal of Control*, vol. 56, pp. 319–346, 1992.

[15] P. J. Brockwell and R. A. Davis, *Time Series: Theory and Methods*, Springer, New York, NY, USA, 2009.

[16] J. Ma and A. Song, "Development of a novel two-axis force sensor for chinese massage robot," *Applied Mechanics and Materials*, vol. 103, pp. 299–304, 2012.

[17] N. Hu, H. Fukunaga, S. Matsumoto, B. Yan, and X. H. Peng, "An efficient approach for identifying impact force using embedded piezoelectric sensors," *International Journal of Impact Engineering*, vol. 34, no. 7, pp. 1258–1271, 2007.

[18] M. Tracy and F.-K. Chang, "Identifying impact load in composite plates based on distributed piezoelectric sensor measurements," in *Smart Structures and Materials: Smart Structures and Integrated Systems*, vol. 2717 of *Proceedings of SPIE*, pp. 231–236, February 1996.

[19] M. P. Schoen, "Dynamic compensation of intelligent sensors," *IEEE Transactions on Instrumentation and Measurement*, vol. 56, no. 5, pp. 1992–2001, 2007.

21

Role of the Material Electrodes on Resistive Behaviour of Carbon Nanotube-Based Gas Sensors for H_2S Detection

M. Lucci,[1] F. Toschi,[2] V. Guglielmotti,[2] S. Orlanducci,[2] and M. L. Terranova[2]

[1] *Dipartimento di Fisica, Università di Roma "Tor Vergata", Via della Ricerca Scientifica, 00133 Roma, Italy*
[2] *Dipartimento di Scienze e Tecnologie Chimiche, Minimalab Università di Roma "Tor Vergata", Via della Ricerca Scientifica, 00133 Roma, Italy*

Correspondence should be addressed to F. Toschi, francesco.toschi@uniroma2.it

Academic Editor: Michele Penza

Miniaturized gas-sensing devices that use single-walled carbon nanotubes as active material have been fabricated using two different electrode materials, namely, Au/Cr and NbN. The resistive sensors have been assembled aligning by dielectrophoresis the nanotube bundles between 40 μm spaced Au/Cr or NbN multifinger electrodes. The sensing devices have been tested for detection of the H_2S gas, in the concentration range 10–100 ppm, using N_2 as carrier gas. No resistance changes were detected using sensor fabricated with NbN electrodes, whereas the response of the sensor fabricated with Au/Cr electrodes was characterized by an increase of the resistance upon gas exposure. The main performances of this sensor are a detection limit for H_2S of 10 ppm and a recovery time of few minutes. The present study suggests that the mechanism involved in H_2S gas detection is not a direct charge transfer between molecules and nanotubes. The hypothesis is that detection occurs through passivation of the Au surfaces by H_2S molecules and modification of the contact resistance at the Au/nanotube interface.

1. Introduction

Among the many new technology opportunities and scientific challenges provided by carbon nanotubes (CNT), the sensing represents one of the most important topic. In particular, the design and fabrication of gas sensors assembled with CNT is a burgeoning research field still under development. The gas-sensing devices based on carbon nanomaterials can operate by a variety of different mechanism, such as resistivity change, capacitive effects, field-effect, and gas ionization [1, 2]. The resistive sensors, based on the property of some materials to modify their conductivity when in contact with chemical species, represent nowadays the most widespread class of gas sensors [3].

The intrinsic properties of CNTs meet indeed totally the requirements for a chemical sensor in terms of sensitivity, reproducibility, and durability [4, 5], whereas chemical/physical modifications of the nanotube walls can enhance the molecular specificity, making it possible to achieve a good selectivity [6].

Moreover the use of the carbon nanotubes as active sensing material for gas detection enables to minimize size/weight of the device, with reduction of power consumption and of manufacturing costs.

In our labs researches related to gas detection are carried out using resistive sensors assembled with single wall carbon nanotubes (SWCNTs) [7–9]. The experiments performed up to now demonstrated that such sensors are able to detect at room temperature sub-ppm levels of gas molecules (NO, NO_2, and NH_3) in some hundreds of μs. Moreover, after only few seconds, the devices are ready for a new operation. To reach such good performances, the main challenges to be managed are the relative orientation of the nanotubes and the application of a back-gate voltage [7–9].

The task of optimizing the organization of the SWCNT between the conductive stripes has been faced and successfully achieved by means of an dielectrophoresis process (DEP) that induces alignment of the SWCNT bundles between the electrodes [10]. A series of experiments confirmed that this kind of nanotube organization is indeed strictly needed in order to increase the sensitivity of the device. The second task regards a proper tuning of the voltages applied to the back-gate contact during the various phases of the nanotube/gas interaction processes. The voltage tuning was found to strongly improve the response times and to allow sensor self-calibration.

In resistive devices the p-type semiconducting behaviour of SWCNT is exploited to discriminate species with different electronic properties; an electron-donor gas affects the conductivity in opposite way with respect to an electron-acceptor gas. In this contest the material of the electrode was assumed not to influence the sensor response.

However, it was noted [11] that the response of the SWCNT-based sensor was influenced by the nature of the electrode metallic material, and it was suggested that the interface between the nanotubes and the metal could play an important role in the response of the resistive sensors. This finding could open the way to a selective sensing, obtained by the use of different electrode materials.

In this context we felt it worthwhile to fabricate nanotube-based devices with electrodes made by different materials and to test their detecting performances for the H_2S gas. In the present study devices with electrodes made by Au/Cr and NbN were used.

The choice of NbN, a rather unconventional material for a gas sensor, was made because we asked for a second material with a work function (WF) value quite similar to that of Au (WF = 5,1 eV), but unable to chemically interact with S-containing species. Based on these requirements, the preference went to NbN (WF = 4.95 eV) [12], that is, chemically stable, also against oxidation, [13] and mechanically resistant.

H_2S is a flammable, dangerous, and reactive gas originated from soils and from human activities, such as coal combustions and petroleum refining. The levels of personal exposure that workers may receive range between 100 and 250 ppm [14].

Accurate and rapid detection of hydrogen sulphide is needed for safety reasons and environmental pollution control [15]. The sensing devices presently in use are rather bulky and expensive and moreover tend easily to saturate. There is therefore a pressing need of technological improvements for portable detectors with fast response for the identification of such a gas.

2. Experimental

The sensor consists in a field emission transistor (FET)-like 3-pin multifinger element having an infinite electrical resistance, that starts to drop when the nanotubes are deposited between the interdigitated electrode stripes. The resistance is measured between drain-source pins, and the gate is used to improve the desorption rate.

Two kinds of materials, namely, Au/Cr and NbN, were employed to fabricate the microelectrodes. Au/Cr (Cr is the adhesive layer) or NbN films were deposited on the SiO_2-insulating layers grown on a p-doped Si substrate. The Au and Cr deposits (thickness: 80 nm and 20 nm, resp.) were produced by thermal evaporation of solid targets, the NbN deposits by DC-reactive sputtering, following an already established procedure [16].

Figure 1: Optical image of the tested sensor.

The interdigitated microelectrodes, 40 μm spaced, were patterned by a lift-off technique. The dimensions of the sensor are about 1 × 1 cm (Figure 1).

The sensing element is realized using controlled amounts of commercial SWCNT (Nanocyl: purity > 90% wt, diameter: 0.8–1.6 nm, length 5–30 μm), purified following previously settled procedures [17] and dispersed in $CHCl_3$. The dispersions are sonicated in order to assure a good dispersion of the nanotubes and deposited by casting on the electrode platform within the interdigitated Au/Cr or NbN electrodes.

The alignment of the nanotubes between the electrodes is carried out by means of a dielectrophoretic method. An AC field having a frequency of 1 MHz and 12 V_{pp} is applied up to the complete evaporation of the solvent ($CHCl_3$). Details have been reported in [9].

The experiments have been performed using a H_2S/N_2 gas mixture purchased by Rivoira S.p.A. in a tank with a certificated concentration of 100 ppm of H_2S in N_2. Further dilutions have been obtained by means of a flow meter system. The final H_2S concentrations used for the experiments were 10, 20, 40, 60, 80, and 100 ppm. The flowing of the gaseous mixtures occurred at 200 sccm for 300 sec under standard conditions (25°C, 1 atm). The desorption processes were carried out applying a back bias of 6 V for 20 sec, coupled with a thermal shock at the temperature of 90°C for 15 sec.

3. Results and Discussion

As shown in Figures 2(a) and 2(b), the DEP process provides a useful way to align SWCNTs between electrodes establishing a good electrical connection to the external measurement circuit.

The alignment of the SWCNTs amplifies the conductance changes of the material when exposed to the gas [7, 9]. We have firstly tested a multifinger device fabricated with Au/Cr stripes.

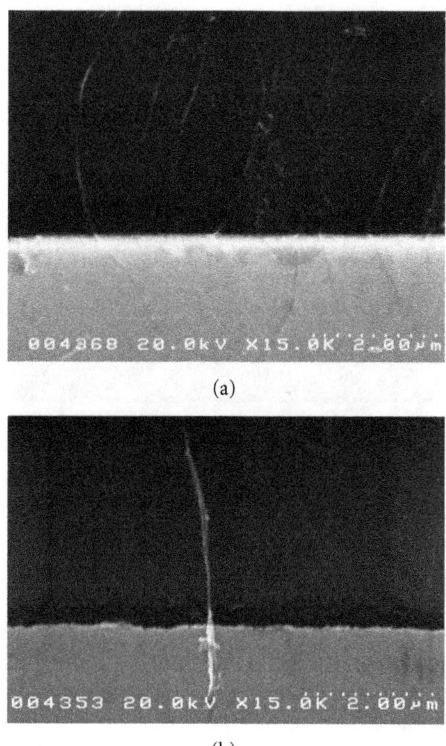

(a)

(b)

Figure 2: (a) and (b) FE-SEM images of SWCNT bundles aligned between the electrodes of the multifinger device.

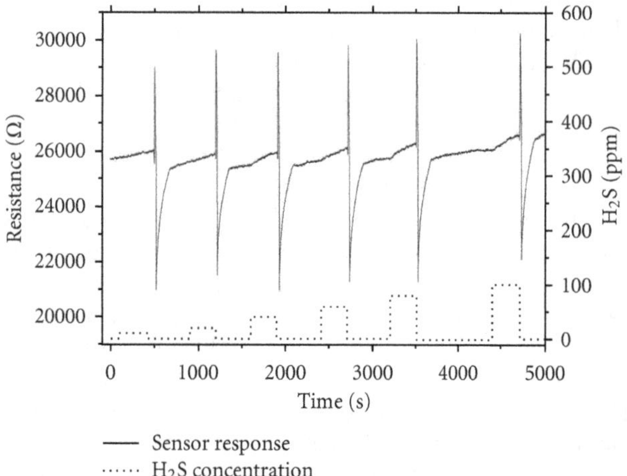

Figure 3: Resistance curves as a function of time for H_2S concentrations of 10, 20, 40, 60, 80, and 100 ppm (in N_2), under standard conditions. The narrow peaks are due to the effect of the applied gate voltage coupled with a thermal shock (90°C), required for the complete desorption of the gas.

Figure 3 reports the resistance values as a function of time and shows the behaviour of the sensor for H_2S concentrations of 10, 20, 40, 60, 80, and 100 ppm (in N_2), measured under standard conditions. The narrow peaks are due to the effect of the applied gate voltage coupled with a thermal shock (90°C), required for the complete desorption of the gas. Using this approach the recovery time of the sensor was at the longest 5 min for the higher H_2S concentrations.

In order to better visualize the sensor response, we report in Figure 4 for each concentration the relative curve of resistance variation ($\Delta R = R_f - R_0$), where R_0 is the resistance of the sensor exposed to pure N_2, and R_f is the maximum value of the resistance. The resistance of the device increases upon exposure to the H_2S, and the curve is Languimir type. As seen from the graphs of Figure 4, very short response times are obtained, the longest being about 5 min for the H_2S concentration of 100 ppm.

The $\Delta R/R_0$ data, reported in Figure 5 as a function of the different H_2S concentrations, demonstrate a very high precision of the sensor response (see error bars) with a linear trend for concentrations of H_2S ranging from 10 to about 60 ppm.

For each concentration 3 measurements have been carried out. The electrical resistance's data and the error bars are obtained by averaging over the 3 measurements. The sensitivity of the sensor is reported in terms of relative resistance change, $\Delta R/R_0$.

A different situation occurs using the same sensing material and the same procedures for the assembling of multifinger devices fabricated with NbN stripes. In this case, independently from the H_2S concentration, no signals related to gas adsorption could be detected (Figure 6). The resistance (R_{NbN}) of about 13.6 kΩ shown in Figure 6 is referred to the entire sensor device, that is, the multifingers NbN electrodes interconnected by means of SWCNTs. This R_{NbN} value obtained enables us to exclude the presence of Nb_2O_5 on the metal contacts because, in the case of Nb oxide formation, a higher value of resistance would be detected. The absence of signals in the presence of H_2S deserves some comments.

In general, the resistive response of SWCNT to gaseous species has been interpreted in terms of charge transfer between the p-type semiconducting SWCNT and electron-donors or electron acceptors species [18–21]. In their study, however, Suehiro et al. [11] noted that different electrode materials could modify the response of resistive SWCNT-based sensor and ascribed such behaviour to the occurrence of effects different from the conventional direct gas/nanotube interaction. In the present case too, the absence of resistance variation when using the NbN-based electrodes suggests that no H_2S uptake by the nanotubes occurs under our experimental conditions. The fact that the exposure to H_2S does not result in a resistance variation when using NbN electrodes means that the detection of H_2S is driven by a mechanism intrinsically different from a electron transfer between nanotubes and adsorbed molecules.

On the basis of the results obtained using different materials for electrode fabrication, it can be argued that the H_2S is not directly detected by the Au/Cr-based electrodes through a mechanism of charge transfer between nanotube and gas molecules, as in the case of other species, such as NOx or NH_3, but rather through the modification of the Au-SWCNT contact resistance.

A modulation of the Schottky barrier is thought to take place between the gold electrode and SWCNTs when exposed

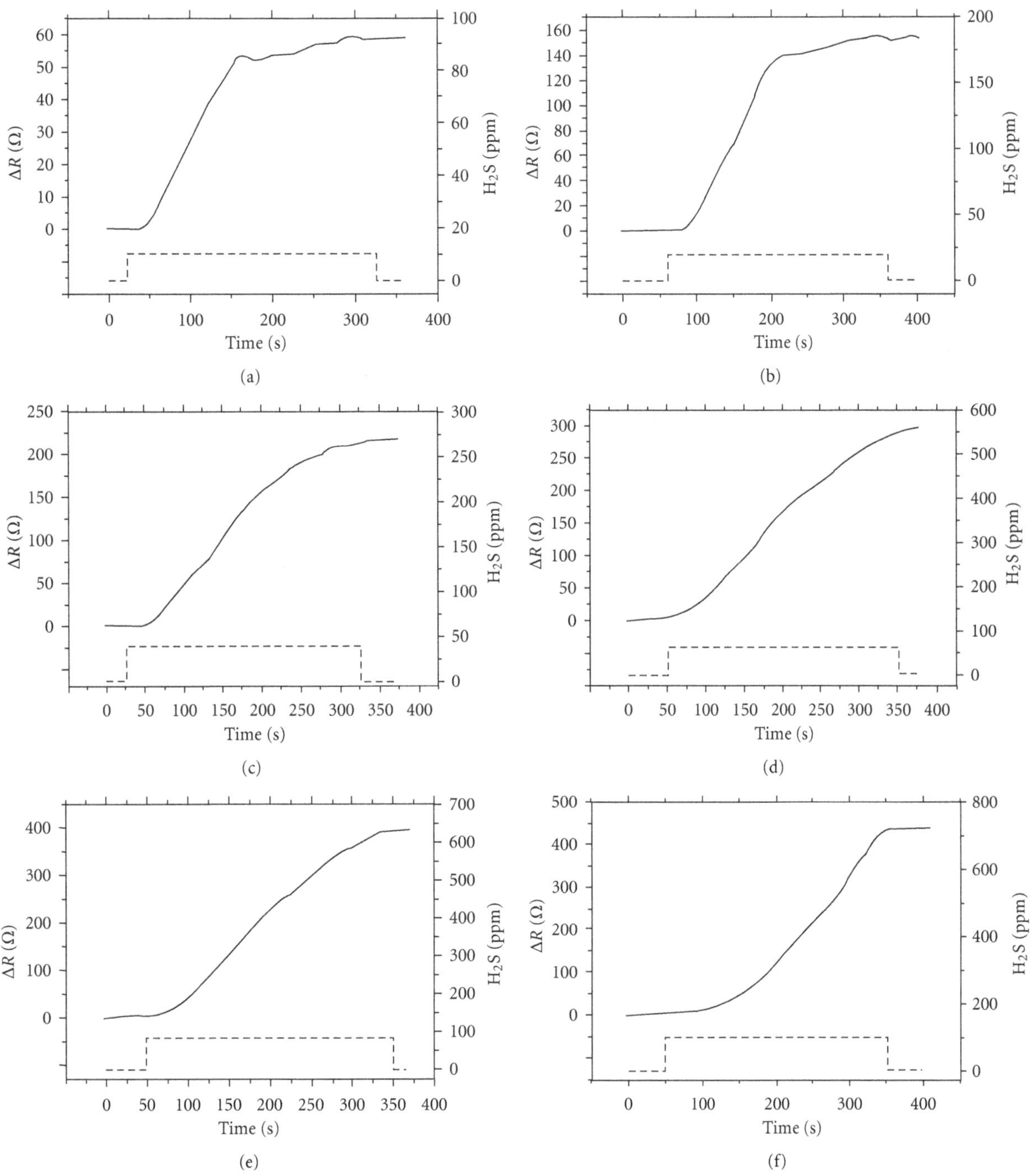

FIGURE 4: Curve of the resistance variation ($\Delta R = R_f - R_0$) versus time where R_0 is the resistance of the sensor exposed to pure N_2, and R_f is the maximum value of resistance. The H_2S concentrations (in N_2) are: (a) 10, (b) 20, (c) 40, (d) 60, (e) 80, and (f) 100 ppm, respectively.

to H_2S. The adsorbed gas molecules modify the Au/SWCNT Schottky contact areas and induce instantaneous modulations of the metal contact work function, which changes the device's resistance.

In other words, selective chemisorption of the molecules on Au and passivation of the Au/SWCNT interface are responsible for the measured increase of the contact resistance between Au and nanotubes.

Considering the strong bonding forces acting between Au and H_2S [22], this interpretation is also able to explain the need of a simultaneous thermal and electrostatic shock (voltage gate) to obtain complete desorption of the gas and recovery of the sensor.

It is to be noted that some papers [22–27] reported about gas sensors assembled with nanotubes decorated by metal nanoparticles. In these systems the metallic species act as

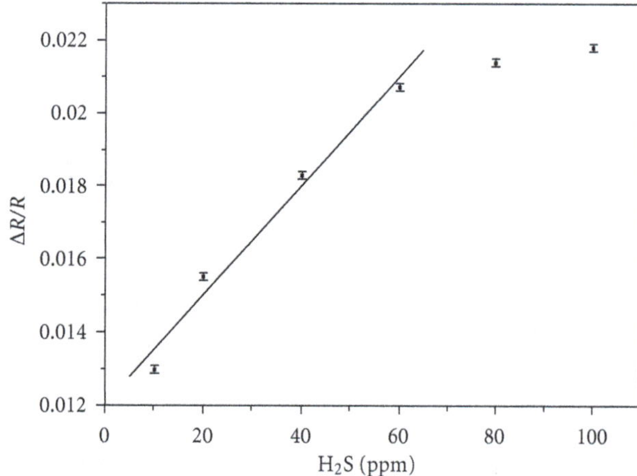

Figure 5: Relative resistance change ($\Delta R/R_0$) for the different H_2S concentrations.

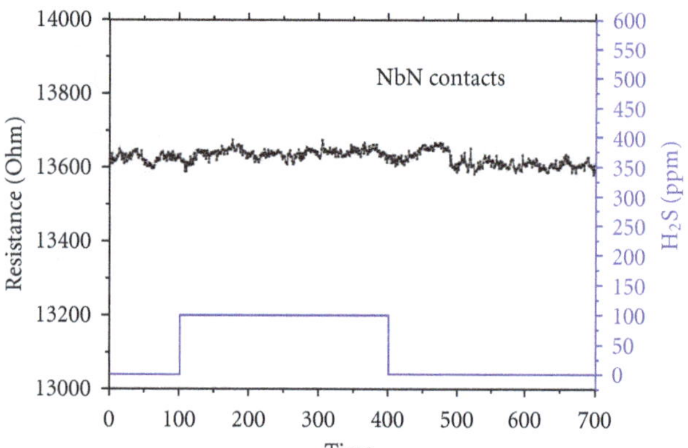

Figure 6: Sensing device fabricated using NbN electrodes: resistance changes upon exposure to 100 ppm of H_2S.

the sensing materials, and a charge transfer occurs between metal clusters and nanotubes. As regards H_2S, due to the large amount of metal nanoparticles, very low concentration of the gas could be detected [24–26]. On the other hand, such sensors saturate for gas concentration around 10 ppm, a value clearly below the threshold limit that poses serious health risks in workplaces or through accidental exposures.

4. Conclusions

We have fabricated two types of gas detectors using the same SWCNT as active material and either Au/Cr or NbN for the electrodes of the multifinger device. The task of ordering the nanotube bundles perpendicularly to the electrodes of the sensors has been pursued by using DEP. The other focal point, that is, the reduction of the recovery time and the optimization of the response time, has been overcome by the combined use of a back gate potential and of a thermal shock.

In the investigated concentrations range (10–100 ppm), the response of the sensor fabricated using Au/Cr electrodes is characterized by high level performance in terms of reproducibility and response times.

The accurate detection, at room temperature, of H_2S concentrations near to the threshold of health risk makes our devices really useful for in-door and out-door safety applications, including use in coal and natural gas processing, petroleum industries, security monitoring, and biogas production.

The fact that the NbN-based electrodes does not enable to register the presence of the gas suggests the lack of a specific interaction between nanotubes and H_2S, that is, a poor electron-donor specie. The experiments performed using electrodes fabricated with different materials allowed us to highlight that a mechanism different from a charge transfer between gas molecules and nanotube is responsible for H_2S detection. The conductivity changes detected by our circuit are instead likely caused by the Schottky barrier modulation at the metal/SWCNT interface induced by chemisorption of H_2S on the Au electrode [28].

It is to be underlined that the operating mechanism of this kind of sensors could represent an attractive approach towards a selectivity based not only on the conventional nanotube functionalization but also on the combination of different electrode materials for manufacturing a multi gas sensor platform. The perturbations of the conductivity induced in the nanotube-based circuit by the chemical interaction of a gaseous species with a selected electrode material give indeed easy to read out direct information. These sensors can be essential components to interface with conventional electronic architectures in a lab-on-chip hybrid system.

References

[1] N. Sinha, J. Ma, and J. T. W. Yeow, "Carbon nanotube-based sensors," *Journal of Nanoscience and Nanotechnology*, vol. 6, no. 3, pp. 573–590, 2006.

[2] M. Penza, "Carbon nanotubes for gas sensing applications: principles and transducers," in *Carbon Nanomaterials for Gas Adsorbtion*, M. L. Terranova, S. Orlanducci, and M. Rossi, Eds., CRC Press, Taylor Francis Group 2012.

[3] D. W. H. Fam, A. Palaniappan, A. I. Y. Tok, B. Liedberg, and S. M. Moochhala, "A review on technological aspects influencing commercialization of carbon nanotube sensors," *Sensors and Actuators B*, vol. 157, no. 1, pp. 1–7, 2011.

[4] T. Zhang, S. Mubeen, N. V. Myung, and M. A. Deshusses, "Recent progress in carbon nanotube-based gas sensors," *Nanotechnology*, vol. 19, no. 33, Article ID 332001, 2008.

[5] Y. Wang and J. T. W. Yeow, "A review of carbon nanotubes-based gas sensors," *Journal of Sensors*, vol. 2009, Article ID 493904, 24 pages, 2009.

[6] D. R. Kauffman and A. Star, "Carbon nanotube gas and vapor sensors," *Angewandte Chemie - International Edition*, vol. 47, no. 35, pp. 6550–6570, 2008.

[7] M. Lucci, A. Reale, A. Di Carlo et al., "Optimization of a NO_x gas sensor based on single walled carbon nanotubes," *Sensors and Actuators B*, vol. 118, no. 1-2, pp. 226–231, 2006.

[8] M. Lucci, P. Regoliosi, A. Reale et al., "Gas sensing using single wall carbon nanotubes ordered with dielectrophoresis," *Sensors and Actuators B*, vol. 111-112, pp. 181–186, 2005.

[9] M. L. Terranova, M. Lucci, S. Orlanducci et al., "Carbon nanotubes for gas detection: materials preparation and device assembly," *Journal of Physics Condensed Matter*, vol. 19, no. 22, Article ID 225004, 2007.

[10] M. Salvato, M. Lucci, M. Cirillo et al., "Low temperature conductivity of carbon nanotube aggregates," *Journal of Physics Condensed Matter*, vol. 23, no. 47, 2011.

[11] J. Suehiro, H. Imakiire, S. I. Hidaka et al., "Schottky-type response of carbon nanotube NO_2 gas sensor fabricated onto aluminum electrodes by dielectrophoresis," *Sensors and Actuators B*, vol. 114, no. 2, pp. 943–949, 2006.

[12] R. Fujii, Y. Gotoh, M. Y. Liao, H. Tsuji, and J. Ishikawa, "Work function measurement of transition metal nitride and carbide thin films," *Vacuum*, vol. 80, no. 7, pp. 832–835, 2006.

[13] R. Schneider, B. Freitag, D. Gerthsen, K. S. Ilin, and M. Siegel, "Structural, microchemical and superconducting properties of ultrathin NbN films on silicon," *Crystal Research and Technology*, vol. 44, no. 10, pp. 1115–1121, 2009.

[14] Iowa State University, Department of Chemistry MSDS, Hydrogen Sulfide Material Safety Data Sheet, http://avogadro.chem.iastate.edu/MSDS/hydrogen_sulfide.pdf.

[15] N. Izadi, A. M. Rashidi, S. Golzardi, Z. Talaei, A. R. Mahjoub, and M. H. Aghili, "Hydrogen sulfide sensing properties of multi walled carbon nanotubes," *Ceramics International*, vol. 38, no. 1, pp. 65–75, 2012.

[16] M. Lucci, S. Sanna, G. Contini et al., "Electron spectroscopy study in the NbN growth for NbN/AlN interfaces," *Surface Science*, vol. 601, no. 13, pp. 2647–2650, 2007.

[17] F. Valentini, A. Amine, S. Orlanducci, M. L. Terranova, and G. Palleschi, "Carbon nanotube purification: preparation and characterization of carbon nanotube paste electrodes," *Analytical Chemistry*, vol. 75, no. 20, pp. 5413–5421, 2003.

[18] J. Kong, N. R. Franklin, C. Zhou et al., "Nanotube molecular wires as chemical sensors," *Science*, vol. 287, no. 5453, pp. 622–625, 2000.

[19] E. S. Snow, J. P. Novak, P. M. Campbell, and D. Park, "Random networks of carbon nanotubes as an electronic material," *Applied Physics Letters*, vol. 82, no. 13, pp. 2145–2147, 2003.

[20] J. P. Novak, E. S. Snow, E. J. Houser, D. Park, J. L. Stepnowski, and R. A. McGill, "Nerve agent detection using networks of single-walled carbon nanotubes," *Applied Physics Letters*, vol. 83, no. 19, pp. 4026–4028, 2003.

[21] L. C. Wang, K. T. Tang, I. J. Teng et al., "A single-walled carbon nanotube network gas sensing device," *Sensors*, vol. 11, no. 8, pp. 7763–7772, 2011.

[22] D. R. Kauffman, D. C. Sorescu, D. P. Schofield, B. L. Allen, K. D. Jordan, and A. Star, "Understanding the sensor response of metal-decorated carbon nanotubes," *Nano Letters*, vol. 10, no. 3, pp. 958–963, 2010.

[23] A. Star, V. Joshi, S. Skarupo, D. Thomas, and J. C. P. Gabriel, "Gas sensor array based on metal-decorated carbon nanotubes," *Journal of Physical Chemistry B*, vol. 110, no. 42, pp. 21014–21020, 2006.

[24] M. Penza, R. Rossi, M. Alvisi et al., "Pt- and Pd-nanoclusters functionalized carbon nanotubes networked films for sub-ppm gas sensors," *Sensors and Actuators B*, vol. 135, no. 1, pp. 289–297, 2008.

[25] M. Penza, R. Rossi, M. Alvisi, G. Cassano, and E. Serra, "Functional characterization of carbon nanotube networked films functionalized with tuned loading of Au nanoclusters for gas sensing applications," *Sensors and Actuators B*, vol. 140, no. 1, pp. 176–184, 2009.

[26] S. Mubeen, T. Zhang, N. Chartuprayoon et al., "Sensitive detection of H_2S using gold nanoparticle decorated single-walled carbon nanotubes," *Analytical Chemistry*, vol. 82, no. 1, pp. 250–257, 2010.

[27] S. Mubeen, J. H. Lim, A. Srirangarajan, A. Mulchandani, M. A. Deshusses, and N. V. Myung, "Gas sensing mechanism of gold nanoparticles decorated single-walled carbon nanotubes," *Electroanalysis*, vol. 23, no. 11, pp. 2687–2692, 2011.

[28] A. J. Leavitt and T. P. Beebe, "Chemical reactivity studies of hydrogen sulfide on Au(111)," *Surface Science*, vol. 314, no. 1, pp. 23–33, 1994.

22

Development of a Fiber-Optic Sensing System for Train Vibration and Train Weight Measurements in Hong Kong

C. C. Lai,[1] Jacob C. P. Kam,[2] David C. C. Leung,[2] Tony K. Y. Lee,[2] Aiken Y. M. Tam,[2] S. L. Ho,[1] H. Y. Tam,[1] and Michael S. Y. Liu[1]

[1] *Department of Electrical Engineering, Hong Kong Polytechnic University, Kowloon, Hong Kong*
[2] *Mass Transit Railway, Fo Tan Railway House, Fo Tan, Shatin, NT, Hong Kong*

Correspondence should be addressed to Tony K. Y. Lee, tkylee@mtr.com.hk

Academic Editor: Yu-Lung Lo

A novel operation system to detect train vibration and train weight using FBG sensing network has been designed and tested in Hong Kong. The purpose of the system is for real time condition monitoring of trains. Because of the fast response of optical systems, the trains can be monitored in real-time during its normal service without any special arrangement. Hence, the condition checking can be realized without any disruption on the operating condition of the railway system.

1. Introduction

The health condition monitoring of trains is becoming increasingly important in modern cities like Hong Kong that rely heavily on mass transportation systems. Due to the heavy reliance of the general public on public transportation, it is inconceivable to have long delays in any of the public transportation systems, not to mention the suspension of services for many days. In the past, it is common to use the conventional approach of planning the maintenance work using either time-interval-based or mileage-based scheduling. However, it is well known that not all systems are exactly the same and hence the queuing approach may not be the best arrangement for the railway industry. In other words, there may be trains which should be serviced earlier than being planned, and there are also trains which do not need to be looked at even though they have been scheduled for routine checkup. To make the best use of the limited maintenance resources available, a reliable train health conditioning monitoring system is crucial.

This paper describes a novel health monitoring system using optical fibres. The benefit of optical fiber sensor is electromagnetic immunity (EMI), reliability, and durability. Indeed EMI is crucial because it is common to have 25 kV overhead lines in mainline trains and there are also hundreds, if not thousands of amperes of current flowing in the return rail. The use of optical fiber as both the sensing element and the transmission media can minimize the interference of electromagnetic waves from the traction motors and from the power lines of mainline systems.

The optical sensor being exploited in this study is based on fibre Bragg grating (FBG) technology [1–4]. The FBG sensor network is installed on a 2.5 km long track in Hong Kong and it is close to the Siu Ho Wan depot (SHD) between Tung Chung Station and Yam O station in the Tung Chung Line.

To validate the measured data, a trainset "TCL 08" with cars carrying sand bags with known weights was tested at around 11 : 41 : 42.6 am on 20th September, 2012. The data being recorded are then analysed offline carefully. An antiderailment system based on the calibrated results as described in this paper are now in operation.

2. Fibre Bragg Grating Sensors Network

Fiber Bragg grating (FBG) is an optical sensor in an optical fiber for sensing changes in either strain or temperature. FBG sensor has been selected for the study being carried out

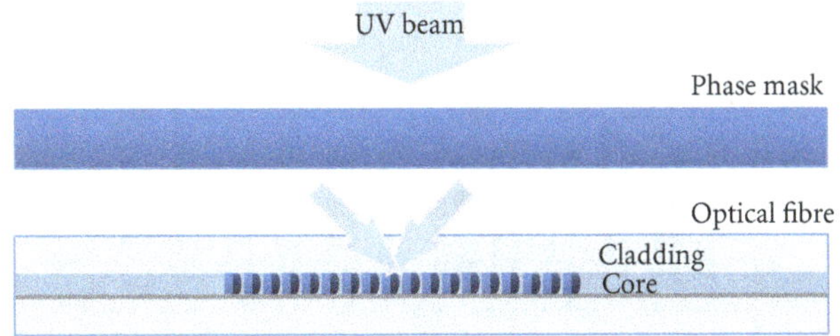

FIGURE 1: Fabrication of optical sensors using Ultraviolet lights and masks.

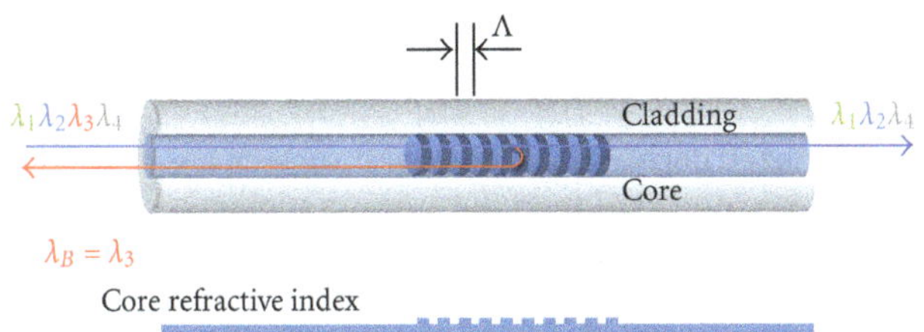

FIGURE 2: Changes in core reflective index after FBG has been inscribed on the optical fibre.

because of its fast and quick time response as well as its EMI property.

2.1. The Working Principle of FBG. An Fibre Bragg Grating (FBG) sensor is an optical device that measures strains by means of detecting changes in the reflected wavelength of light. A FBG sensor can be fabricated on a short-length (≈1 cm) of single-mode optical fibre, which is as thin as human hair, by exposing the sensors to ultraviolet light through a phase mask [5] as shown in Figure 1.

The FBG sensor consists of a short length of periodic refractive-index [6] changes as shown in Figure 2 inside the optical fibre. A light source is used to pass a band of light spectrum into the sending end of the optical fibre. Without the FBG sensor, the light passes through the optical fibre unobstructed. When there is an FBG sensor, a narrowband of wavelength of the light spectrum is reflected back to the sending end and these reflected wavelengths are analyzed by an optical interrogator. The light being reflected back has a spectrum that characterizes the pitch (e.g., the separation between two periodic marks) of the periodic refractive-index variation. It can be seen in Figure 2 that if the input light with a band of wavelengths 1, 2, 3, and 4 is fed into the optical fibre, it is only wavelength $B = 3$ which will be reflected back because the grating satisfies the Bragg condition of $B = 2n\Lambda$, where B is the Bragg wavelength, n is the refractive index of the core material, Λ is the pitch of the grating which is defined by the phase mask during FBG inscription. The rest of the wavelengths 1, 2, and 4 are transmitted through the grating. As the pitch is changed when the FBG is subjected to strain (i.e., the pitch becomes longer, when the FBG sensor is being extended, and becomes shorter when the FBG sensor is being compressed), there will be corresponding changes in the reflected wavelength.

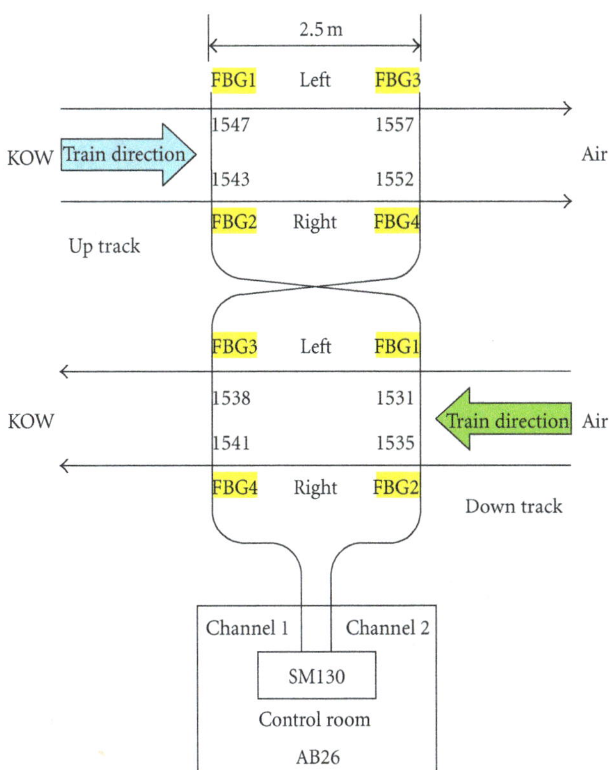

FIGURE 3: Schematic of Train Load balance index monitoring system in SHD. The sensor group 1543, 1547, 1552, 1557 are installed on up track while sensor group 1531, 1535, 1538, 1540 are installed on down track.

Since the parameter of measurement is the wavelength of light which is not affected by electromagnetic fields, the process is immune to electromagnetic interference and hence is intrinsically more stable than any electrical monitoring system in an electr4magnetically noisy environment which is typical in an electrified railway.

2.2. FBG Sensor Network. The FBG sensor network being installed includes a number of FBGs in an optical fiber. The length of the optical fibre can be as long as 100 km and many FBGs with different grating characteristics can be inscribed on the same optical fibre which is connected to a commercially available interrogator system as shown in Figure 3 in the control room. This interrogator system is further connected to a computer with an application program which is developed in-house for train vibration monitoring and train weighing.

Figure 4 shows a typical pickup from a sensor installed on the track with the passage of a 12-car train. It is noted that there are 4 wheels in one car and hence the first 4 peaks on the left-hand side of the upper figure correspond to the first car. It can be seen that there are more vibrations in Cars 6 and 8 (when counting from the left hand side) and such observation would allow the engineer to pay attention to cars with abnormal or excessive vibrations. Subsequent investigation revealed that there were roundness problems in

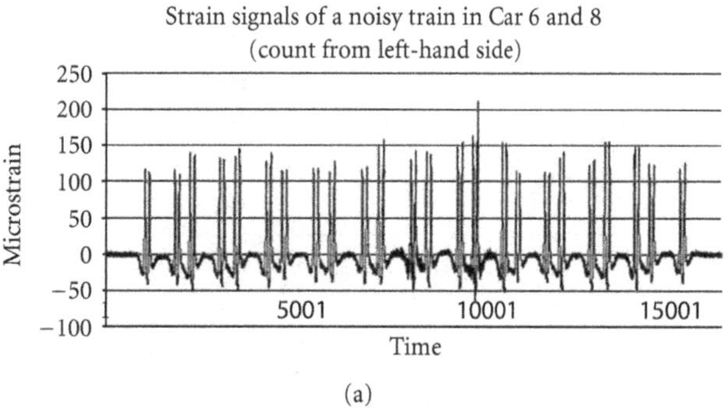

(a)

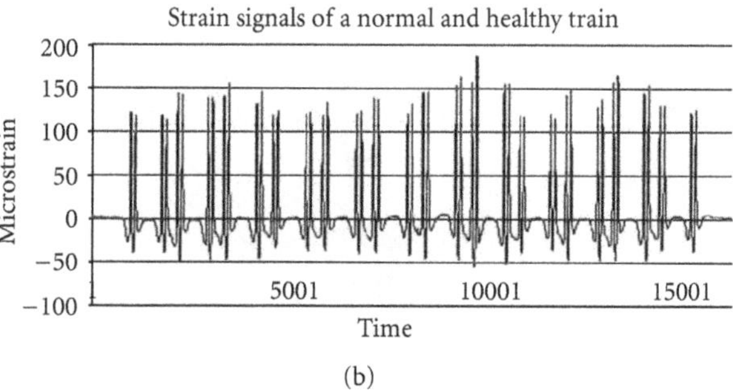

(b)

FIGURE 4: Strains picked up by optical sensors mounted on the track with the passage of a 12-car train.

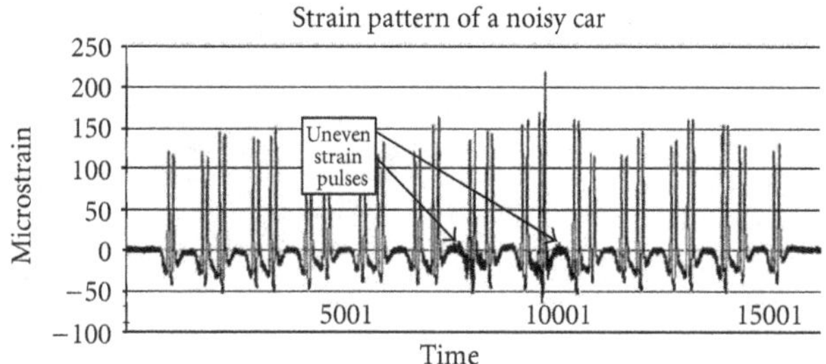

FIGURE 5: Strains picked up on the track due to the passage of a noisy train.

the wheels of Cars 6 and 8 and this will be described in more details in subsequent sections of this paper.

3. Applications

3.1. Monitoring of Imperfections in Train Wheels. With the installation of the sensors on the tracks, a wealth of investigations could be carried out readily. For example, from the strain measurements at the track, it was confirmed that there were some noisy trains. It was suspected that these noisy trains might have relatively imperfect wheels. Hence, a series of tests were carried out. The first test is to identify a noisy train (a train producing strains as shown similar to the upper half of Figure 4) using an FBG sensor installed on the rail track as described in the previous section.

The principle of wheel imperfection detection by FBG strain [4] sensors installed at track is based on the fact that wheel defects such as flange pits, wheel flats, and particularly out-of-round wheels which are also known as polygonal wheels, will exert periodic impact force on the track. In this work, it was found that an imperfect wheel will produce an uneven strain impulse on the track. In contrary, a newly turned wheel will produce a symmetrical strain impulse. In another approach, one attempts to relate the interaxial vibrations as shown in Figure 5 with a wheel that has out-of-roundness. In order to compare parameters from two different systems, the interaxial vibrations were quantified by a vibration index which is obtained arithmetically by considering the train speed, vibration frequency, and magnitude as a whole (Figure 6).

The detail derivation of the vibration index, is however, a proprietary information of the authors and hence cannot be described in too much details in this paper. Figure 7 shows the vibration indexes obtained from a noisy train.

The local vibrations such as those in Cars 6 and 8 (counting from the left) of Figure 7 have a vibration index of 1.5+ which are higher than those of the other cars in the same train with a vibration index lower than 1.1.

The wheel of the train cars with out-of-roundness (Cars number 6 and 8) [7] are measured.

The wheels on Car number 6 were turned right after the out-of-roundness measurement while the wheels on Car number 8 were kept unturned as a control. The wheels out-of-roundness after turning were between 0.05 to 0.07 mm. The strain pattern and vibration indexes of the serviced train are shown in Figure 8 from which one can observe that the vibration at Car number 6 has been eliminated after wheel turning while the vibration at Car number 8 persists. By comparing Figure 7 with Figure 9, it can be seen that the vibration index of Car number 8 (unturned wheels) remained high at around 1.9 while the index of Car number 6 (turned wheels) was greatly reduced from 1.8 to 0.8. This shows that the vibration index, which is deduced from the FBG strain sensor measurement results, is an effective means to distinguish wheels with out-of-roundness from wheels which are healthy.

3.2. Train Weight Measurement. The Bragg wavelengths of the FBGs will change because of deformation of the rail when the train passes over the location at which the FBG is installed. The data are being logged by a computer with a sampling frequency of 1000 Hz. The waveforms are shown as in Figure 10 below.

At the initial stage of the study, it was found intuitively that the amplitude of the peaks such as $P1$ as shown in Figure 10 is broadly corresponding to the weight of the axle. However, it is noted that the amplitude of the peak can be found in different ways. Therefore, four approaches are proposed as given below. Here, $W1$ and $W2$ are defined as the respective weight of axles 1 and 2 of the same bogie. $P1$ and $P2$ are the peaks of the signal excited by the load being applied to axle 1 and axle 2 respectively. $V1$, $V2$, and $V3$ are the respective valleys which are generated when the train car

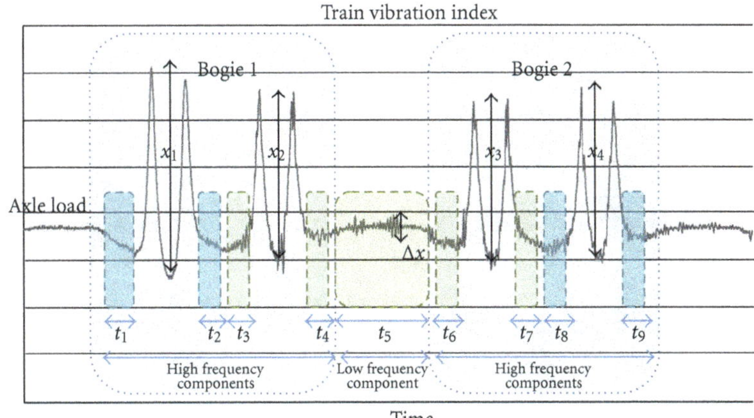

Vibration index $= \sum_{n=1\ldots9}^{n} K_n f(\Delta X_n/X_n, t_n)$, where K_n is the weighting factor of each component

FIGURE 6: Graphical explanation of how the vibration index is being derived after the vibration waveforms are sectionalized and weighted in order to generate the vibration index.

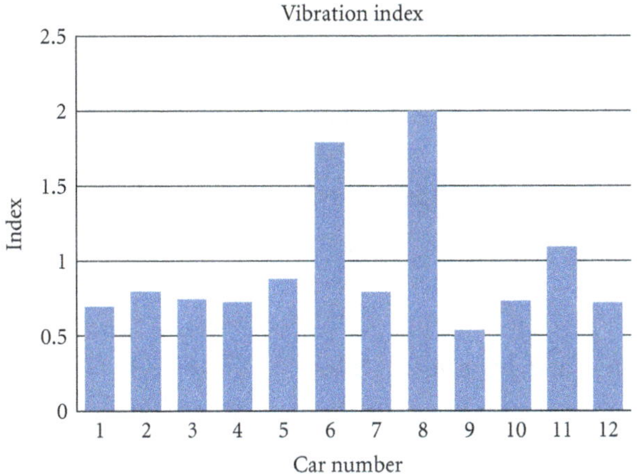

FIGURE 7: Vibration index of the 12 cars.

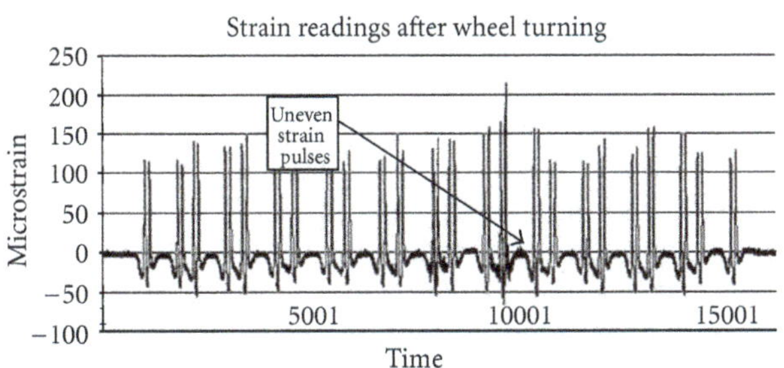

FIGURE 8: The strain readings after wheel turning.

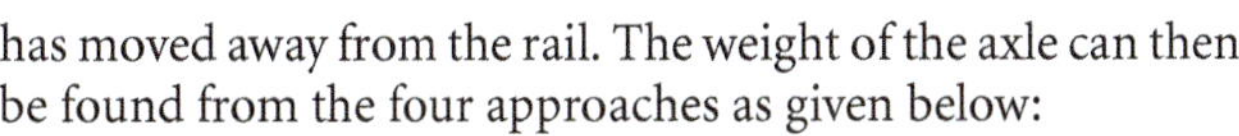

has moved away from the rail. The weight of the axle can then be found from the four approaches as given below:

Approach 1: $W1 = (P1 - V1)$; $W2 = (P2 - V2)$,

Approach 2: $W1 = (P1 - V2)$; $W2 = (P2 - V3)$,

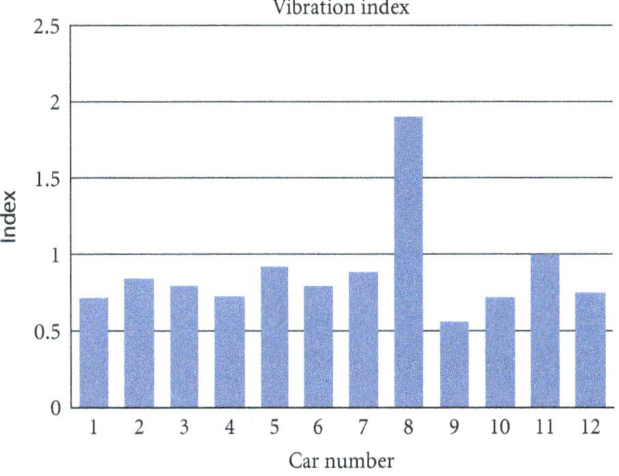

FIGURE 9: Vibration index of the train with the wheels of car 10 turned.

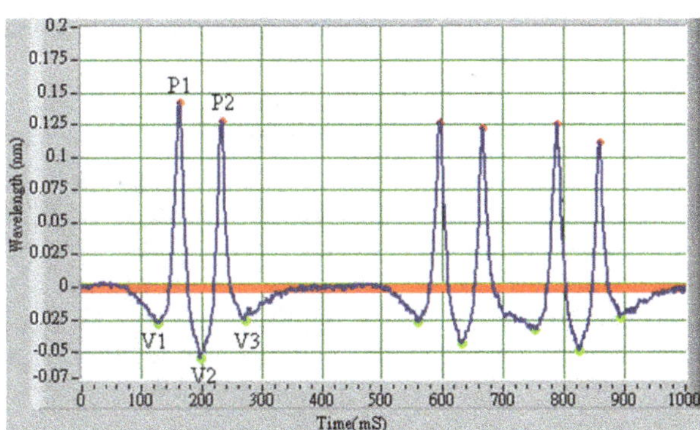

FIGURE 10: Strain measurement using FBG when a wheel of a train is passing through.

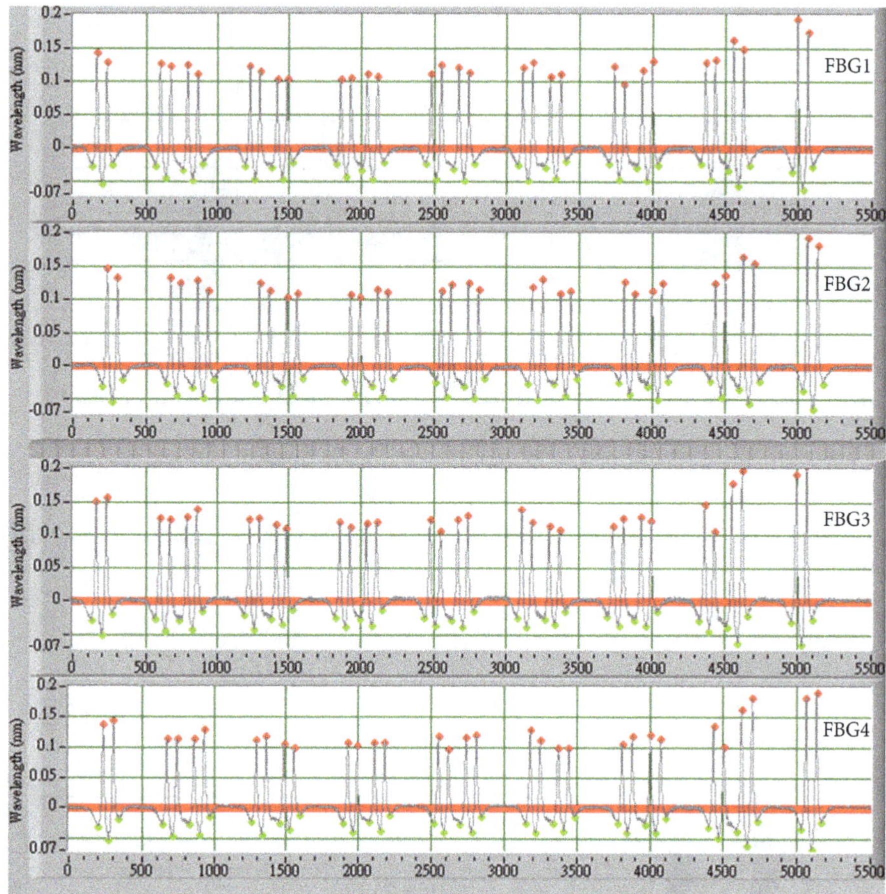

FIGURE 11: Wavelength signals of the trainset "TCL 08" as measured in the down track.

Approach 3: $W1 = (P1 - (V1 + V2)/2)$; $W2 = (P2 - (V2 + V3)/2)$,

Approach 4: $W1 = P1$; $W2 = P2$.

3.2.1. The Static Weight Data. The trainset "TCL 08" with the weights of all the wheels were measured and the static data are then used to compare with the strain measured by using FBGs.

3.2.2. Measurement Results. The wavelengths signal of the trainset "TCL 08" are shown as in Figure 11. These signals are measured in the down track.

The strains as derived from the shifts in wavelengths are then feed to the application program and then compared with static weight data. In order to calibrate the system, the static weight of one axle was equated to the amplitude of the FBG signal when the train was moving very slowly over the track at which an FBG sensor was installed. The ratio, which is equal to the weight of axle/amplitude of the FBG signal, is then used to convert the dynamic amplitude of each FBG signal in train weight. The results of the four approaches in finding the amplitude of the FBG signals are as shown in Table 1. The best scenario is Approach 4 which has the minimum error percentage. The average error from the four FBGs are 10.39% and, bearing in mind that the "dynamic" weight of trains is normally considered to be 10% higher than their static weight, the measured data are considered to be in excellent agreement with the static data.

3.3. Train Antiderailment System. One of the most important safety assessments for running train on rail is on the possibility of derailment, particularly in negotiating curve and track twist. The likelihood of derailment is very much dependent on the conditioning of track and the vehicle. The former refers to the deviation of the maintenance limits on the track geometries, while the latter refers to the conditions of the vehicle suspensions system together with their wheel profiles. Checking of both conditions will have to be done during off-traffic hours under prescheduled intervals, unless there are means for online in-service checking.

Derailment caused by track twist is related to the combination of the horizontal guiding force and the reduction of the vertical wheel-load of the leading wheel. The resultant of the two force vectors may cause the guiding wheel to climb the rail.

Nadal's limit (Y/Q) on the lateral force (Y) on the wheel being evaluated for derailment and its vertical load (Q) is used to measure the probability of flange climbing. It has to be kept to an acceptable limit. Figure 12 indicates the relationship of the forces at the wheel/rail interface.

TABLE 1: The result of the four approaches.

Up track				
Error %	Approach 1	Approach 2	Approach 3	Approach 4
FBG 1,2	11.51	12.51	14.39	12.48
FBG 3,4	11.53	13.10	12.59	9.94
Down track				
Error %	Approach 1	Approach 2	Approach 3	Approach 4
FBG 1,2	14.43	14.37	12.00	9.40
FBG 3,4	12.36	12.82	12.34	9.74

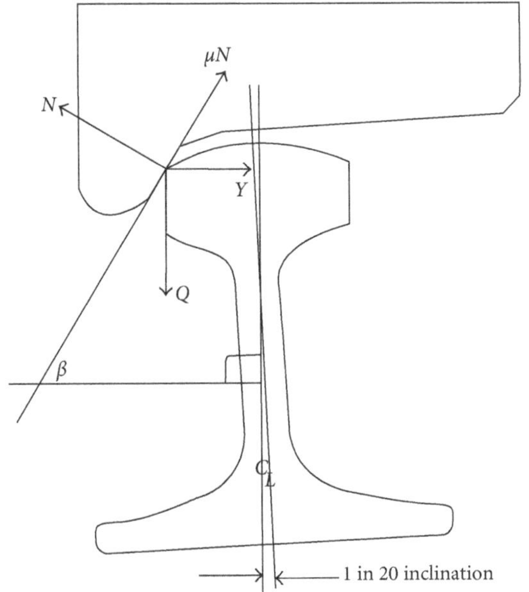

$$\frac{Y}{Q} = \frac{\tan\beta - \mu}{1 + \tan\beta},$$

where

Y: lateral force at wheel angle

Q: vertical load on the wheel

β: wheel ange angle at plane of contact

μ: coefficient of friction at ange/rail contact point

FIGURE 12: Relationship of forces at the wheel/rail interface.

As lateral force [8] is inevitably present for train wheels in negotiating track curve and twist, it is important to ensure that the wheel sets must be designed to operate with adequate vertical loads to keep the Y/Q ratio to fall within limits.

Load transfer between wheel sets is thus critical and must be controlled also to within a reasonable limit, which is no more than 60% of the load from one wheel being transferred to the other as shown in (1), where Q_1 and Q_2 are the vertical loads of two wheels of the same axle acting on rail:

$$\text{off-loading ratio } \frac{\Delta Q}{Q} = \frac{(Q_1 - Q_2)}{(Q_1 + Q_2)} < 0.6. \quad (1)$$

The weight of the train acting on the rail by the wheel would create both tensile strain and stress on the rail beam. By putting FBG on rail, their corresponding changes in Λ can be detected readily. Practically, changes in the FBG grating Λ sensed would be presented as changes in λ and, hence, according to the principle described, the weight of the train will trigger a shape change in wavelength ($\Delta\lambda$). Once it is captured, it will represent the weight of the wheel acting on the rail Q.

As the off-loading ratio $\Delta Q/Q$ concerns only on the relativity ratio of the loadings and not on the absolute value of the vehicle weights, thus, the temperature effect and the need of fine tuning and resolving the transfer function of the $\Delta\lambda$ with respect to the weight Q are not a matter of concern as they all will be canceling out in deriving the off-loading ratio $\Delta Q/Q$.

It is noted that because one is more concerned about the ratio of the axle weight on the two sides of the axle when determining the potential derailment danger, there is no real need to measure the axle weight very accurately. The fact that the sensors were measuring the axle weight very accurately is, therefore, an additional bonus which might provide further intelligence into the development of an even more comprehensive smart railway conditioning monitoring system in future.

4. Summary

An optical system using Fibre Brag Grating has been designed and tested in a commercial railway testbed. Electromagnetic interferences could cause serious problems in main line railways operating with a 25 kV supply. Hence, this research is focused on the use of optical sensors which are immune of EMI. In essence, this study reports the use of FBG sensors for detecting vibrations from noisy trains as well as train weigh, thereby paving the way for the development of an antiderailment system which is an essential requirement for a modern railway.

The response signal obtained by the FBG sensors is positive and clear enough for further development. Apart from the objectives mentioned above, the FBG system could be tailor made for many additional railway applications including the following:

(a) counting the numbers of axles in and out of a given section (track circuit occupation), that is, for axle counter,

(b) train identification by tracking the differences of axle bases, weights, and car numbers of various train stocks,

(c) speed detection.

In summary, a small piece of FBG sensor can generate lots of train running information with good data integrity. It can be applied to monitor the train running performances of the railway by building a virtual system on top of the conventional ones as a standby system both on operational and safety aspects.

Particularly, FBG axle counter system can be built on top of existing conventional axle counter system of a moving block signal system to allow a virtual track circuit to remain operational in case of emergency or for running engineering train during nontraffic hours.

Further analysis and measurements together with a fully blown development on its mounting on rail is necessary before the maturity of the FBG sensor technology can be put into practical usage in the highly conservative railway industry.

Acknowledgments

The authors would like to acknowledge the support from the MTR for allowing a number of tests to be carried out in an operational rail line.

References

[1] P. F. Weston, C. J. Goodman, P. Li, C. S. Ling, R. M. Goodall, and C. Roberts, "Track and vehicle condition monitoring during normal operation using reduced sensor sets," *Transactions Hong Kong Institution of Engineers*, vol. 13, no. 1, pp. 47–53, 2006, Special Issue on Railway Development in the 21st Century.

[2] K. Y. Lee, K. K. Lee, and S. L. Ho, "Exploration of using FBG sensor for axle counter in railway engineering," *WSEAS Transactions on Systems*, vol. 3, no. 6, pp. 2440–2447, 2004.

[3] K. K. Lee and S. L. Ho, "Unconventional method of train detection using fibre optic sensors," *Transactions Hong Kong Institution of Engineers*, pp. 16–21, 2006, Special Issue on Railway Development in the 21st Century.

[4] A. D. Kersey, M. A. Davis, H. J. Patrick et al., "Fiber grating sensors," *Journal of Lightwave Technology*, vol. 15, no. 8, pp. 1442–1462, 1997.

[5] K. O. Hill, B. Malo, F. Bilodeau, D. C. Johnson, and J. Albert, "Bragg gratings fabricated in monomode photosensitive optical fiber by UV exposure through a phase mask," *Applied Physics Letters*, vol. 62, no. 10, pp. 1035–1037, 1993.

[6] T. Erdogan, "Fiber grating spectra," *Journal of Lightwave Technology*, vol. 15, no. 8, pp. 1277–1294, 1997.

[7] A. Johansson, "Out-of-round railway wheels-assessment of wheel tread irregularities in train traffic," *Journal of Sound and Vibration*, vol. 293, no. 3-5, pp. 795–806, 2006.

[8] P. Remington and J. Webb, "Estimation of wheel/rail interaction forces in the contact area due to roughness," *Journal of Sound and Vibration*, vol. 193, no. 1, pp. 83–102, 1996.

Industrial Qualification Process for Optical Fibers Distributed Strain and Temperature Sensing in Nuclear Waste Repositories

S. Delepine-Lesoille,[1] X. Phéron,[1] J. Bertrand,[1] G. Pilorget,[1] G. Hermand,[1] R. Farhoud,[1] Y. Ouerdane,[2] A. Boukenter,[2] S. Girard,[2] L. Lablonde,[3] D. Sporea,[4] and V. Lanticq[5]

[1] *National Radioactive Waste Management Agency (Andra), 1-7 rue Jean Monnet, Parc de la Croix blanche, 92298 Chatenay-Malabry, France*
[2] *LabHC, UMR CNRS 5516, Université de Saint-Etienne, 42023 Saint-Etienne, France*
[3] *iXFIber, 22300 Lannion, France*
[4] *National Institute for Laser, Plasma and Radiation Physics, Magurele, Romania*
[5] *Cementys, 27 Villa Daviel, 75013 Paris, France*

Correspondence should be addressed to S. Delepine-Lesoille, sylvie.lesoille@andra.fr

Academic Editor: Qiang Wu

Temperature and strain monitoring will be implemented in the envisioned French geological repository for high- and intermediate-level long-lived nuclear wastes. Raman and Brillouin scatterings in optical fibers are efficient industrial methods to provide distributed temperature and strain measurements. Gamma radiation and hydrogen release from nuclear wastes can however affect the measurements. An industrial qualification process is successfully proposed and implemented. Induced measurement uncertainties and their physical origins are quantified. The optical fiber composition influence is assessed. Based on radiation-hard fibers and carbon-primary coatings, we showed that the proposed system can provide accurate temperature and strain measurements up to 0.5 MGy and 100% hydrogen concentration in the atmosphere, over 200 m distance range. The selected system was successfully implemented in the Andra underground laboratory, in one-to-one scale mockup of future cells, into concrete liners. We demonstrated the efficiency of simultaneous Raman and Brillouin scattering measurements to provide both strain and temperature distributed measurements. We showed that 1.3 μm working wavelength is in favor of hazardous environment monitoring.

1. Introduction

Distributed optical fiber sensors (OFSs) [1–3] are a key technology for the monitoring of the planned French deep geological repository for long-lived high-level and intermediate-level wastes, called Cigéo. Temperature and strain distributed sensing based on Raman, Rayleigh, and Brillouin scatterings offer exceptional advantages over traditional electronic sensors, especially as they provide distributed data over the entire structure and thus overcome limitations of traditional sensors, whose information is restricted to local effects.

This paper focuses on temperature and strain distributed sensing based on Raman and Brillouin scatterings in optical fibers for structural health monitoring, more precisely for nuclear industry. Although commercial off-the-shelf sensors and interrogation units are numerous, the global measuring chain may provide disappointing monitoring results to the end-users, unless a number of considerations specific to nuclear environments are taken into account. These are further developed within this paper, with an emphasis on environmental conditions influence, especially (i) temperature, (ii) gamma rays, and (iii) hydrogen influences.

Andra's (French National Radioactive Waste Management Agency) potential applications include surface and deep geological radioactive waste disposal structure monitoring, for instance within the future geological repository that would contain highly instrumented disposal cells. Intermediate-level long-lived waste cells are presently designed as 400 m long tunnels, with a 1 m thick concrete liner, placed 500 m deep in a clay rock called Callovian-Oxfordian formation.

Monitoring aims at preserving retrievability of nuclear wastes, assessing long-term safety, enabling optimization of structures all along the exploitation which is expected to last a century. A major specification of the geological repository monitoring system is durability, required to last up to a century, despite hazardous conditions: gamma rays and hydrogen release. Because of high gamma radiation doses, disposal cells are not accessible as soon as exploitation starts, and the first nuclear waste package is placed inside the disposal cell. This implies the monitoring system to be robust for decades without any maintenance. OFS are a robust technology known to handle radiations quite well. Durability is also ensured thanks to their ability to perform remote sensing, which enables maintenance of optoelectronic instruments. Finally, OFSs are very attractive for their small size and varied external coatings, which limits the preferential flow paths and reduces invasiveness, a highly important aspect to avoid affecting the long-term safety of the geological repository for nuclear wastes.

For these attractive advantages, Andra drives many research studies on distributed temperature and strain sensing with optical fiber sensors. To ensure measurement quality and lifetime of the global monitoring system, Andra has implemented a qualification procedure, which will be presented in the first part of the paper. Previous reported results gathered within this qualification process will be recalled. The second part focuses on recent results: field test where temperature compensation of strain measurements, based on combined Brillouin and Raman scatterings, was successfully implemented. The third part presents research results on hazardous condition influences on both Brillouin and Raman scatterings as a function of optical fiber types. We conclude with a recommendation on the optimized sensing system for our application.

2. Qualification Procedure and Previous Results

2.1. Qualification Procedure. The outstanding properties of optical fiber sensors drove major interest in structural health monitoring applications. Nevertheless, it is important to notice that optical fiber and optical installation practices used in telecommunications and other industries are significantly different from nuclear constraints and applications, which may be far less tolerant.

Moreover, optical fiber sensing systems presently suffer from a lack of standardization of claimed performances. Dedicated qualification processes are not defined yet. Andra has implemented a multistage qualification procedure for each selected measurement chain. For OFS, it was chosen to study both sensing cables and optoelectronic instruments separately before pairing such elements and focusing on data processing.

The described overall process is inspired from [4]. Global test sequence includes four stages.

Stage one consists in acquiring in-depth knowledge of the sensing technology, engineering solutions, and practical implementation constraints. It aims at selecting the technologies best suited to the specific requirements of monitoring the geological repositories for long-lived nuclear wastes. When off-the-shelf sensing chain performances do not fulfil requirements, Andra initiates research programs. It has been the case for distributed optical fiber sensing system whose results are presented in this paper.

Stage two consists in carrying out laboratory tests, under fully supervised and/or controlled environmental conditions, to qualify the sensitive component and assess the complete measurement chain performances. Sensors are tested alone, then embedded in the host material of interest.

Stage three consists in outdoor tests, to evaluate field implementation influence. At this stage, the sensing chain is preserved from hazardous conditions, extreme temperature, or gamma rays. Unexpected influence of various parameters might thus be revealed.

The fourth stage involves hardening in view of the application environmental conditions. In the envisioned French geological repository, temperature would range from 20°C to 90°C. Gamma radiation rates reach 1 Gy/h, total dose 10^7 Gy. Hydrogen release is also expected; its maximum levels could approach 100% hydrogen content in the atmosphere.

2.2. Previous Results. This qualification methodology has been implemented for distributed temperature and strain sensing.

Andra selected Raman scattering to perform distributed temperature measurements, as it is the most advanced technology with superior temperature sensitivity, better than 0.1°C [5]. Andra selected Brillouin scattering for distributed strain sensing, since interferometric measurements are in favor of durability. The first three stages of the qualification procedure, laboratory tests, and preliminary outdoor tests were reported in [6]. Two remaining questions were to be addressed to end the qualification procedure.

First, how to compensate for Brillouin scattering temperature sensitivity? The Brillouin frequency shifts are known to be proportional to temperature (ΔT) and strain (ε) variations as in (1) [7]:

$$\Delta\nu_B = C_T \Delta T + C_\varepsilon \varepsilon. \quad (1)$$

C_T and C_ε are characteristics of the optical fiber type. At the operating wavelength (1550 nm), for standard G652 single-mode fiber, C_T and C_ε are in the order of 1 MHz/°C and 0.05 MHz/$\mu\varepsilon$ [8]. Instruments based on Brillouin scattering would perform either temperature or strain measurements; strain is 20 times less influent than temperature. A solution to decorrelate strain from temperature is detailed in Section 3.

Second, what are hydrogen and gamma rays influences on Raman and Brillouin scatterings? Quantitative evaluation of such influence as a function of optical fiber types is presented in Section 4. Based on these various evaluations, a last paragraph predicts lifetime of the optical fiber geological repository monitoring system, anticipating strain and temperature measurements uncertainties and distance range.

3. Tunnel Liner Instrumentation: Strain Measurement Temperature Compensation

Andra has created an Underground Research Laboratory to evaluate the constructability, safety, and reversibility of the potential radioactive waste disposal in the Callovian-Oxfordian clay stones. A multidisciplinary program is implemented in this 500 m underground structure. The "retaining and covering observation experiment" is taking place into a dedicated gallery with a concrete liner (Figure 1), similar with intermediate-level long-lived waste disposal cells. Its purpose is to evaluate several monitoring solutions. This experiment integrates onfield constrains (drilling environment, dust, and operational phases with limited intervention time) and a representative experiment in view of future geological repository (one-to-one scale, specific rock in its natural location).

3.1. Instrumented Structure. The gallery is 5 m in diameter with a 50 cm thick concrete liner. Many parameters were monitored in both clay stones and covering concrete: temperature, strain, water content, and interstitial fluid pressure.

One section of the concrete liner was instrumented with collocated fiber optic sensing cables whose picture is illustrated in Figure 2.

Andra had previously tested Raman temperature sensing into single-mode fibers in a surface building slab. This test highlighted the great sensitivity of Raman scattering in single-mode fiber to curvature, which can poorly be avoided in civil engineering structures [9]. This is why two different sensing lines were implemented: multimode fibers were used for Raman temperature monitoring and single-mode fibers were installed for Brillouin strain sensing. More precisely, we used three different sensing cables, two for Brillouin measurements, composed of G652 and G657 fiber types to evaluate curvature sensitivity, and one for Raman sensing. The three cables were placed redundantly so that, finally, six sensing arches were embedded inside the concrete liner.

In order to ensure accurate positioning and maintaining during concrete pouring, sensing cables were attached to a thin piece of wire mesh, spitted on the retaining concrete.

Measurements were remotely performed every 15 minutes by commercially available instruments, a Brillouin OTDA (optical time-domain analyser) and a Raman distributed temperature sensing (DTS). Both devices were set at 0.5 m spatial resolution.

Instruments were located in another gallery of the underground laboratory. The full sensing line is approximately 500 m, with only 20 m embedded inside the concrete liner.

Electronics sensors, such as vibrating wire extensometers and platinium probes, were colocated with optical fiber sensors.

3.2. Results of Early Age Monitoring. The gallery liner was constructed in 2 steps. First, the bottom part (inverted arch) concrete was poured. One month after, the second arch completed the liner.

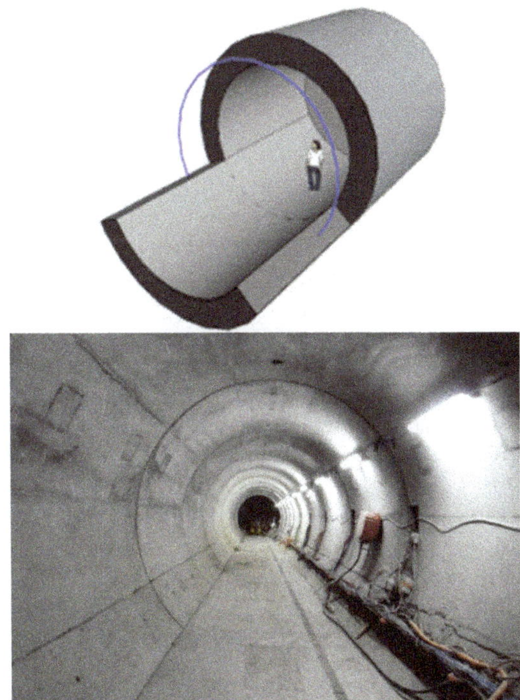

FIGURE 1: Scheme of gallery instrumentation by distributed sensing and picture of the gallery 6 months after liner casting.

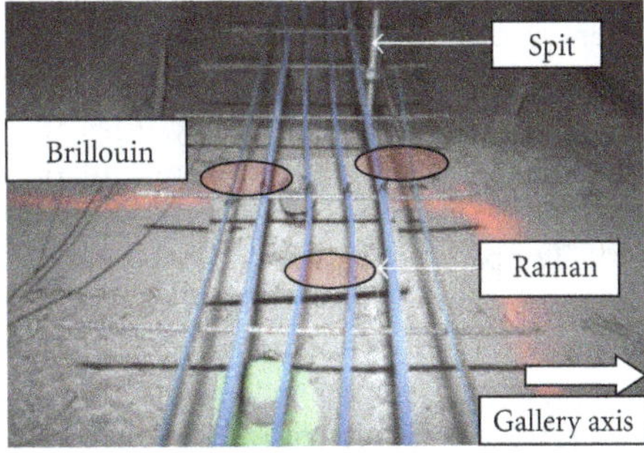

FIGURE 2: Picture of optical fiber cables installed on a gallery circumference, before concrete casting.

One from the six sensing lines got damaged during construction. We did not observe any advantage of G652 compared with G657 on the strain measurement quality.

During concrete hardening, an exothermal chemical reaction takes place. Consequently, temperature increases and Brillouin frequency shifts. Thermal expansion of concrete induces strain on the optical fiber cable, which also increases Brillouin frequency. Temperature measurement acquired by Raman sensing lines (left y axis) and raw Brillouin frequency (right y axis) are illustrated in Figure 3.

To provide useful information, concrete thermal expansion must be differentiated from the parameters of interest: strain induced by stress, creep, and shrinkage, noted $\varepsilon_{\text{comp}}$ (comp. stands for "compensated"). If α_{concrete} is the concrete

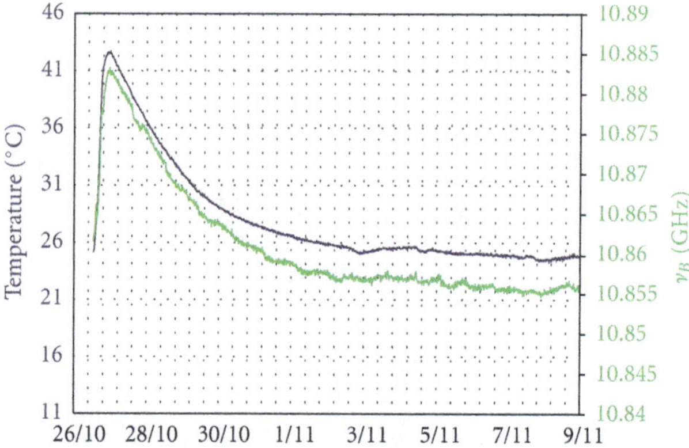

FIGURE 3: Raman temperature measurement and Brillouin frequency shift during concrete hardening.

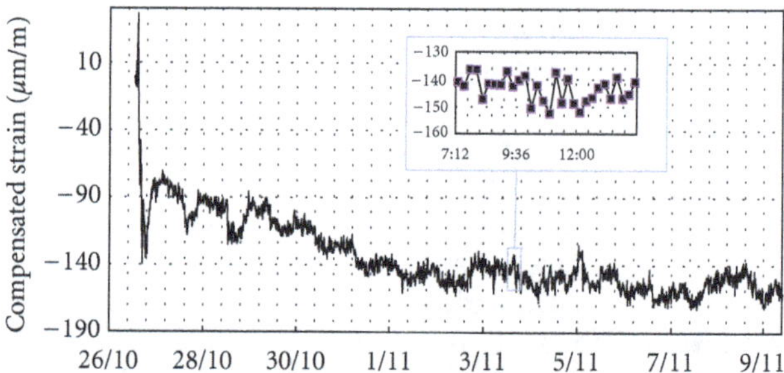

FIGURE 4: Processed data: temperature compensated strain acquired during concrete pouring (x axis is dates, from October to November 2011).

thermal expansion coefficient, temperature-compensated strain can be obtained following equation (2):

$$\varepsilon_{\text{comp}} = \frac{\Delta \nu_B}{C_\varepsilon} - \frac{(C_T + \alpha_{\text{concrete}} C_\varepsilon) \cdot \Delta T}{C_\varepsilon}. \quad (2)$$

Embedded and instrumented concrete samples [10] enabled in situ measurement of $\alpha_{\text{concrete}} = 10\,\mu\varepsilon/^\circ\text{C}$. Assuming $C_T = 1\,\text{MHz}/^\circ\text{C}$ and $C_\varepsilon = 0.05\,\text{MHz}/\mu\varepsilon$, we obtained the compensated strain measurements plotted in Figure 4.

The 150 μm/m compressive strain measured is the consequence of the early-age shrinkage of concrete. This value is fully consistent with the one measured by the vibrating wire extensometers placed nearby the optical fibers.

Uncertainties were evaluated analyzing 24 repeated measurements acquired during 6 h (inset Figure 4). Such analysis has been repeated. We obtain uncertainty in the order of 30 μm/m, which corresponds to the Brillouin instrument performance. Temperature uncertainty was better than 0.2°C in the Raman sensing line (Figure 3). The Raman-Brillouin compensation method does not increase the initial strain and temperature measuring system uncertainties.

Our results were obtained with 500 m distance range and a 0.5 m spatial resolution. Similar experiments of Raman temperature compensated Brillouin strain measurements reported 3.6°C and 80 μm/m repeatability with 5 m resolution [11].

This method also enables long-term monitoring. Concrete liner strain evolution has been acquired all along construction steps; measurements will go on in the next years to acquire as much information as possible before the repository construction.

An advantage of distributed measurements provided by optical fibers is illustrated in Figure 5. In this gallery, covering concrete is not reinforced (no steel bar). Tensile strain zones were identified and precisely located by distributed measurements at locations where there was no electronic sensor.

As conclusion, a sensing scheme combining Raman and Brillouin instruments with multiple sensing cables is a promising solution for performing simultaneous temperature and strain monitoring.

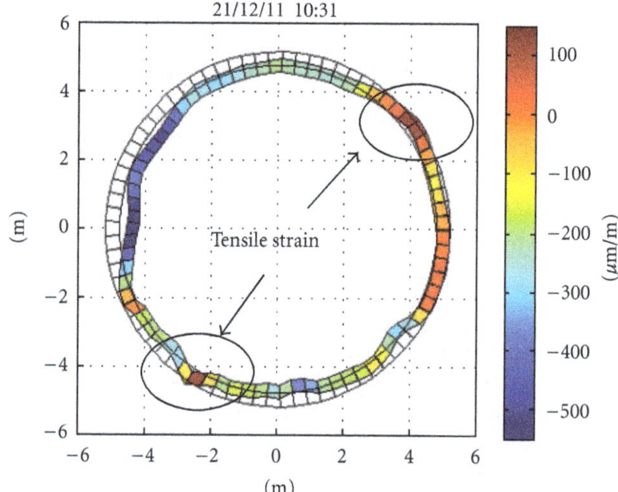

FIGURE 5: Compensated strain reported on section after arch construction.

This real-scale experiment demonstrates that a Raman-Brillouin temperature and strain sensing is very well suited for underground tunnel monitoring. It also appeared that chosen optical fiber cables were robust enough to put up with construction conditions.

4. Compatibility with Future Hazardous Environments

Nuclear waste repository is a challenging environment due to the presence of gamma radiations which is known to degrade the optical properties of fibers through three different phenomena [12]: radiation-induced attenuation (RIA) decreases fiber transmission efficiency, radiation-induced emission (RIE) decreases signal-to-noise ratio and compaction changes refractive index (especially for very high particle fluences or doses).

The amplitudes and kinetics of these changes depend on many parameters, among which are dose rates, total doses, optical fiber type, and operating wavelength.

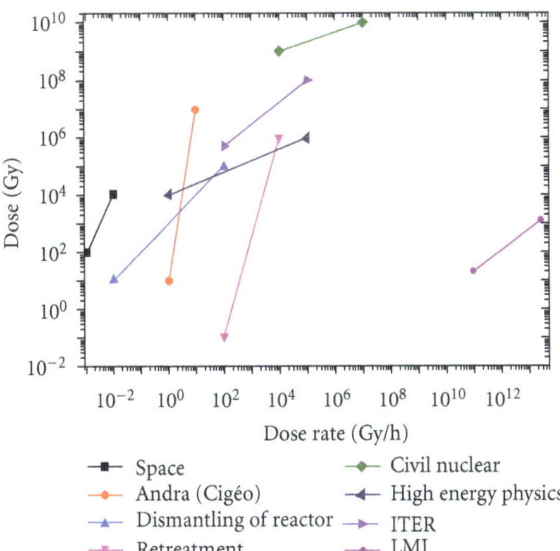

FIGURE 6: Overview of typical applications as a function of total dose and dose rate [13].

As illustrated in Figure 6, in Andra geological repository, doses rates are moderate; however, after 100 years of monitoring in the vicinity of high-level wastes structure cells, total doses will reach 10^7 Gy.

4.1. Gamma Influence on Brillouin Scattering. It has been demonstrated on standard G652 fibers that there is a limited influence of gamma rays on the Brillouin frequency at least up to doses of 100 kGy [14]. To evaluate possible influence at total doses corresponding to our application, we performed in situ [15] and postmortem gamma-ray irradiation tests in different classes of fibers.

Gamma-ray irradiations were performed at room temperature using the Brigitte facility, cobalt-60 source, at SCK-CEN (Belgium) [16], with a dose rate of $\sim$28 kGy$\cdot$h^{-1}.

We selected our samples among the most commonly used fibers, like the SMF28 from Corning, since almost all commercially available Brillouin instruments are designed to operate with this type of fibers. Thus, the first selected fiber was the step-index fiber SMF28 from Corning. In view of the geological repository monitoring requiring high gamma dose tolerance, we selected a fluorine-doped cladding, pure silica-core fiber. The fiber has a 5 μm core radius and a 40 μm cladding radius with a fluorine concentration of 1.25% mol. Finally, to evaluate the influence of high concentrations of dopants, we tested a highly GeO_2-doped core fiber at 28% mol concentration with a core radius of 1.4 μm and a pure silica cladding with radius of 62.5 μm (HGe in the following).

As expected and detailed in Table 1, measured RIA levels are important at the high doses (up to 10^7 Gy), with a factor up to 4 between the RIA measured for the two extreme cases. The most tolerant fiber, the F-doped fiber, quickly saturates, at a moderate dose level of 5 MGy, at RIA level around 50 dB/km. Impact on distance range is discussed later.

For the dose range corresponding to geological repository application, measured central Brillouin frequencies also shifted (BFS) with dose, whatever the considered optical fiber types. Main parameters of the Brillouin spectrum are listed in Table 1 as well as their changes with dose. Unlike SMF28 and HGe fiber, a clear saturation in the Brillouin frequency shift ($\Delta\nu_B$) occurs at $\sim$2 MGy for the F-doped fiber. This saturation effect is of major interest for the target application where expected doses largely exceed this saturation level dose. Moreover, for this fiber, Brillouin frequency shift remains small, in the order of 2 MHz, which corresponds to 2°C measurement error or 40 μm/m. On the opposite, 18°C temperature measurement error would occur if HGe fiber was selected.

Strain and temperature sensing based on Brillouin scattering involves both the central Brillouin frequency measurement and hypothesis on calibration coefficients (C_ε, C_T) (see (1)). These calibration coefficients might also be impacted by radiation. To our knowledge, it is the first time radiation influence on these parameters is evaluated.

No significant effect of radiations on C_T could be observed for the SMF28 and HGe fibers (Figure 7). For the F-doped fiber, a decrease of about 6% was noted in the C_T coefficient after a deposited dose of 10^7 Gy. C_ε strain coefficients do not seem to depend on received gamma dose, whatever the fiber type. As a conclusion, gamma impact on calibration factors would increase measurement uncertainties by few percent only.

As a conclusion, radiation effects on the performances of strain and temperature Brillouin scattering based optical fiber sensors have been deeply investigated for different classes of optical fibers. There is a strong influence of the core composition. Thus, optical fiber type should be carefully chosen to ensure durability of the monitoring system.

The presented tests are relevant since it is well known that higher dose rates increase RIA; thus, results overestimate degradations that will endure the geological repository monitoring system.

Fluorine-doped pure-silica core fibers are able to handle dose rates and total doses during one century of monitoring within the geological repository. Strain and temperature measurement uncertainties would slightly degrade. The most impacting radiation effect is RIA. Since optoelectronic instrument optical budget is in the order of 10 dB, distance range would be reduced down to 200 m for the F-doped fiber after a century of monitoring. It suits requirements as (i) high-level long-lived waste (HLW) disposal cell will be 40 to 100 m long and (ii) in 400 m intermediate-level long-lived waste storage cell, doses are greatly reduced.

On the opposite, with SMF28 fiber, distributed sensing would be compromised after only a decade of monitoring along 100 m of HLW disposal cell, instead of the required century.

We focused our study on the response at the 1550 nm wavelength since it is the commercial instrument working wavelength. However, a global optimization of the entire system must also consider this parameter as a possible variable. Initially, optical losses are slightly more important in pristine optical fibers at the 1310 nm wavelength compared to the ones reported at 1550 nm. Yet, after a 10 MGy dose, the fibers exhibit lower propagation losses at 1310 nm; for instance, 23 dB/km versus 56 dB/km at 1550 nm for the fluorine fiber.

TABLE 1: Summary of radiation-induced attenuations and Brillouin frequency shifts.

	ν_{B0} (MHz)	Dose (MGy)	1.1	3	5.5	7.82	10
SMF 28	10843	RIA (dB/km)	84.8	147	186	214	253
		$\Delta\nu_B$ (MHz)	1.0	1.5	2.0	3.0	4.0
HGe fiber	9276	RIA (dB/km)		270		324	406
		$\Delta\nu_B$ (MHz)		9.6		13.3	17.8
F-fiber	11050	RIA (dB/km)	25	38	51	54	56
		$\Delta\nu_B$ (MHz)	0.8	2.0	2.1	2.2	2.3

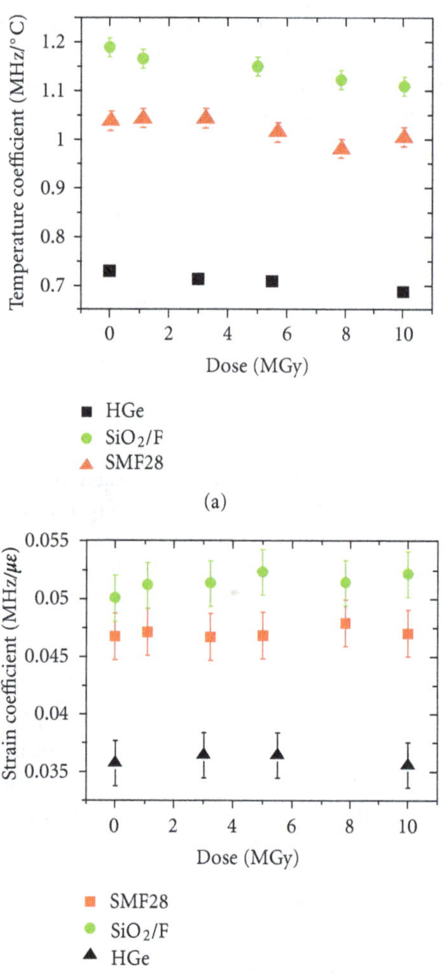

FIGURE 7: Temperature (C_T-up) and strain (C_ε-down) Brillouin calibration coefficients, with received gamma dose for the three optical fiber types.

4.2. Hydrogen Influence on Brillouin Scattering. As introduced previously, for nuclear waste repository instrumentation, hydrogen influence must also be quantified.

Hydrogen originates from (i) nuclear waste release and (ii) anoxic corrosion of metallic materials. Although hydrogen releases are expected small (in the order of 430 mmol/hour release for each intermediate-level nuclear waste), when ventilation stops with cell closure, concentrations would slowly yet regularly increase. Its maximum levels could approach 100% hydrogen content in the atmosphere in few months.

The chosen samples were the same F-doped and HGe fibers used for radiation studies. A G652 from iXFiber company was chosen as a reference. It is composed of a GeO_2-doped core fiber at 3.4 mol% concentration with a 4.6 μm radius and a pure silica cladding with radius 62.5 μm.

The optical fibers were placed inside autoclave chambers, where molecular hydrogen pressure was maintained at 150 bars and temperature was regulated at 25°C. With such conditions, hydrogen concentrations into the optical fiber core reach more than 95% of the saturation level in the fiber core after 330 h (around 13 days).

Selected pressure condition accelerates hydrogen diffusion. We also performed measurement during natural hydrogen release at the end of the experiment: 16 days (versus 13 days under pressure) were required for exposed optical fibers to retrieve their original characteristics.

During hydrogen loading, samples were removed from autoclave regularly to perform measurements: distributed Brillouin scattering and absorption losses on a large wavelength span, from 1.1 μm to 1.56 μm.

Measured spectral attenuations confirmed three absorption bands appear in the attenuation spectra of exposed optical fibers, the most important at 1245 nm and two smaller at 1165 nm and 1130 nm, consistently with [17]. Losses reached 70 dB/km (resp. 50 dB/km) at saturation at 1550 nm (resp. 1310 nm). Standard Brillouin instruments have optical budget in the order of 10 dB. As a result, in instrumented disposal cells for long-lived nuclear wastes, maximal distance range would significantly diminish with the increase of hydrogen content in the atmosphere, from the kilometer range down to one hundred meters. Improvement could be expected if working wavelength was tuned down to 1.3 μm.

Brillouin spectra before and after hydrogen exposure are illustrated in Figure 8 for the G652 fiber.

Brillouin scattering is modified by hydrogen content in optical fibers. On top of reduced amplitude, Brillouin scattering shifts towards high frequencies. This shift is somewhat linear and reaches 21 MHz at saturation for the G652 and the F-doped fibers, 18 MHz for the HGe fiber. Assuming standard coefficients (C_T = 1 MHz/°C, C_ε = 0.05 MHz/$\mu\varepsilon$), such a shift would induce an error in temperature (resp. strain) measurement in the order of 21°C (resp. 420 μm/m).

Fluorine fiber is revealed to be more sensitive to small and moderate hydrogen contents than other fibers.

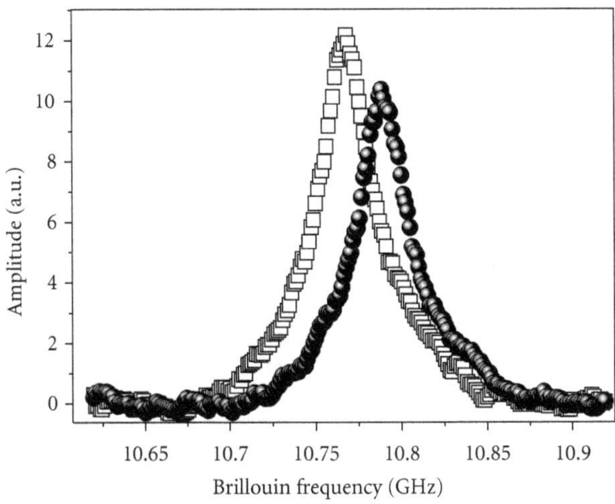

Figure 8: Measured Brillouin spectra in the G652 fiber before and after 13 days exposure to hydrogenated atmosphere (at saturation level).

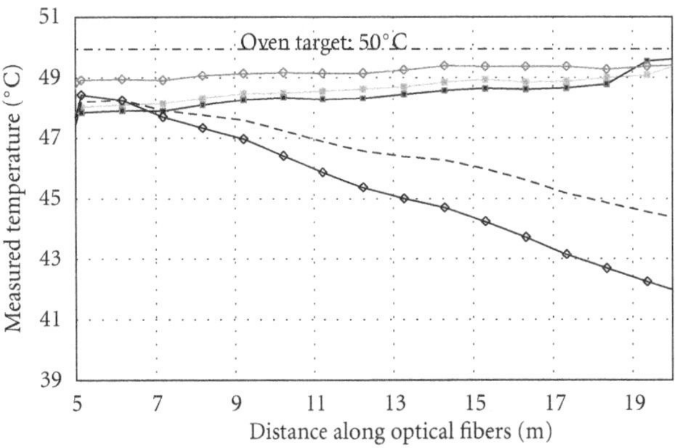

Figure 9: Temperature measurements based on Raman scattering in OM4 and radiation-hard fibers, pristine, and irradiated samples placed in an oven at 50°C.

At saturation, F-doped fiber and G652 are similar; HGe fiber is slightly less impacted. Further work is required to conclude if dopant type, internal stress, or defect concentrations are the major cause, or if differences are induced by different diffusion rates.

This result is not in favor of the application requirements since the F-doped fiber was the only sample which could endure radiations. In view of our application, we tested carbon coating which is known to prevent hydrogen migration into silica [18]. Samples endured the same hydrogen exposure. As detailed in [19], carbon coating of G652 fibers proved to be fully efficient since neither variation of propagation losses nor Brillouin frequency shift could be detected along these samples.

As a conclusion, an F-doped fiber with carbon coating is mandatory for the monitoring of the future geological repository where there will be both gamma radiation and hydrogen release.

4.3. Gamma Influence on Raman Sensing. Unlike Brillouin sensing which is wavelength encoded, for Raman sensing phenomena, RIA not only degrades distance range, but also induces measurement error. Indeed, temperature values are obtained by taking the ration of the two Raman components (Antistokes and Stokes) as detailed in (3) [20, 21]:

$$\frac{I_{AS}}{I_S} = \left(\frac{\lambda_S}{\lambda_{AS}}\right)^4 \exp\left(\frac{-h * \Delta\nu}{k * T(z)} - \int_0^z [\alpha_{AS}(\xi) - \alpha_S(\xi)] d\xi\right), \quad (3)$$

where λ_{AS} and λ_S are Raman Antistokes and Stokes wavelengths, $\Delta\upsilon = 13.2$ THz for silica, h is the Planck's constant, c is the speed of light in vacuum, k is Boltzmann's constant, T is the temperature of the optical fiber, and α refers to propagation losses.

Gamma radiation induces differential losses at Stokes and Antistokes wavelengths, which then induce temperature measurement errors, consistently with (3), as detailed in [20, 21]. Promoted solutions were to use double-ended configuration (closed loop).

However, single-ended arrangement (open loop) suits geological repository specification much better since (i) waste disposal cells are not always accessible at both ends and (ii) ability to perform measurements up to breaking points is important for durability in the order of a century. This is why we turned towards single-ended configuration and chose to focus on radiation-hard optical fibers.

Multimode silica-based fibers with 50 μm core diameters were selected: standard (OM4 type) and radiation-hard fibers. Two samples were extracted from each optical fiber coil; one has been left pristine, the second sample has been irradiated.

We used a Co source at a dose rate of 0.65 kGy/h. A 67 kGy total dose was deposited in two steps. Postmortem temperature measurements were performed on pristine and irradiated samples of the two optical fiber types, placed inside a climatic chamber. Results are illustrated in Figure 9 for the 50°C temperature step.

Temperature measurements performed on standard pristine fiber matches imposed oven temperature (50°C) with a 1°C shift. Once irradiated, large measurement errors appear; error increases with distance, up to 10°C error cumulated in 20 m. OTDR (optical time-domain reflectometer) measurements revealed 80 dB/km at 850 nm and 21 dB/km at 1300 nm. As shown in Figure 9, raw data may be processed to take into account this 0.06 dB/m differential loss value, applying (3). It limits significantly temperature error, but linear drift still remains. Since Raman instrument works at 1064 nm, propagation losses measurements should rather be performed at the real Stokes and Antistokes wavelengths, 1110 nm and 1020 nm. It will be performed in a near future.

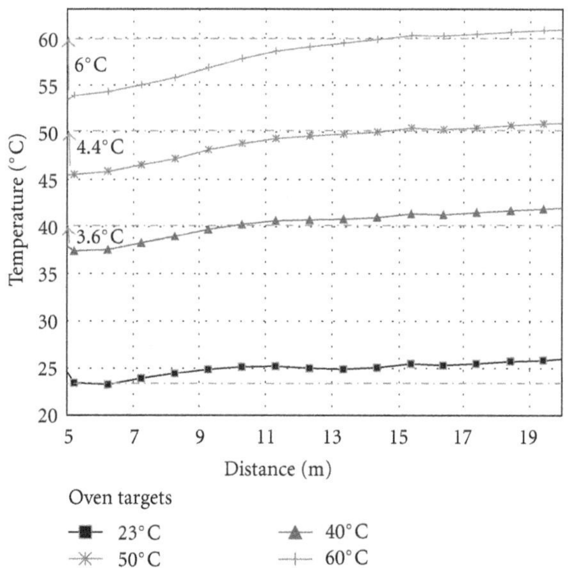

Figure 10: Temperature measurements based on Raman scattering in the radiation-hard fiber irradiated at 618 kGy total dose.

Pristine as well as irradiated radiation hard-fiber samples provide successful temperature measurements (Figure 9), however, with poor accuracy, in the order of 3°C.

We increased irradiation dose. Second irradiation run was performed in MOL facilities, simultaneously with single-mode fibers. It consisted in gamma rays, with a dose of 1,5 kGy/h during 341 hours. Total dose of 551,5 kGy was thus deposited on top of the previous 67 kGy, which provided 618 kGy. Sample was submitted to temperature changes, 23°C, 40°C, 50°C, and 60°C. As illustrated in Figure 10, temperature sensing remains possible even after $0.6 * 10^7$ Gy value that approaches a hundred years of monitoring in the geological repository.

Similarly, with the pristine radiation-hard fiber, temperature measurement quality is poor, degraded by (i) a bias to the oven temperature value and (ii) a linear drift that increases with distance. This linear error is positive, whereas it was negative for standard fiber. Indeed, measured losses at 850 nm and 1300 nm were, respectively, 51 dB/km and 74 dB/km (larger at 1300 nm); thus, differential losses are opposite to previous case.

4.4. Hydrogen Influence on Raman Scattering. Hydrogen influence on Raman distributed temperature sensing has been described in the literature [22] and is expected to induce differential losses, thus temperature reading errors similar with radiation effects. Based on hydrogen tests performed with single-mode fibers (Section 4.2), Andra will supply carbon-coated multimode radiation-hard fibers in the future.

4.5. Discussion. In order to predict distributed temperature and strain measurement performances in the future geological repository for long-lived nuclear waste, we considered separately gamma radiation and hydrogen influences on Brillouin and Raman scatterings in various optical fiber types. However, combined influence of hydrogen and radiation is expected to be in favor of measuring performances [23]. Presented results are worst cases.

Carbon coating should also be in favor of radiation tolerance.

Temperature, in the order of 90°C in the vicinity of high-level waste disposal cell, will accelerate hydrogen diffusion in optical fibers [24]; for ease of manipulation, pressure, instead of temperature, was used in the test. Equivalence will need to be checked.

Tests on radiation influence on Raman sensing were limited to postmortem measurements. These preliminary results will be augmented with on-line measurements on hydrogen-loaded carbon-coated radiation-hard fiber. Spectral transmission measurements will also be included to take full advantage of the compensation model.

5. Conclusion

We developed a strategy to design a durable monitoring system for nuclear structures, based on truly-distributed optical fiber sensors.

We demonstrated Brillouin strain measurement can be efficiently compensated from temperature influence installing a parallel sensing line paired with a Raman instrument.

State-of-the-art results on gamma and hydrogen influences on Rayleigh, Raman, and Brillouin scatterings were obtained. Fluorine-doped pure-silica-core fibers coated with carbon will be mandatory to handle high gamma dose (10^7 Gy) and hydrogen release in the vicinity of high-level long-lived waste disposal cells. After a century of monitoring based on Brillouin scattering, maximal range would decrease down to a hundred meter but strain measurement uncertainty should remain stable. With other fiber types or primary coatings, large strain measurement errors would ruin the monitoring system reliability.

To reduce hydrogen and radiation influences, a significant future improvement could be achieved by choosing the operating wavelength at 1.3 μm instead of 1.55 μm.

We showed that radiation-hard fibers enabled single-end Raman measurements up to 0.6 MGy; however, temperature measurement uncertainty, presently in the order of few degrees, must be improved. A postprocessing to compensate for differential loss induced by radiation has been proposed and will, in the future, take benefit of spectral measurements.

Next step of Andra qualification procedure is to incorporate these special fibers into strain sensing cables and run (i) combined hydrogen-gamma influence test and (ii) another outdoor test, to evaluate if curvatures or splicing may get more sensitive than with standard fibers.

Acknowledgments

Part of this work has been done in the frame of the COST Action TD1001: Novel and Reliable Optical Fibre Sensor Systems for Future Security and Safety Applications (OFSeSa). D. Sporea acknowledges that this work was partially supported by the grant of the Romanian National

Authority for Scientific Research, CNDI-UEFISCDI, project 8/2012 "Sensor Systems for Secure Operation of Critical Installations."

References

[1] J. M. Lopez-Higuera, *Handbook of Optical Fiber Sensing Technology*, Wiley, New York, NY, USA, 2002.

[2] A. Rogers, "Distributed optical-fibre sensing," *Measurement Science and Technology*, vol. 10, no. 8, pp. R75–R99, 1999.

[3] V. Lanticq, R. Gabet, F. Taillade, and S. Delepine-Lesoille, "Distributed optical fibre sensors for structural health monitoring: upcoming challenges," in *Optical Fibre, New Developments*, L. Christophe, Ed., chapter 9, pp. 177–199, InTech, Olajnica, 2010.

[4] "Qualification guide FD CEN/TR, 14748," Non-destructive testing—methodology for qualification of non-destructive tests, 2005.

[5] S. W. Tyler, J. S. Selker, M. B. Hausner et al., "Environmental temperature sensing using Raman spectra DTS fiber-optic methods," *Water Resources Research*, vol. 45, Article ID W00D23, 11 pages, 2009.

[6] S. Delepine-Lesoille, J. M. Henault, G. Moreau et al., "Truly distributed optical fiber sensors for structural health monitoring: From the telecommunication optical fiber drawling tower to water leakage detection in dikes and concrete structure strain monitoring," *Advances in Civil Engineering*, vol. 2010, Article ID 930796, 13 pages, 2010.

[7] M. Niklès, L. Thévenaz, and P. A. Robert, "Simple distributed fiber sensor based on Brillouin gain spectrum analysis," *Optics Letters*, vol. 21, no. 10, pp. 758–760, 1996.

[8] M. Niklès, L. Thévenaz, and P. A. Robert, "Brillouin gain spectrum characterization in single-mode optical fibers," *Journal of Lightwave Technology*, vol. 15, no. 10, pp. 1842–1851, 1997.

[9] J. P. Dubois, S. Delepine-Lesoille, V. H. Tran et al., "Raman versus Brillouin optical fiber distributed temperature sensing: an outdoor comparison (metallic beam and concrete slab)," in *Proceedings of the 4th International Society for Structural Health Monitoring of Intelligent Infrastructures Conference (SHMII '09)*, July 2009.

[10] V. Lamour, A. Haouas, J. P. Dubois, and R. Poisson, "Long term monitoring of large massive concrete structures: cumulative effects of thermal gradients," in *Proceedings of 7th International Symposium on Nondestructive Testing in Civil Engineering*, Nantes, France, July 2009.

[11] M. N. Alahbabi, Y. T. Cho, and T. P. Newson, "Simultaneous temperature and strain measurement with combined spontaneous Raman and Brillouin scattering," *Optics Letters*, vol. 30, no. 11, pp. 1276–1278, 2005.

[12] D. Sporea, A. Sporea, S. O'Keeffe, D. McCarthy, and E. Lewis, "Optical fibers and optical fiber sensors used in radiation monitoring," in *Selected Topics on Optical Fiber Technology*, Y. Moh, S. W. Harun, and H. Arof, Eds., Intech, Vienna, Austria, 2012.

[13] M. Van Uffelen, *Modélisation de systèmes d'acquisition et de transmission à fibres optiques destinés à fonctionner en environnement nucléaire [Ph.D. thesis]*, Université de Paris, Orsay, France, 2001.

[14] D. Alasia, A. Fernandez Fernandez, L. Abrardi, B. Brichard, and L. Thévenaz, "The effects of gamma-radiation on the properties of Brillouin scattering in standard Ge-doped optical fibres," *Measurement Science and Technology*, vol. 17, no. 5, pp. 1091–1094, 2006.

[15] X. Pheron, Y. Ouerdane, S. Girard et al., "In situ radiation influence on strain measurement performance of Brillouin sensors," in *Proceedings of the 21st International Conference on Optical Fiber Sensors*, May 2011.

[16] A. Fernandez-Fernandez, H. Ooms, B. Brichard et al., "SCK-CEN gamma irradiation facilities for radiation tolerance assessment," in *Proceedings of the NSREC Data Workshop*, pp. 171–176, 2002, 02HT8631.

[17] J. Stone, "Interactions of hydrogen and deuterium with silica optical fibers," *Journal of Lightwave Technology*, vol. 5, no. 5, pp. 712–733, 1987.

[18] P. J. Lemaire and E. A. Lindholm, "Hermetic optical fibers: carbon coated fibers," in *Specialty Optical Fibers Handbook*, A. Mendez and T. F. Morse, Eds., pp. 453–490, New York Academic, 2007.

[19] S. Delepine-Lesoille, J. Bertrand, L. Lablonde, and X. Phéron, "Distributed hydrogen sensing with Brillouin scattering in optical fibers," *Photonics Technology Letters*, vol. 24, no. 17, pp. 1475–1477, 2012.

[20] F. B. H. Jensen, E. Takada, M. Nakazawa, T. Kakuta, and S. Yamamoto, "Consequences of radiation effects on pure-silica-core optical fibers used for raman-scattering-based temperature measurements," *IEEE Transactions on Nuclear Science*, vol. 45, no. 1, pp. 50–58, 1998.

[21] A. Fernandez Fernandez, P. Rodeghiero, B. Brichard et al., "Radiation-tolerant Raman Distributed Temperature monitoring system for large nuclear infrastructures," *IEEE Transactions on Nuclear Science*, vol. 52, no. 6, pp. 2689–2694, 2005.

[22] C. Martelli, A. L. C. Triques, A. Braga et al., "Operation of optical fiber sensors in hydrogen-rich atmosphere," in *Proceedings of the 4th European Workshop on Optical Fibre Sensors*, September 2010.

[23] B. Brichard, A. L. Tomashuk, V. A. Bogatyrjov et al., "Reduction of the radiation-induced absorption in hydrogenated pure silica core fibres irradiated in situ with γ-rays," *Journal of Non-Crystalline Solids*, vol. 353, no. 5–7, pp. 466–472, 2007.

[24] P. J. Lemaire, "Reliability of optical fibers exposed to hydrogen: prediction of long-term loss increases," *Optical Engineering*, vol. 30, no. 6, pp. 780–789, 1991.

Permissions

The contributors of this book come from diverse backgrounds, making this book a truly international effort. This book will bring forth new frontiers with its revolutionizing research information and detailed analysis of the nascent developments around the world.

We would like to thank all the contributing authors for lending their expertise to make the book truly unique. They have played a crucial role in the development of this book. Without their invaluable contributions this book wouldn't have been possible. They have made vital efforts to compile up to date information on the varied aspects of this subject to make this book a valuable addition to the collection of many professionals and students.

This book was conceptualized with the vision of imparting up-to-date information and advanced data in this field. To ensure the same, a matchless editorial board was set up. Every individual on the board went through rigorous rounds of assessment to prove their worth. After which they invested a large part of their time researching and compiling the most relevant data for our readers. Conferences and sessions were held from time to time between the editorial board and the contributing authors to present the data in the most comprehensible form. The editorial team has worked tirelessly to provide valuable and valid information to help people across the globe.

Every chapter published in this book has been scrutinized by our experts. Their significance has been extensively debated. The topics covered herein carry significant findings which will fuel the growth of the discipline. They may even be implemented as practical applications or may be referred to as a beginning point for another development. Chapters in this book were first published by Hindawi Publishing Corporation; hereby published with permission under the Creative Commons Attribution License or equivalent.

The editorial board has been involved in producing this book since its inception. They have spent rigorous hours researching and exploring the diverse topics which have resulted in the successful publishing of this book. They have passed on their knowledge of decades through this book. To expedite this challenging task, the publisher supported the team at every step. A small team of assistant editors was also appointed to further simplify the editing procedure and attain best results for the readers.

Our editorial team has been hand-picked from every corner of the world. Their multi-ethnicity adds dynamic inputs to the discussions which result in innovative outcomes. These outcomes are then further discussed with the researchers and contributors who give their valuable feedback and opinion regarding the same. The feedback is then collaborated with the researches and they are edited in a comprehensive manner to aid the understanding of the subject.

Apart from the editorial board, the designing team has also invested a significant amount of their time in understanding the subject and creating the most relevant covers. They scrutinized every image to scout for the most suitable representation of the subject and create an appropriate cover for the book.

The publishing team has been involved in this book since its early stages. They were actively engaged in every process, be it collecting the data, connecting with the contributors or procuring relevant information. The team has been an ardent support to the editorial, designing and production team. Their endless efforts to recruit the best for this project, has resulted in the accomplishment of this book. They are a veteran in the field of academics and their pool of knowledge is as vast as their experience in printing. Their expertise and guidance has proved useful at every step. Their uncompromising quality standards have made this book an exceptional effort. Their encouragement from time to time has been an inspiration for everyone.

The publisher and the editorial board hope that this book will prove to be a valuable piece of knowledge for researchers, students, practitioners and scholars across the globe.

List of Contributors

M. Tamazin and M. J. Korenberg
Electrical and Computer Engineering Department, Queen's University, Kingston, ON, Canada K7L 3N6

A. Noureldin
Electrical and Computer Engineering Department, Queen's University, Kingston, ON, Canada K7L 3N6
Electrical and Computer Engineering Department, Royal Military College of Canada, Kingston, ON, Canada K7K 7B4

Hirokazu Kobayashi, Toshimasa Tsuzuki, Toshitake Onishi, Yuhei Masaoka and Koji Nonaka
Department of Electronic and Photonic Systems Engineering, Kochi University of Technology, Kochi 782-8502, Japan

Xunjian Xu
Key Laboratory, State Grid Corporation of China, Beijing 100031, China

W. Sharatchandra Singh, B. P. C. Rao, S. Thirunavukkarasu and T. Jayakumar
NDE Division, Indira Gandhi Centre for Atomic Research, Kalpakkam 603 102, India

Yasutaka Kishima, Kentarou Kurashige and Toshihisa Kimura
Muroran Institute of Technology, 27-1 Mizumoto, Hokkaido, Muroran 0508585, Japan

Daehee Kim and Baek-Young Choi
Department of Computer Science Electrical Engineering, School of Computing and Engineering, University of Missouri, Kansas City, MO 64110, USA

Sejun Song
Department of Engineering Technology Industrial Distribution, Dwight Look College of Engineering, Texas A&M University, College Station, TX 77843, USA

Mikel Bravo and Manuel López-Amo
Departamento de Ingenier´ıa El´ectrica y Electr´onica, Universidad P´ublica de Navarra, Campus Arrosadia S/N, Navarra, 31006 Pamplona, Spain

Ping Wang and Yan Yan
College of Automation Engineering, Nanjing University of Aeronautics and Astronautics, Yudaojie Road 29, Jiangsu 210016, China

Gui Yun Tian, Omar Bouzid and Zhiguo Ding
School of Electrical and Electronic Engineering, Newcastle University, Newcastle upon Tyne NE1 7RU, UK

Shutao Xing
Deep Sea (US) Inc., Houston, TX 77042, USA
Department of Civil and Environmental Engineering, Utah State University, Logan, UT 84322, USA

Marvin W. Halling and Paul J. Barr
Department of Civil and Environmental Engineering, Utah State University, Logan, UT 84322, USA

Abdelrahman Ali and Naser El-Sheimy
Department of Geomatics Engineering, The University of Calgary, 2500 University Drive N.W., Calgary, AB, Canada T2N 1N4

C. A. Galindez-Jamioy and J. M. López-Higuera
Photonic Engineering Group of the University of Cantabria, R&D&i Telecommunication Building, Avenida Los Castros, 39005 Santander, Spain

Hua Xia
Photonics Laboratory, Micro and Nano Structures Technologies, GE Global Research, 1 Research Circle, Niskayuna, NY 12309, USA
College of Nanoscale Science and Engineering, State University of New York at Albany, NY 12222, USA

Sachin Dekate and Boon Lee
Photonics Laboratory, Micro and Nano Structures Technologies, GE Global Research, 1 Research Circle, Niskayuna, NY 12309, USA

Doug Byrd
Engineering Division, GE Energy, 300 Garlington Road, GTTC 200D, Greenville, SC 29615, USA

Peter Noy, Roger Steiner, Joerg Voelkle and Christof Fattinger
F. Hoffmann-La Roche Ltd., Pharma Research and Early Development, Discovery Technologies, 4070 Basel, Switzerland

Martin Hegner
CRANN—The Naughton Institute, School of Physics, Trinity College Dublin, Dublin 2, Ireland

Sangkyu Lee, Jongyun Cho and Wonkyu Moon
Department of Mechanical Engineering, Pohang University of Science and Technology, Pohang 790-784, Republic of Korea

Yeolho Lee
Corporate Technology Operations SAIT, Samsung Electronics Co., Ltd., Yongin 446-712, Republic of Korea

Sangmin Jeon and Hyung Joon Cha
Department of Chemical Engineering, Pohang University of Science and Technology, Pohang 790-784, Republic of Korea

Ronan Daly, Shishir Kumar, Kangho Lee, Anne Weidlich and Georg S. Duesberg
CRANN and School of Chemistry, Trinity College Dublin, Dublin 2, Ireland

Gyongyi Lukacs and Martin Hegner
CRANN and School of Physics, Trinity College Dublin, Dublin 2, Ireland

Masayuki Morisawa and Shinzo Muto
Department of Information and Communication System, Graduate School of Medicine and Engineering, University of Yamanashi, 4-3-11 Takeda, Kofu 400-8511, Japan

Gabriele Rescio, Alessandro Leone and Pietro Siciliano
Institute for Microelectronics and Microsystems, Italian National Research Council (CNR), Via Monteroni, c/o Campus Universit`a del Salento, Palazzina A3, 73100 Lecce, Italy

Jason Jensen and Martin Hegner
Centre for Research on Adaptive Nanostructures and Nanodevices (CRANN) and School of Physics, Trinity College Dublin, Dublin 2, Ireland

Thomas Lehmann and Mahdi Tavakoli
Department of Electrical and Computer Engineering, University of Alberta, 9107-116 Street, Edmonton, AB, Canada T6G 2V4

Nawaid Usmani and Ronald Sloboda
Cross Cancer Institute, University of Alberta, 11560 University Avenue, Edmonton, AB, Canada T6G 1Z2

J.Mirapeix, O.M. Conde, J. M. Lopez-Higuera and A. Cobo
Photonics Engineering Group, University of Cantabria, 39005 Santander, Spain

P. B. Garc´ıa-Allende
Hamlyn Centre for Robotic Surgery, Institute of Global Health Innovation, and Department of Surgery and Cancer, Imperial College London, London SW7 2AZ, UK

Junqing Ma, Aiguo Song and Dongcheng Pan
Jiangsu Key Lab of Remote Measurement and Control, School of Instrument Science and Engineering, Southeast University, Nanjing 210096, China

M. Lucci
Dipartimento di Fisica, Universit`a di Roma "Tor Vergata", Via della Ricerca Scientifica, 00133 Roma, Italy

F. Toschi, V. Guglielmotti, S. Orlanducci and M. L. Terranova
Dipartimento di Scienze e Tecnologie Chimiche, Minimalab Universit`a di Roma "Tor Vergata", Via della Ricerca Scientifica, 00133 Roma, Italy

C. C. Lai, S. L. Ho, H. Y. Tam and Michael S. Y. Liu
Department of Electrical Engineering, Hong Kong Polytechnic University, Kowloon, Hong Kong

Jacob C. P. Kam, David C. C. Leung, Tony K. Y. Lee and Aiken Y.M. Tam
Mass Transit Railway, Fo Tan Railway House, Fo Tan, Shatin, NT, Hong Kong

S. Delepine-Lesoille, X. Ph´eron, J. Bertrand, G. Pilorget, G. Hermand and R. Farhoud
National Radioactive Waste Management Agency (Andra), 1-7 rue Jean Monnet, Parc de la Croix blanche, 92298 Chatenay-Malabry, France

Y. Ouerdane, A. Boukenter and S. Girard
LabHC, UMR CNRS 5516, Universit´e de Saint-Etienne, 42023 Saint-Etienne, France

L. Lablonde
iXFIber, 22300 Lannion, France

D. Sporea
National Institute for Laser, Plasma and Radiation Physics, Magurele, Romania

V. Lanticq
Cementys, 27 Villa Daviel, 75013 Paris, France

www.ingramcontent.com/pod-product-compliance
Lightning Source LLC
Chambersburg PA
CBHW080702200326
41458CB00013B/4932